# OXFORD STUDIES IN PROBABILITY

SERIES EDITORS

L. C. G. ROGERS

*with*

P. BAXENDALE   P. GREENWOOD   F. P. KELLY
J.-F. LE GALL   E. PARDOUX   W. VERVAAT
D. WILLIAMS

OXFORD STUDIES IN PROBABILITY · 1

# Foundations of the Prediction Process

FRANK B. KNIGHT

*Department of Mathematics*
*University of Illinois at*
*Urbana-Champaign*

CLARENDON PRESS · OXFORD
1992

*Oxford University Press, Walton Street, Oxford OX2 6DP*
*Oxford New York Toronto*
*Delhi Bombay Calcutta Madras Karachi*
*Petaling Jaya Singapore Hong Kong Tokyo*
*Nairobi Dar es Salaam Cape Town*
*Melbourne Auckland*
*and associated companies in*
*Berlin Ibadan*

*Oxford is a trade mark of Oxford University Press*

*Published in the United States*
*by Oxford University Press, New York*

© *Frank B. Knight, 1992*

*All rights reserved. No part of this publication may be reproduced, stored in a retrieval system, or transmitted, in any form or by any means, electronic, mechanical, photocopying, recording, or otherwise, without the prior permission of Oxford University Press*

*A catalogue record for this book is available from the British Library*

*Library of Congress Cataloging in Publication Data*
Knight, Frank B.
Foundations of the prediction process / by Frank B. Knight.
Includes bibliographical references and index.
1. Stochastic processes. I. Title.
QA274.K57    1991    519.2—dc20    91-20603

*ISBN 0-19-853593-7*

*Typeset by Integral Typesetting, Gorleston, Norfolk NR31 6RG*
*Printed in Great Britain by*
*Biddles Ltd, Guildford & King's Lynn*

# Preface

The subject of continuous time stochastic processes currently finds itself in a state which, however suitable it may be for practitioners of the subject, leaves much to be desired from an aesthetic standpoint. It is a little reminiscent of certain bulky plant-eating dinosaurs who, we are told, had very large and well-developed appendages but proportionately very small brains. In short, there seems to be little overall focus or unifying principle to guide the creature. The analogy can be carried even further if we recall that some dinosaurs developed rudimentary extra brains (or, at least, control centres) for the more remote appendages. One can easily bring to mind branches of stochastic theory which have developed an analogous degree of independence. Some, indeed, seem about ready to split off from the rest of the subject. Here, however, our analogy breaks down.

We do not, in the present work, propose to supply any guide to the development of stochastic processes. Rather, what we propose is a guide to the understanding of that development. In this way, our goal is like a brain for the dinosaur which can understand what the creature is doing in a unified way, but does not exercise any actual control. Such a guide, besides providing aesthetic unity, can be an interesting subject of study in its own right. We develop below a unifying principle of understanding, which we call the prediction processes.

The prediction processes (or perhaps simply 'process,' as P.-A. Meyer was the first to emphasize) were introduced by Knight (1975) and rapidly consolidated by Meyer (1976b) and Meyer and Yor (1976). Thereafter little activity followed (excepting Yor (1977) and Getoor (1978)) until the essays of Knight (1981a). These attempted to give an overview of the subject by examining some rather specific areas of application. It was anticipated that more papers would soon follow, but they did not materialize. In fact, apart from an unpublished work of D. Aldous, the present work seems to be the first real successor to Knight (1981a).

It is beyond the present writer to explain this lack of development but at least it has the advantage of leaving a clear field for the present book. In the following, we try to give a comprehensive account of the subject. However, except for the first two sections of Essay I of Knight (1981a), the previous work is not superseded. Rather, we focus here on the predictive aspects of the subject, which were somehow not emphasized in earlier treatments, and on a direct development, as distinct from the scattered approach taken before.

On the other hand, we do not go into the convergence of sequences of processes, although the prediction process may provide a natural approach.

Nor has the theory been completed with regard to the range of processes considered. For example, a promising possibility not considered here would seem to be replacement of sample paths by sample measures or distributions (which would entail, however, a narrowing of the state spaces). In another direction, we do not explicitly demonstrate the connections of our theory with the (Strasbourg) general theory of processes. The reader wishing to learn how the optional and previsible projections relate to the prediction process is referred to Meyer (1976*b*) and Meyer and Yor (1976). A nice brief account of the 'general theory' itself, for purposes of relating it to general Markov processes, is found in Appendix A.5 of Sharpe (1988). In this book, the emphasis is on how the theory can be avoided rather than on how it fits in. In spite of these omissions, however, we do reach a natural ending point, providing a nice well-rounded theory which is also very general.

For a rapid idea of what the present work is really about, the reader should look over the first six pages of the Introduction, followed by the appendix to Chapter 6. Granting that this is no substitute for the book, it may nevertheless give the general gist of the subject. Chapters 1–3 were first developed in lectures at the Universities of Illinois, Strasbourg, and Vancouver. Special thanks are due to these institutions, as well as to Professor P.-A. Meyer (Strasbourg) and Professors E. Perkins and J. Walsh (Vancouver) who helped and encouraged this work. Chapter 1 contains the core of the method, while Chapters 2 and 3 are rather distinct applications of it. Section 2.4 is not used in subsequent chapters.

Chapters 4–6 largely represent quite new developments, although parts are anticipated in Knight (1986*a,b*). Chapters 4 and 5 are quite separate, and the reader interested only in prediction might omit Chapter 4. Chapter 6 is more comprehensive; it contains not only illustrations of prediction (Section 6.1), but also the enlargements of the theory to cover the set-up of filtering and the general theory of processes (Section 6.2). Section 6.3 contains the application of the method of the Introduction to decomposition of semimartingales, with a new result on the stationary Gaussian case. Section 6.4 treats the modification of processes, ending with the application of modification of the prediction process to obtain information about a general (non-Markovian) process. Lastly, the Appendix contains material of general theoretical interest about the method itself.

Final thanks are due to Professor Jerry Sacks, who enabled me to have a reduced teaching load while working on Chapters 4 and 5, and to Hilda Britt and the rest of the secretarial staff of the Department of Mathematics of the University of Illinois at Urbana-Champaign, for their unflagging patience in typing a manuscript which must occasionally have looked like a forest after the tornado.

*Urbana*  F.B.K.
January 1991

# Contents

| | | |
|---|---|---|
| **List of symbols** | | ix |
| **Introduction** | | 1 |
| **1.** | **The measure-theoretic prediction process** | 15 |
| **2.** | **Topological considerations and prediction space: application to Markov processes** | 41 |
| | 2.1 The prediction topology | 41 |
| | 2.2 Application to Markov processes | 51 |
| | 2.3 Prediction space: Ray processes and right processes | 62 |
| | 2.4 Compactifications and the Ray space | 80 |
| | Appendix | 89 |
| **3.** | **Gaussian processes and the prediction process in the wide sense** | 91 |
| | 3.1 The structure of a measurable Gaussian process | 91 |
| | 3.2 Specialization to continuous covariances | 123 |
| | Appendix | 141 |
| **4.** | **A classification of measurable processes** | 144 |
| | 4.1 Introduction | 144 |
| | 4.2 The continuous prediction processes: intrinsic diffusion | 149 |
| | 4.3 Classification of discontinuities | 151 |
| **5.** | **Prediction of measurable processes** | 165 |
| | 5.1 Introduction | 165 |
| | 5.2 Strict-sense prediction of square-integrable processes | 171 |
| **6.** | **Application to concrete examples, and ramifications of the theory** | 195 |
| | 6.1 Application to explicit prediction | 195 |
| | 6.2 Adaptation to a general filtration | 212 |
| | 6.3 Decomposition of semimartingales, and Gaussian processes | 218 |
| | 6.4 Modifications of the future | 229 |
| | Appendix | 238 |
| **References** | | 244 |
| **Index** | | 247 |

# List of symbols

| | | |
|---|---|---|
| $\mathscr{A}$ | weak infinitesimal generator | 234 |
| $b(\cdot)$ | bounded measurable functions | 16 |
| $C(\Omega')$ | continuous functions | 21 |
| $d_E$ | metric on $\bar{E}$ | 44 |
| $d'$ | weak metric on $\Omega'$ | 20 |
| $d_n, d_\infty$ | increasing metrics on $U$ | 75, 80 |
| $D$ | non-branch points | 38 |
| $D_\lambda(z)$ | | 234 |
| $\Delta_t h$ | jump of $h$ at $t$ | 155 |
| $(E, \mathscr{E})$ | Lusin space | 16 |
| $E(X|\mathscr{F})$ | conditional expectation | 11 |
| $\bar{E}$ | compact metric space | 69 |
| $(E_p, \mathscr{E}_p)$ | state space of $X_t^e$ | 56 |
| $\bar{\bar{E}}$ | closure of $E$, inherited from $\overline{\phi(E)}$ | 57 |
| $E(t)$ | mean function | 91 |
| $\mathscr{E}(X|H)$ | Hilbert space projection | 10 |
| $F_n$ | projector of $X_z^c$ onto $Y_n$ | 123 |
| $F^*, F^\#$ | projectors of $X^+$ onto $Y$ | 185 |
| $\mathscr{F}_t$ | general filtration | 7 |
| $\mathscr{F}', \mathscr{F}_t'$ | canonical filtration for $\Omega'$ | 17 |
| $\mathscr{F}_t^z$ | augmented filtration | 24 |
| $\Gamma_z^*(u; v_1, v_2)$ | inner product of conditional expectations | 171 |
| $\Gamma(s, t), \Gamma(s - t)$ | covariances | 91, 133 |
| $G_\lambda^*, G_\lambda^\#$ | projectors of $M_\lambda$ onto $Y$ | 185 |
| $G_t^z(S)$ | prediction process given one component | 214 |
| $\mathscr{G}_T'$ | filtration of one component | 214 |
| $G_t^{X,W}$ | prediction process relative to observed $W$ | 239 |
| $H_z(t+)$ | Hilbert space of given process | 107, 167 |
| $H_z^*(t)$ | Hilbert space of conditional expectations | 171 |
| $H_z^\#(t), H_z'(t)$ | Hilbert space of $(M_n)$ | 181, 228 |
| $J$ | times of discontinuity | 151 |

| | | |
|---|---|---|
| $J_D, J_{D^c}$ | unforseeable and foreseeable times of discontinuity | 152 |
| $K_{\lambda,k}$ | projector of $M_\lambda$ onto $Y_k$ | 124 |
| l.c.r.l. | left continuous with right limits | 8 |
| LMR | linear martingale representation | 171 |
| $L^2(\Omega, \mathcal{F}, P)$ | Hilbert space | 10 |
| $(M', \mathcal{M}')$ | space of probability measure | 19 |
| $(M_E, \mathcal{M}_E)$ | space of probability measures | 28 |
| $M_{\lambda,n}$ | r.c.l.l., $\lambda$-supermartingales | 52 |
| $M_p$ | enlarged $\phi(E)$ | 55 |
| $M_\lambda^A(t)$ | martingales of absorbed process | 234 |
| $M_G, M_{G(c)}$ | Gaussian demesnes | 93, 95, 228 |
| $M_r$ | demesne of finite $r$th moments | 95 |
| $M_\lambda(t)$ | fundamental martingales | 4, 109 |
| $\langle M_\lambda \rangle$ | previsible quadratic variation | 173 |
| $M_\infty, M_0$ | limit martingales | 197 |
| $M_{2,E_i}$ | demesne of square integrable components | 216 |
| $(m,n)_z^t$ | inner products of $R_\lambda^Z \hat{\rho}$ | 114 |
| $(m,n)_z^*(t)$ | inner products of conditional expectations | 172 |
| $M_c$ | demesne of intrinsic diffusions | 149 |
| $M(J_D), M(J_{D^c})$ | demesnes of jumps | 157 |
| $M_d$ | demesne of intrinsic jump processes | 159 |
| $M_{\text{WOJ}}$ | demesne of well-ordered jumps | 161 |
| $M_2^+$ | processes right-continuous in $L^2$ | 166 |
| $\hat{M}_\lambda$ | projection of $M_\lambda$ onto $H_z(t+)$ | 168 |
| $M_\lambda^f$ | Kunita–Watanabe martingale | 145 |
| $\hat{\mathcal{M}}_\lambda$ | wide sense $M_\lambda$ | 113 |
| $\mu(\lambda)$ | mean transform | 91 |
| $\Omega', \Omega'_E$ | canonical coordinate space | 16, 17 |
| $\Omega_Z$ | canonical prediction space | 62, 63 |
| $\Omega_U$ | canonical space of $U$ | 64 |
| $\bar{\Omega}_E$ | compactification of $\Omega'_E$ | 47 |
| $p(t, x, B)$ | Markov transition function | 51 |
| $\mathcal{P}$ | previsible (predictable) $\sigma$-field | 183 |
| $P^z$ | probability (equivalent to $z$) | 19, 62 |
| PMSIR | previsible martingale stochastic integrable representation | 171 |

List of symbols | xi

| Symbol | Description | Page |
|---|---|---|
| $q_E^*(t, z, A)$ | transition function of $G_t^z$ | 214 |
| $q(t, z, A)$ | prediction process transiton function | 31 |
| r.c.l.l., {r.c.l.l.}, {r.c.l.l.}$_z$ | right continuous with left limits | 8, 69, 233 |
| {r.c.} | right continuous paths | 67 |
| $\tilde{\rho}, \tilde{\rho}_E, \tilde{\rho}_{E_i}$ | generalized a.e. coordinate function | 18, 49, 216 |
| $\hat{\rho}, \hat{\rho}_E, \hat{\rho}_{E_i}$ | process function of prediction process | 4, 39, 50, 216 |
| $R_\lambda$ | resolvent of Markov process | 52 |
| $R_\lambda^Z$ | resolvent of $Z_t$ | 4, 42 |
| $R_\lambda^Z(X(t))$ | Z-resolvent of $X(t)$ | 196, 218 |
| $\Sigma_n^Y(t)$ | summation notation | 176 |
| $\sigma(\cdot)$ | generated $\sigma$-field | 7 |
| $\sigma^2(\lambda_1, \lambda_2)$ | covariance transform | 91 |
| $\Theta_t, \Theta_t^Z$ | translation operators | 18, 49, 62 |
| $\hat{\phi}$ | Kuratowski isomorphism | 17 |
| $\phi(x)$ | map from $E$ to $M_E$ | 53 |
| $\phi_z^*(t)$ | strict sense multiplicity | 181 |
| $\phi_z^\#(t)$ | orthogonal martingale multiplicity | 182 |
| $\Phi_z$ | wide sense index of multiplicity | 118, 122 |
| $t_k^n$ | partition | 173 |
| $T_t$ | Markov semigroup | 51 |
| $U$ | prediction demesne | 64 |
| $U^*$ | completion of $U$ | 73 |
| $\bar{U}_\infty$ | Ray compactification of $U$ | 75, 80 |
| $U_e$ | entrance space | 82 |
| $U_R$ | Ray space | 80 |
| $V_\lambda(z)$ | variance of transform | 175 |
| $X_t'$ | measurable standard modification | 52 |
| $\tilde{X}_t$ | generalized coordinate process | 18, 49 |
| $X_t^x, X_t^e$ | regularized version of $X_t'$ | 57 |
| $X_z^c$ | Gaussian process continuous in $L^2$ | 123 |
| $X_z^+$ | process right-continuous in $L^2$ | 167 |
| $X_B$ | absorbed process | 233 |
| $w_z(t)$ | canonical prediction path | 63 |
| $Y_n(t)$ | ordered orthogonalization of $M_n$ | 118 |
| $\hat{\mathcal{Y}}_n$ | wide sense $Y_n$ | 122 |
| $Y_n^\#$ | orthogonal martingales | 181 |

## xii | List of symbols

| | | |
|---|---|---|
| $\hat{Y}_n$ | projections of $Y_n$ onto $H_z(t+)$ | 168 |
| $Z_t^z$ | prediction process of $z$ | 23, 24 |
| $Z_t$ | canonical prediction process | 63 |
| $\mathscr{L}_{t+}^0$ | canonical filtration on prediction space | 62–3 |
| $\mathscr{L}_t^z$ | augmentation of $\mathscr{L}_{t+}^0$ for $P^z$ | 171 |

# Introduction

The theory of stochastic processes, if measured by the volume of publications devoted thereto, has undergone an enormous expansion in the years since 1948 and 1953 when the well-known books of P. Lévy and J. L. Doob on this subject were published. Setting aside all but the case of a single continuous parameter $t$, which represents time in our development, we would still find too many books, and far too many papers, to list in a manageable bibliography. Nevertheless, the expansion is limited in some respects. With regard to the basic methods used, and the types of dependence treated among the random variables constituting the processes, there is comparatively little that was not already present (or at least clearly foreshadowed) in these early works. To mention a few examples, we already had quite general theories of processes with independent increments, martingales, Markov processes, and stationary processes, both in the strict and the $L^2$ senses. Stochastic integrals were also familiar (Doob 1953, IX, Section 5)), and indeed the stochastic calculus had been started by K. Ito long before (Ito 1942).

Since that time, the huge expansion of the field may be said to be one of more and more detailed application of the concepts and methods that were recognized by 1953. In recent years, indeed, the field has increasingly turned to applying these ideas in more abstract and complicated settings, such as using Banach spaces and algebraic structures for state spaces, and using arbitrary filtrations of $\sigma$-algebras on the probability space. The connections with potential theory, moreover, have been explored to a degree which outweighs much of the subject.

The present work, by contrast, seeks to develop a single theme which applies to all of these situations, but which involves a different approach. It suffices to discuss only a single type of example, chosen from among many possibilities, to illustrate both the fact that the traditional theory of real-valued processes $X_t$, $0 \leq t < \infty$, is still very far from exhausting the subject, and at the same time to indicate the general approach to be taken here.

Suppose that we begin with an ordinary real-valued Brownian motion process $B_t$, $0 \leq t$, starting at $x$, defined on the probability space $(C, \mathscr{G}_\infty, P^x)$, where $C = C[0, \infty)$, $\mathscr{G}_t = \sigma(B_s, 0 \leq s \leq t)$, and $P^x$ are the natural objects generated by the coordinate functions $B_t$. We assume familiarity with this object, at least for the purposes of the example. Now let $0 \leq A \in \mathscr{G}_\infty$ be an arbitrary finite positive random variable, and, letting $\wedge$ denote the minimum, consider the process $X_t = B_{t \wedge A}$, $0 \leq t$. What can be said in general about

## 2 | Foundations of the prediction process

such a process, and how is it to be approached? The ingredients and operations involved are all quite familiar, so it might seem that the process $X_t$ also should be subject to some familiar methods of investigation But a few specific choices for $A$ will indicate that the process $X_t$ may be quite deep and inaccessible to the usual methods of treatment. Let us introduce first of all, for introductory purposes, the following provisional definition.

**Definition 0.1** The process $X_t = B_{t \wedge A}$ on $(C, G_\infty, P^x)$ is called an arrested Brownian motion, with starting point $x$ and arresting time $A$. For any $S \in G_\infty$ with $P^x(S) > 0$, the process $X_t$ restricted to the conditional probability space $(S, (\mathcal{G}_\infty|S), (P^x|S))$ is called a conditional arrested Brownian motion, conditioned by $S$.

**Remark** It is easy to see that we can alternatively first condition by $S$ and then arrest by $A$ on $S$. The result is the same in either order.

To give a first example, let $x = 0$ and $A = |B(10)|$. (We are choosing 10 in place of 1 in the hope that it will be ten times clearer.) What can we say about $X_t$? The first basic principle in approaching the process is to consider the behaviour of $X_{t+s}$, $0 \leq s$, conditional on $\sigma(X_s, s \leq t)$. We denote this last $\sigma$-field by $\mathscr{F}_t$. If $\mathscr{F}_t$ is given, then so is $\{A < t\}$, and on this set $X_{t+s} = X_t$ for all $s$. This is an arrested Brownian motion with $A \equiv 0$. On $\{A \geq t\}$ we claim that $X_{t+s}$ has the probability law of a conditional arrested Brownian motion, provided that $P\{A \geq t\} \neq 0$ (as may be assumed). Indeed, since $\mathcal{G}_\infty = \sigma(B_s, s \leq t) \otimes \sigma(B_s, s > t)$, for fixed $w_0 \in \{A \geq t\}$, if we restrict $A$ to the section $\{w: w(s) = w_0(s), 0 \leq s \leq t\}$ then $A - t$ is $\sigma(B_s, s \geq t)$-measurable over the section, and if $\mathscr{F}_t$ is given, then $B_s \circ \theta_t$ has the law of a conditional Brownian motion for the conditional probabilities $(P^x|\mathscr{F}_t)$ over $\sigma(B_s, s \geq t)$, given $\{A \geq t\}$, and $A - t$ is an arresting time. The conditional process $X_{t+s}$ given $\mathscr{F}_t$, for $w \in \{A \geq t\}$, thus has the law of the a conditional arrested Brownian motion starting at $X_t$, with respect to the Brownian motion $B_s \circ \theta_t$ $(= B_{t+s})$, the arresting time $A - t$, and as conditioning set the section of $S_t = \{A \geq t\}$ at given $w(s)$, $s \leq t$. Its probabilities are determined by a regular conditional probability $(P^x|\mathscr{F}_t)$ over $\sigma(B_s, s \geq t)$.

Now none of these assertions depends on the particular choice $A = |B(10)|$. They are true for any arresting time $A$. More than this, we could begin with a conditional Brownian motion (given some set $S$ with $P(S) > 0$) in place of $B(t)$. It is easy to see that the future $X_{t+s}$ given $\mathscr{F}_t$ would again be a conditional arrested Brownian motion, where the conditioning set is now the section of $S \cap \{A \geq t\}$ at a given initial segment $(w(s), s \leq t)$. In short, the property of having the probability law of a conditional arrested Brownian motion propagates itself in time to the future under conditioning by $\mathscr{F}_t$.

This type of property will have a special place in our development. The reader is invited to observe that many other familiar properties of processes, including independent increments, the Markov property, the jointly Gaussian property, the martingale property, continuity of path, absence of accessible jumps, etc. also propagate in this way.

Returning to the case $A = |B(10)|$, for $t \geq 10$, either $A < t$ and the process is already arrested, or else $A \in \mathscr{F}_t$ is a given constant and the conditioning set is the whole space, so that $X_{t+s}$ is an ordinary Brownian motion stopped at a fixed time $A - t$. On the other hand, for $t < 10$, conditional on $A \geq t$ the conditioning set for the Brownian motion becomes $\{|B(10)| \geq t\}$, so that when $|B(t)| < 5$ (for example) the process has a drift away from 0 given $\{A \geq t\}$, and at time $t = 10$ we must have $|B(10)| \geq 10$. Evidently for $t < 10$ the arrested Brownian motion $X_t$ does not behave much like a true Brownian motion prior to $A$, but may exhibit strong drifts away from 0. Nevertheless, it is clear that, conditional on $\{A \geq t\}$, $X_t$ is Markovian in the sense that $I_{\{A \geq t\}}$ and $X_t$ alone determine the law of $X_{t+s}$. The determination is inhomogeneous in time, so that for $t \leq 10$ the pair $(I_{\{A \geq t\}}, X_t)$ is an inhomogeneous Markov process. This Markov property, however, is not present for arrested Brownian motion in general but happens because of our particular choice of $A$.

There is, nevertheless, a basic Markov property that does hold in general, and it is a homogeneous Markov property as well. Namely, we consider the new process whose state at time $t$ is the conditional probability law of $X_{t+s}$, $0 \leq s$, given $\mathscr{F}_t$. One can see from the above description that this law determines its own law of evolution homogeneously, independently of the past. Thus we may replace $X_t$ by a process whose states are the probability laws of conditional arrested Brownian motions, and this new process is a homogeneous Markov process. This observation is the basic clue to the method which we are going to pursue, and it turns out to be essentially unrestricted in applicability.

Let us see what may be the dividends in the particular case at hand, where $A = |B(10)|$. Let us denote by $Z_t$ this process of conditional futures. (The exact definitions are postponed until Chapter 1.) It is clear that $Z_t$ determines $X_t$, in fact $X_t = \hat{\rho}(Z_t)$ for a fixed function $\hat{\rho}$, hence we can study $X_t$ in terms of $Z_t$. But for purposes of applying or predicting $X_t$ it would often be $Z_t$ which would occupy our main attention in any case. There is a natural sense in which $Z_t$ has continuous paths. (It turns out to be a true diffusion process in our example.) However, for other choices of $A$, say if the distribution of $A$ has an atom at $t_0 > 0$ ($P\{A = t_0\} > 0$ and $P\{A > t_0\} > 0$), then, generally speaking, $Z_t$ will have a jump discontinuity at $t = t_0$ on $\{A \geq t_0\}$, registering the fact that the event $\{A = t_0\}$ is decided at time $t_0$. The discontinuities of $Z_t$ may also be 'totally inaccessible' (in the language of the general theory of processes), as illustrated by the case when $A$ is the starting time of the first excursion away from 0 of $B(t)$ having duration greater than 1. It is evident that, for general $A$, the Markov process $Z_t$ may exhibit a variety of

## 4 | Foundations of the prediction process

properties other than those observed directly from the process $X_t$, and these properties of $Z_t$ are of interest for the understanding and prediction of $X_t$.

Returning once more to the example $A = |B(10)|$, the principle difficulty in understanding $X_t$ is traceable to the fact that $(X_t, \mathscr{F}_t)$ is not a martingale. Indeed, it is not even obvious that it is a semimartingale, although this will follow below. Once we identify the Markov process $Z_t$, however, there are always many martingales available. For example, if $f(z)$ is any real-valued bounded measurable function on the space of futures $z$, and if

$$R_\lambda^Z f(z) = E^z \int_0^\infty e^{-\lambda t} f(Z_t)\, dt$$

is the resolvent of $Z_t$, then the prescription

$$M_\lambda^f(t) = R_\lambda^Z f(Z_t) - R_\lambda^Z f(Z_0) + \int_0^t f(Z_s) - \lambda R_\lambda^Z f(Z_s)\, ds \qquad (0.1)$$

always defines a martingale if $\lambda > 0$. One should think of the integrand as $-\mathscr{A} R_\lambda^Z f(Z_s)$, where $\mathscr{A}$ is the infinitesimal generator of the process $Z_t$. It is not quite as familiar, but true nevertheless, that if $\int_0^\infty e^{-\lambda t} EX^2(t)\, dt < \infty$ (as it is here) we may apply the same principle to $f = \hat\rho$ where $\hat\rho$ is the function such that $X_t = \hat\rho(Z_t)$ for all $t$. We write $M_\lambda(t)$ for $M_\lambda^{\hat\rho}(t)$ in this case, which will be of fundamental importance.

In simple examples such as the above, $M_\lambda(t)$ may be calculated explicitly. It suffices to consider the case $t \le 10$, because, for $t \ge 10$, $X_t$ is already a martingale and it is easy to show that $M_\lambda(t) - M_\lambda(10) = \lambda^{-1}(X(t) - X(10))$. According to the meaning of $Z_t$ as a conditional future, we have

$$R_\lambda^Z \hat\rho(Z_t) = \int_0^\infty e^{-\lambda s} E(X_{t+s}|\mathscr{F}_t)\, ds$$

$$= \int_0^{10-t} e^{-\lambda s} E(X_{t+s}|\mathscr{F}_t)\, ds + \frac{e^{-(10-t)\lambda}}{\lambda} E(X_{10}|\mathscr{F}_t).$$

Now clearly $R_\lambda^Z \hat\rho(Z_t) = R_\lambda^Z \hat\rho(Z_{t \wedge A})$, so we may assume $t \le A$. Then, according to the meaning of $X_{t+s}$ given $\mathscr{F}_t$, as an arrested Brownian motion starting at $X_t = B_t$ with arresting time $|B(10-t)| - t$ (for the Brownian motion $B(t+s)$), we have for $s \le 10 - t$ and $x = X_t$,

$$E(X_{t+s}|\mathscr{F}_t) = E^x(B(s \wedge (|B(10-t)| - t)|\,|B(10-t)| > t).$$

On the right we may condition by $|B(10-t)|$, which has probability density

$$p_x(y, 10-t) = (2\pi(10-t))^{-\frac{1}{2}}\left\{\exp -\frac{1}{2}\frac{(y-x)^2}{(10-t)} + \exp -\frac{1}{2}\frac{(y+x)^2}{(10-t)}\right\}. \quad (0.2)$$

When $y = |B(10-t)|$ is given, the path of $B(s)$ from $x$ to $y$ becomes a Brownian bridge, whose expectation at time $s$ is $x + s(y-x)/(10-t)$. Then the term $x$ may be brought out, and we get

$$E(X_{t+s}|\mathscr{F}_t) = x + (P^x\{|B(10-t)| > t\})^{-1}$$
$$\times \left\{ \int_{t<|y|<t+s} \left(\frac{y-t}{10-t}\right)(y-x)p_x(y, 10-t)\,dy \right.$$
$$\left. + \int_{t+s\leq |y|} \left(\frac{s}{10-t}\right)(y-x)p_x(y, 10-t)\,dy \right\}.$$

We can thus write down and simplify the expression (0.1) for $M_\lambda(t)$, but the details are a bit tedious and will be omitted. An important simplification occurs when we consider $\lim_{\lambda \to \infty} \lambda R_\lambda^Z \rho(Z_t) = X_t$. Hence we can expect to derive a semimartingale representation of $X_t$ if the integral term in $\lambda M_\lambda(t)$ converges pathwise and in $L^1$. It is not difficult to show that this is the case, and we are led to the following result.

**Result 0.1** *The process given by*

$$Y_t = X_t - \int_0^{t \wedge A \wedge 10} P^{X_u}\{|B(10-u)| > u\}^{-1}$$
$$\times \left\{ p_{X_u}(u, 10-u) + 2(2\pi(10-u))^{-\frac{1}{2}} \right.$$
$$\left. \times \exp\frac{-(u+X_u)^2}{2(10-u)}\left(\frac{u(u+X_u)}{10-u} - 1\right) \right\} du$$

*with $X_t = B_{t \wedge A}$, $A = |B(10)|$, and $p_x$ from (0.1), is a martingale for $(P^0, \mathscr{F}_t)$. Its law is that of a Brownian motion stopped at $A$ (i.e. $Y_t$ is Brownian relative to $\mathscr{F}_t$, stopped at $A$ which is a stopping time). Thus $X_t$ is an $\mathscr{F}_t$-semimartingale. The paths of $Y_t$ determine those of $X_t$ uniquely for $t \leq 10$.*

**Remark** A slightly simpler expression for $Y_t$ is obtained in terms of the density of $B$ absorbed at 0, instead of reflected as in $p_x$.

We do not suppose that we have hereby exhausted the study of $X_t$ even in this particular case, but it is clear that the above is a key step. More importantly, the methods we have used are not restricted either to the choice of $A$, or for that matter to arrested Brownian motion. They provide a natural means of investigation of any reasonable stochastic process.

Here, however, the means seem to outweigh the ends. Because of its generality, the theory transcends its application to examples. We will be content, therefore, to indicate one more example here, which the reader will have no trouble completing in all details. We will return to the above type of example, as well as to the theory behind it, in Section 6.3.

## 6 | Foundations of the prediction process

Let $X_t$ be a process consisting of a single jump at a random time $A$: $X_t = I_{[A, \infty)}(t)$. A theory of such examples is given in Knight (1981a, Essay II), but here, to be specific, let us suppose $A$ has density $x \exp(-x)$ for $0 < x$ (a gamma density, but not exponential). Then, for $t < A$, the conditional future $Z_t$ of $X_t$ may be regarded as the probability law of a process similar to $X_t$ but with $A$ replaced by a random variable $A_t$ having density

$$(x + t) \exp(-(x + t))\left\{\int_t^\infty x \exp(-x)\,dx\right\}^{-1} = \frac{x+t}{1+t} e^{-x}, \quad \text{for } 0 < x.$$

For $t \geq A$, $Z_t$ is the law of the constant process 1, so that $A_t \equiv \infty$. To express the martingales $M_\lambda(t)$ in this case, we need first

$$\lambda R_\lambda^Z \hat{\rho}(Z_t) = E \int_0^\infty e^{-\lambda s} I_{[A_t, \infty)}(s)\,ds$$

$$= \begin{cases} 1 & \text{for } t \geq A \\ \{(1+t)(1+\lambda)^2\}^{-1}\{(1+\lambda)t + 1\} & \text{for } t < A. \end{cases}$$

It follows by (0.1) and a routine calculation that

$$M_\lambda(t) = (1 + \lambda)^{-2}\{-t^2(1+t)^{-1} + \lambda(\ln(1+t) - t)\}, \quad \text{for } t < A.$$

Of course, for $t \geq A$, we have $M_\lambda(t) = M_\lambda(A-) + \Delta R_\lambda^Z \hat{\rho}(A)$.

The basis for our method of prediction is now to express $X_t - EX_t$ (where in this case $EX_t = 1 - (1+t)e^{-t}$) in terms of the martingales $M_\lambda(s)$, $s \leq t$, in such a way as to facilitate calculation of $E(X_t|\mathscr{F}_s)$, $s < t$. It is easiest to consider again the limit martingale

$$M_\infty(t) = \lim_{\lambda \to \infty} \lambda M_\lambda(t) = \begin{cases} \ln(1+t) - t & \text{for } t < A \\ \ln(1+A) - A + 1 & \text{for } t \geq A \end{cases}$$

(using the fact that $\lim_{\lambda \to \infty} \lambda \Delta R_\lambda^Z \hat{\rho}(A) = 1$, where $\Delta$ denotes the jump). We wish to determine a function $f_t(s)$ such that $X_t - EX_t = \int_0^t f_t(s)\,dM_\infty(s)$. Now for $t \geq A$ this becomes

$$(1+t)e^{-t} = -\int_0^A f_t(s) \frac{s}{1+s}\,ds + f_t(A)$$

and differentiating with respect to $A$ we easily obtain

$$f_t(s) = (1+t)(1+s)^{-1} \exp(s-t), \quad s \leq t.$$

Thus we have

$$E(X_t|\mathscr{F}_s) = \int_0^s f_t(u)\,dM_\infty(u) + EX_t, \quad s \leq t$$

and the prediction problem is solved in terms of the martingale $M_\infty$.

Introduction | 7

Having now completed our introduction to the prediction processes, we turn to introducing our general set-up for studying them. Let $(\Omega, \mathscr{F}, P)$ be a filtered probability space. Much of the general theory of processes is done in such a setting, but we generally need to assume a little more, as follows.

**Definition 0.2** $\mathscr{F}(=V_t\mathscr{F}_t)$ is said to be essentially countably generated if there is a sequence $(A_k; 1 \leq k < N + 1)$, $N \leq \infty$, $A_k \in \mathscr{F}$, such that $\mathscr{F}$ and $\sigma(A_k; k < N + 1)$ differ only by $P$-null sets. That is, if for $B \in \mathscr{F}$ there exists $A \in \sigma(A_k; k < N + 1)$ with $P(A \triangle B) = 0$.

Let us pause to set some notation. For any collection $S$ of sets or random variables on $(\Omega, \mathscr{F}, P)$, $\sigma(S)$ denotes the least $\sigma$-field with respect to which they are measurable, and $\bar{\sigma}$ denotes the completion of $\sigma$. As usual, $A \triangle B$ denotes the symmetric difference $(A \cap B^c) \cup (B \cap A^c)$.

For $\mathscr{F}$ to be essentially countably generated, it suffices that

$$\bar{\mathscr{F}} = \bar{\sigma}(A_k; k < N + 1), \qquad A_k \in \mathscr{F}.$$

Indeed, every $\bar{A} \in \bar{\sigma}$ satisfies $P(\bar{A} \triangle A) = 0$, for some $A \in \sigma$, by the monotone class argument. A condition equivalent to that of Definition 0.2 is that $L^2(\Omega, \mathscr{F}, P)$ be separable. Suppose $\mathscr{F}$ and $\sigma(A_k; k < \infty)$ differ only by $P$-null sets. Then, for

$$f \in L^2(\mathscr{F}), E(f|\sigma(A_1, \ldots, A_n)) \xrightarrow{L^2} E(f|\sigma(A_k; k < \infty)) = f \quad \text{a.s.}$$

(as usual, 'a.s.' means with probability 1) by martingale convergence, and $L^2(\sigma(A_1, \ldots, A_n))$ has finite dimension. Here the converse assertion is routine. Consequently, if $\mathscr{F}$ is essentially countably generated, so is each $\mathscr{F}_t$, since of course $L^2(\Omega, \mathscr{F}_t, P) \subset L^2(\Omega, \mathscr{F}, P)$.

Now something is drastically lacking from this initial set-up, namely, there are no 'futures'. Even if all $\mathscr{F}_t$ are known, one does not know how to go from $\mathscr{F}_{t_1}$ to $\mathscr{F}_{t_2}$, $t_1 < t_2$. What $\sigma$-field of additional events is to be added? There are many choices in general. We have to identify $\sigma$-fields $G_{(t_1, t_2]}$ such that $\mathscr{F}_{t_1} \vee G_{(t_1, t_2]} = \mathscr{F}_{t_2}$. In practice this is done by deciding what can be 'observed' between $t_1$ and $t_2$. We introduce a *process* (or processes) $X_t$ so that $\mathscr{F}_t = \sigma(X_s, s \leq t)$, $G_{(t_1, t_2]} \equiv \sigma(X_s, t_1 < s \leq t_2)$ where $\equiv$ denotes equality up to $P$-null sets. In fact, this can always be done for given $(\Omega, \mathscr{F}_t, P)$. We can take $(X_n)$ dense in $L^2(\Omega, \mathscr{F}, P)$, and set $M_n(t) = E(X_n|\mathscr{F}_t)$, $1 \leq n$. (We can even choose the martingales $M_n$ to be measurable processes.) Then

$$\mathscr{F}_t \equiv \sigma(M_n(s), s \leq t).$$

Indeed, for any $A \in \mathscr{F}_t$, choose $X_{n_k} \xrightarrow{L^2} I_A$. Then by Jensen's inequality

$$E(M_{n_k}(t) - E(I_A|\mathscr{F}_t))^2 \leq E(M_{n_k}(t) - I_{A_k})^2 \to 0,$$

and so we have $A \in \bar{\sigma}(M_n(s), s \leq t, 1 \leq n)$.

We can say more. Introduce the right-continuous (resp. left-continuous) filtrations

$$\mathscr{F}_{t+} = \bigcap_{\varepsilon > 0} \mathscr{F}_{t+\varepsilon} \quad \left(\mathscr{F}_{t-} = \bigvee_{\varepsilon > 0} \mathscr{F}_{t-\varepsilon}, t > 0\right).$$

## 8 | Foundations of the prediction process

Then (although $\mathscr{F}_{t+}$ is generally not countably generated) we have

$$\mathscr{F}_{t+} \equiv \sigma(M_n(s+), s \leq t), \qquad \mathscr{F}_{t-} \equiv \sigma(M_n(s-), s \leq t),$$

where $M_n(t\pm)$ are, respectively, the right-continuous with left limits (abbreviated r.c.l.l.) martingale, and the left-continuous with right limits (l.c.r.l.) martingale, defined $P$-a.s. for all $t$ by $M_n(t\pm) = \lim_{\tau \to t\pm} M_n(\tau)$, respectively, with $\tau \in \mathbb{Q}$. Therefore, if we use such $M_n(t+)$ we can apply Doob's representation theory (Doob 1953, I, Section 6) to transfer $(\Omega, \mathscr{F}_{t+}, P)$ to $(\Omega^\circ, \mathscr{F}_t^\circ, P^\circ)$, where $\Omega^\circ = \{(w_n(t)) \text{ r.c.l.l.}\}$, $\mathscr{F}_t^\circ = \sigma(w_n(s), s \leq t, 1 \leq n)$, and $P^\circ(S^\circ) = P\{M_n(\cdot +) \in S^\circ\}$. Every $S \in \mathscr{F}$ will have a counterpart $S^\circ \in \mathscr{F}^\circ (= V_t \mathscr{F}_t^\circ)$ for which $P^\circ(S^\circ) = P(S)$ is known. Thus $P^\circ$ suffices (in theory) to calculate probabilities in $\mathscr{F}_{t+}$.

However, we should ask from the beginning, 'Where do such $(\Omega, \mathscr{F}_t, P)$ come from?' Presumably it is generated by some given processes, and these are not necessarily martingales. Indeed, martingales are conditional expectations; they are computed, not observed, as a rule. It is more realistic and natural to be given only a process $X_t$ with Borel measurable paths to generate $\mathscr{F}_t$. Then we do not presume any topology on the state space. This added generality has some advantages, and it also leads to some difficulties. It does not seem to be known, for example, whether every $P$ on the coordinate $\sigma$-field of measurable paths is essentially countably generated. (Probably it is false, even when $X_t$ is real valued.) However, we shall evade this by not using the coordinate $\sigma$-field. Our replacement will be countably generated (not just essentially so for each $P$) which the coordinate $\sigma$-field clearly is not. Like the coordinate $\sigma$-field, it will be invariant under translations $\Theta_t w(s) = w(t+s)$ when we permit $-\infty < t < \infty$. The formal definition is postponed until Chapter 1 (Definition 1.1).

Here, let us bring up a question in the context of a general $(\mathscr{F}_{t+}, P)$ with $\mathscr{F}$ essentially countably generated. We have seen that it is possible to essentially generate $\mathscr{F}_{t+}$ by a sequence of r.c.l.l. martingales. A question which may be asked is whether it can also be generated in a useful way by a homogeneous r.c.l.l. strong Markov process, for some initial distribution. A consequence of our theory is that the answer is yes, but the strong Markov process is by no means unique.

So much for the preliminaries on processes. We turn next to some preliminaries on prediction. We will indicate that prediction, as it will be understood here, is more or less synonomous with conditional expectation. But it will involve a particular slant on conditional expectation of a rather more applied type than that of probabilistic potential theory, for example, or even of classical probability in the sense of the axiomatic (Kolmogorov) approach. We will extend from a well-known characterization of expectation in the case of an indicator random variable

$$I_A(w) = \begin{cases} 1 & w \in A \\ 0 & w \notin A \end{cases} \quad A \in \mathscr{F},$$

for which $EI_A = P(A)$ is just the probability itself.

## Introduction | 9

**Lemma 0.2** *$P(A)$ is the unique number $c$ which minimizes the $L^2$-norm*
$$E^{\frac{1}{2}}(c - I)^2.$$

**Proof** We have by familiar rules of expectation
$$\begin{aligned} E(c - I)^2 &= c^2 - 2cEI_A + EI_A \\ &= c^2 - 2cP(A) + P(A) \\ &= (c - P(A))^2 + P(A) - (P(A))^2, \end{aligned}$$
so the minimum is attained at $c = P(A)$ and the minimum is $(P(A)(1 - P(A)))^{\frac{1}{2}}$.

In general terms this tells us that we can interpret $P(A)$ as an *approximation* (or prediction) of the random variable $I_A$ (or event $A$). More specifically it is (a) the best approximation of $I_A$ within the class of all *constant* random variables, and (b) the best approximation in the sense of $L^2(\Omega, \mathscr{F}, P)$, i.e. in the Hilbert space norm. There is no ambiguity here involving sets of measure 0 because we limit ourselves to constants as probabilities; two constants that are equal a.s. are, of course, identical. In short, this suggests that we consider probabilities as random variables of a special kind, that is, as constant random variables, approximating more general ones, namely indicators.

Before going on, we pause to consider whether this 'explains' what we mean by $P(A)$. Suppose we had not observed the above. Clearly we might, even so, admit the usefulness of probability on other grounds. We mention two of these in passing.

1. *The law of large numbers, or 'frequency' explanation*. This brings to mind, or requires, other ideas. We need to consider a sequence of random variables satisfying some further conditions, such as independence or orthogonality. We will not pursue this interpretation in the present work, but it should not be overlooked that orthogonality means orthogonality in $L^2(\Omega, \mathscr{F}, P)$, so the $L^2$-norm makes its appearance again.

2. *The martingale, or 'betting', explanation*. In our context, this comes down to the assertion that if you want to place odds on the event $A$ the 'fair price' is $P(A) = EI_A$. In other words, the pair of random variables $0$, $I_A - P(A)$ is a martingale. (We begin with 0 because nothing can be won or lost prior to the game.) In this interpretation the essential thing is that $E(I_A - P(A)) = 0$. In contrast to Lemma 0.2, there is no appearance of the $L^2$-norm, but rather of the $L^1$-norm $EI_A$. However, if we replace $I_A$ by a more general random variable (say $X$) the condition becomes $E(X - EX) = 0$, so that it is not really a norm condition, but only the integral $EX$ is involved. This is a more primitive notion of probablity than the others, relying not as much on quantitative considerations as on striking a kind of balance between $X$ and $EX$.

# 10 | Foundations of the prediction process

We certainly do not intend to choose sides among the different interpretations. Indeed, they all lead to the same conclusion for our purposes, namely that one should use $P(A)$ to approximate (or predict) $I_A$ (or $A$) with a constant. But in line with the most applied meaning of prediction, we will consider the first alternative ('least squares') in somewhat more detail.

Suppose we have a closed subspace $H \subset L^2(\Omega, \mathscr{F}, P)$, and wish to approximate some $X \in L^2(\Omega, \mathscr{F}, P)$ by elements of $H$. Since we are assuming $\mathscr{F}$ is essentially countably generated, $H$ is separable, and so by a familiar procedure there are $\varphi_1, \varphi_2, \ldots, \varphi_n, \ldots \in H$, $E(\varphi_i \varphi_j) = \delta_{ij}$ which form an orthonormal basis of $H$. Now we write, for integer $N$, the $L^2$-approximation $X \equiv \sum_{n=1}^{N} E(X\varphi_n)\varphi_n + R_N$, and $E(\varphi_n R_N) = 0$, $1 \le n \le N$. Then we have

$$EX^2 = \sum_{n=1}^{N} E^2(X\varphi_n) + ER_N^2 \ge \sum_{n=1}^{N} E^2(X\varphi_n)$$

(Bessel's inequality) and it follows by the Riesz–Fisher theorem that there exists a unique $g \in H$ such that $g = L^2 - \lim_{N \to \infty} \sum_{n=1}^{N} E(X\varphi_n)\varphi_n$. Then, for every $n$, we have

$$E((X - g)\varphi_n) = E(X\varphi_n) - E(X\varphi_n) = 0,$$

and by a similar argument $E((X - g)h) = 0$ for any $h \in H$. Now for any $h \in H$ we can write

$$E(X - h)^2 = E(X - g + g - h)^2$$
$$= E(X - g)^2 + 2E((X - g)(g - h)) + E(g - h)^2.$$

But $g - h \in H$, so this becomes

$$E(X - h)^2 = E(X - g)^2 + E(g - h)^2.$$

Obviously this is minimized over $h \in H$ by setting $h = g$, so we have the following theorem.

**Theorem 0.3** *For any $X \in L^2(\Omega, \mathscr{F}, P)$, the best $L^2$-approximation to $X$ in $H$ is*

$$g = \sum_{n=1}^{\infty} E(X\varphi_n)\varphi_n,$$

*where the series converges in $L^2$. Moreover, the mean square error is*

$$E(X - g)^2 = EX^2 - \sum_{n=1}^{\infty} E^2(X\varphi_n),$$

*and $g$ is characterized by $E((X - g)h) = 0$, $h \in H$.*

**Notation 0.3** We write $g = \mathscr{E}(X|H)$, where $g$ is unique up to a $P$-null set if $\mathscr{F}$. We will not regard $g$ as an equivalence class unless stated.

Now let us specialize this to the case $H = L^2(\Omega, \mathscr{F}_1, P)$ where $\mathscr{F}_1$ is a subfield of $\mathscr{F}$ containing all P-null sets. Then $I_B \in H$ for $B \in \mathscr{F}_1$, and we have

$$0 = E((X - \mathscr{E}(X|H))I_B)$$
$$= \int_B X \, dP - \int_B \mathscr{E}(X|H) \, dP.$$

But we have $\mathscr{E}(X|H) \in H$, hence it is measurable over $\mathscr{F}_1$ (for any choice of representative). The next theorem follows by a familiar characterization of conditional expectation.

**Theorem 0.4** *If $H = L^2(\Omega, \mathscr{F}_1, P)$ is separable, and $\mathscr{F}_1$ contains all P-null sets in $\mathscr{F}$, then*

$$\mathscr{E}(X|H) \equiv E(X|\mathscr{F}_1), \quad \text{for } X \in L^2(\Omega, \mathscr{F}, P).$$

**Remarks**

1. If $\mathscr{F}_1$ does not contain all P-null sets, we can always augment it so as to contain them, without altering the structure of $H$ It follows that the theorem holds without this requirement when we restrict our choice of $\mathscr{E}(X|H)$ to a representative measurable over $\mathscr{F}_1$.

2. Separability of $H$ is also unnecessary, but as noted above it follows from our assumptions.

3. Since $L^2(\Omega, \mathscr{F}, P)$ is dense in $L^1(\Omega, \mathscr{F}, P)$ in the $L^1$-norm, and the conditional expectation operators $E(X|\mathscr{F}_1)$ are norm decreasing by Jensen's inequality

$$E|E(X|\mathscr{F}_1)| \leq E(E(|X| \, | \, \mathscr{F}_1)) = E|X|,$$

we can use the above construction to determine $E(X|\mathscr{F}_1)$ for all $X \in L^1(\Omega, \mathscr{F}, P)$.

Let us give an example for Theorems 0.3 and 0.4.

**Example 1** Let $(B_n, 2 \leq n < N + 1)$, $N \leq \infty$, be independent events of positive probability, and let $H$ be the Hilbert subspace generated by $\{1, I_{B_n}; 2 \leq n < N + 1\}$ and their finite linear combinations. For any $A \in \mathscr{F}$, let us find the best approximation of $I_A$ in $H$. An orthonormal basis of $H$ is given by $\varphi_1 = 1$, $\varphi_2 = (I_{B_2} - P(B_2))/\sqrt{P(B_2)(1 - P(B_2))}$, and in general $\varphi_n = (I_{B_n} - p_n)/\sqrt{p_n(1 - p_n)}$, where $p_n = P(B_n)$, $2 \leq n < N + 1$. Of course, this is just a sequence of independent Bernoulli random variables, normalized in the usual way, without assuming them to be identically distributed.

Now for $A \in \mathscr{F}$,

$$E(I_A \varphi_n) = \begin{cases} P(A) & n = 1 \\ \dfrac{P(A \cap B_n) - P(A)p_n}{\sqrt{p_n(1-p_n)}}; & n > 1 \end{cases}$$

and so

$$\mathscr{E}(I_A | H) = P(A) + \sum_{n=2}^{N} (P(A \cap B_n) - P(A)p_n) \frac{(I_{B_n} - p_n)}{p_n(1-p_n)}.$$

But $P(A \cap B_n)/p_n = P(A|B_n)$ and $(P(A \cap B_n) - P(A))/(1 - p_n) = -P(A|B^c)$, so the series terms become $P(A|B_n) - P(A)$ for $w \in B_n$, or $P(A|B_n^c) - P(A)$ for $w \in B_n^c$. Therefore we start with $P(A)$, and add terms which are non-negative if $P(A|B_n) \geq P(A)$ and $B_n$ occurs, or if $P(A|B_n^c) \geq P(A)$ and $B_n^c$ occurs, and negative otherwise. All of this is intuitively reasonable, and in line with what a prediction should be.

At the same time, one must carefully distinguish this $\mathscr{E}(I_A|H)$ from $E(I_A|\sigma\{B_n, 2 \leq b < N+1\})$ where, for any sets $B_\alpha$, we again use the notation $\sigma\{B_\alpha, \alpha \in S\}$ as the smallest $\sigma$-field containing all the $B_\alpha$'s. For example, if $N = 3$, then, on the set $B_2 \cap B_3$, we have

$$E(I_A | \sigma(B_2, B_3)) = \frac{P(A \cap B_2 \cap B_3)}{P(B_2)P(B_3)}$$

and

$$\mathscr{E}(I_A|H) = -P(A) + P(A|B_2) + P(A|B_3).$$

These are equal to $P(A)$ if $A$ is mutually independent of the $B_2, B_3$. But if, for example, $A = B_2 \cap B_3$, the first becomes 1 while the second is $P(B_2) + P(B_3) - P(B_2)P(B_3)$. Clearly this need not be 1.

Thus the conditional expectation is indeed closer to $I_A$ (on $A$) than is the projection on $H$. Of course, this is always true in the $L^2$-norm, since we are approximating over a larger Hilbert space, namely $L^2(\sigma(B_2, B_3))$, but on the other hand, our derivation of $\mathscr{E}(I_A|H)$ makes no essential use of mutual independence of $\{B_n; n < N+1\}$. It suffices that they be pairwise independent, so that the quantities $(\varphi_n)$ are orthogonal. Thus the smaller Hilbert space requires less knowledge about the joint distributions of the random variables in order to compute the approximation.

It is evident by this time (if it was not so already) that the $L^2$-norm is a very natural one to define approximations (or 'predictions'). But one should ask if it is in some sense uniquely determined by the properties that we want a method of approximation to have. Indeed, one way of prescribing it is by its relation to Gaussian processes. This is postponed until Chapter 3. Another type of determination will be discussed here. Given that we want to have the prediction reduce to $E(X|\mathscr{F}_1)$ whenever the random variable $X$ is, say, bounded, and the 'given' is a Banach subspace of functions measurable over

$\mathscr{F}_1 \subset \mathscr{F}$ and including all bounded random variables measurable over $\mathscr{F}_1$ (where we may assume that it does not distinguish functions equal P-a.s.), does this dictate the $L^2$-norm? It seems hard to answer this question completely, but a short digression here is worthwhile.

In the first place, the answer is 'No, $L^2$ is not unique if the whole probability space is too small'. For example, suppose $\mathscr{F} = \sigma\{A\}$, i.e. $\mathscr{F} = \{A, A^c, \Omega, \Phi\}$, where $P(A) > 0$ and $P(A^c) > 0$. Then the only $\mathscr{F}_1 \subset \mathscr{F}$ are $\mathscr{F}_1 = \mathscr{F}$ or $\mathscr{F}_1 = \{\Phi, \Omega\}$. Our requirement reduces to the sole condition that $\min \|X - c\|$ is achieved for $c = EX$. There are many norms other than $L^2$ which will achieve this. For example, let $p = P(A)$, and note that every random variable $X$ (measurable over $\mathscr{F}$) can be written uniquely

$$X = c_1 + c_2 \frac{(1-p)I_A - pI_{A^c}}{\sqrt{p^2 + (1-p)^2}}.$$

This is just the orthogonal decomposition in $L^2$ relative to the 'axes' of

$$\varphi_1 = 1 \quad \text{and} \quad \varphi_2 = \frac{(1-p)I_A - pI_{A^c}}{\sqrt{p^2 + (1-p)^2}}.$$

Now suppose we define our norm as

$$\|X\| = (|c_1|^{p'} + |c_2|^{p'})^{1/p'}, \quad \text{for } 1 \leq p' < \infty, p' \neq 2.$$

It is obvious that the unique $\min \|X - c\|$ is achieved at $c = EX$, just as for $L^2$, and also $\min \|X - c\varphi_2\|$ at $c_2$, so our approximations agree with those of $L^2$ with respect to the two subspaces generated by the two axes. On the other hand, they cannot agree with $L^2$ for all one-dimensional subspaces. This is because the norm is uniquely determined by its 'unit sphere': it is the space $L^2$ if and only if the unit sphere is a circle. Since we may characterize any point $\mathbf{p}$ of the unit sphere as the best approximation of the origin $O$ by points of the tangent line $l$ to the unit sphere at $\mathbf{p}$, by translation of $\mathbf{p}$ to $O$ this is the same as saying that the best approximation of $\mathbf{p}$ on the line $l_0$ through $O$ parallel to $l$ is given by $O$. As $l_0$ rotates, the association of $l_0$ and the direction to the corresponding $-\mathbf{p}$ determines the shape of the unit sphere (namely, it determines the slope of the tangent at a point of norm 1 in any direction). Thus it determines whether or not the unit sphere is a circle.

The same reasoning evidently extends to higher dimensions: we may assert that if, for every closed linear subspace $\mathscr{L} \subset L^2((\Omega, \mathscr{F}, P), \min\|X - Y\|)$ is achieved for the same $Y \in \mathscr{L}$ as it is in an $L^2$-norm (for every $X \in L^2$) then the norm is an $L^2$-norm. Unfortunately, this is much weaker than the problem we originally posed, which required agreement only for subspaces $\mathscr{L}$ of the particular kind corresponding to subfields $\mathscr{F}_1 \subset \mathscr{F}$.

Meanwhile, let us continue the digression in a slightly different direction. Suppose we do not restrict $\mathscr{L}$, but, instead of specifying the 'answers' (approximations) for all $X$ and $\mathscr{L}$, we merely require that they be linear in $X$ for every $\mathscr{L}$. Does this characterize the $L^2$-norm?

14 | Foundations of the prediction process

In two dimensions, the answer is still 'no'. Indeed, in our example above, all the approximations on lines $l_0$ do remain linear. But in (algebraic) dimension $n \geq 3$ the answer is a qualified 'yes'. We have, in fact, the following basic theorem.

**Theorem 0.5** *In a Banach space of dimension at least 3, if operators $\Omega_l(X)$ exist which associate to $X$ a nearest point to $X$ on a line $l$ through the origin and are linear in $X$, for every such line $l$, then the Banach space is a Hilbert space.*

This result follows immediately from a result of Blaschke (1916, p. 157), and another proof in greater generality is in Phillips (1940). Since we will not use the result, and since it is not directly related to probability, we do not write the proof. The reader who has understood this digression so far should be able to reconstruct the proof from Blaschke's result without difficulty. It suffices to show that the unit sphere of any three-dimensional subspace is an ellipsoid, and this is what the theorem of Blaschke does. The point to emphasize here is the significance of the result. Stated in plain terms, it means that if we want our approximation to be linear, we must use a Hilbert space norm. But an even more intuitively significant statement is the following.

**Corollary 0.6** *If, for every subspace (or even every linear subspace) the prediction errors of approximation of $X$ within the subspace are linear, then the Banach space (of dimension at least 3) is a Hilbert space.*

**Proof** Clearly, the errors are linear if and only if the approximations are linear.

# 1

# The measure-theoretic prediction process

*Την δε Τυχην βιοτοιο χυβερνητειραν εχοντες*
Chance, however, is the governess of life

Palladas, 5th Century A.D. Anthologia Palatina 10.65

Before commencing the 'hard mathematics,' we need to focus a little the principles of the Introduction onto the problem at hand. The path functions of our processes are not assumed to have any continuity properties, but are only assumed to be measurable for a prescribed $\sigma$-field on the state space. This fact leads to some technical difficulties which may make our approach seem a bit obscure at first sight. They are not essential, and can be entirely dispensed with if the given process is assumed to have, let us say, right-continuous paths with left limits. But there are sound reasons for not doing this. In our work, continuity properties follow for the prediction process, whether or not we assume them for the original (given, or physical) process.

In this way, there is a fundamental difference between our approach to pathwise continuity properties, and that of the prevalent literature. There is a lengthy literature, of which an early example is Kinney (1953) and a recent example is Chung (1982), of the use of martingale convergence to deduce continuity properties of the paths of a process (chiefly of Markov processes) from analytical assumptions on the data defining the processes (especially on the transition function). This is antithetical to our approach, both in its objective and in its methodology. Instead of introducing extra analytical assumptions, we concern ourselves with intrinsic continuity properties which hold as well for measurable processes as for any more restricted class. To start with continuity assumptions would, in the end, only confuse the picture.

In effect, we always deal with the two kinds (or levels) of processes. The given (or physical) process is only assumed to have measurable paths, because only the assignment of a $\sigma$-field on the state space is necessary in order to specify which events are taken to be observable. It does not presuppose any notion of distance or convergence in the state space. The superimposed prediction process describes a way of handling the given process, and we are largely concerned with the properties of this method of analysis. Therefore, the properties of the original process which we deal with are generally only those which are derived, at a kind of second level, from those of the prediction process. This latter is always endowed with pathwise

16 | Foundations of the prediction process

continuity properties of a natural (but not quite unique) kind. In the usual situations, where the given process also has such properties, it will be seen in Chapter 2 that they obtain because of a natural identification of the given process and its prediction process.

Until the necessary 'duality' (not in the sense of probabilistic potential theory, but rather in the objective–subjective sense) has been established, it is not possible to go into the details of the method. This fact requires of the reader (assumed to be unfamiliar with this approach) perhaps more than the usual amount of forbearance. One really cannot describe the outcome much ahead of time, because the key theorem (Theorem 1.14) is non-trivial both in its meaning and its proof. We have tried, on the other hand, to a give a prior sketch of this development at the appropriate place (preceding Theorem 1.11).

**Definition 1.1** Let $(E, \mathscr{E})$ be a measurable Lusin space (that is a Borel subset of a compact metrizable topological space), let $\Omega'_E = \{w(\cdot) \in \mathscr{B}^+/\mathscr{E}\}$ where $\mathscr{B}^+$ is the Borel field of $\mathbb{R}^+ = [0, \infty)$, and let

$$\mathscr{F}'_t = \sigma\left\{\int_0^s f(w(u))\,du, \quad s < t, f \in b(\mathscr{E})\right\},$$

(as usual, $b(\mathscr{F})$ denotes the set of bounded, $\mathscr{F}$-measurable functions).

We will work with triples $(\Omega'_E, \mathscr{F}'_t, P')$. In some contexts (that of stationary processes in Chapter 3, for example) it is important to permit $\infty < t \le \infty$. The definitions are entirely analogous, and the same notation will be used. We assume $0 \le t$ unless otherwise specified. Note that $\mathscr{F}'_\infty$ is countably generated. However the atoms are not the $w(\cdot)$, but rather their equivalence classes (see below) $w_1 \equiv w_2$ iff $w_1(t) = w_2(t)$ for Lebesgue a.e. $t$. These are called *pseudo-trajectories* in (Dellacherie and Meyer 1975, IV, Sections 40–46). But why does the role of Lebesgue measure $du$ in $\mathscr{F}'_t$ seem to be special? One reason is that if we want to define translation operators $\Theta_t$ the class of null $t$-sets must be translation invariant. Now it is a simple exercise (Exercise 1.1) to show that if a (non-trivial) Borel measure has translation-invariant null sets, the null sets must be the same as those of Lebesgue measure. Indeed, the measure must be equivalent to Lebesgue measure. If we used it in place of $du$, it is easy to see that $\mathscr{F}'_t$ would be the same.

**Exercise 1.1** Suppose $\mu(-\infty, \infty) < \infty$ and $\mu(dx)$ has translation invariant null sets. For $B \subset [0, 2\pi)$ define $\mu'(B) = \sum_k \mu(B + 2\pi k)$. Then $B$ is $\mu'$-null iff it is $\mu$-null. Now consider $\mu'(d\Theta)$ as a measure on the unit circle. Then for $0 < \Theta < 2\pi$ we have (by a slight abuse of notation)

$$\mu'(B + \Theta) = \mu'((B \cap [0, 2\pi - \Theta)) + \Theta) + \mu'(B \cap [2\pi - \Theta, 2\pi) + (\Theta - 2\pi)),$$

and it is seen that $\mu'$ has rotation-invariant null sets. So let us define $V(B) = \int_0^{2\pi} \mu'(B + \Theta) \, d\Theta$. Then the null sets of $V$ are the same as for $\mu'$, hence the same as for $\mu$ on $[0, 2\pi)$. But $V$ is rotation invariant. Hence $V$ is a constant times Lebesgue measure on the circle. It cannot be 0 since then $\mu$ would be 0. So its null sets are those of Lebesgue measure. Now generalize this to measures on $(\mathbb{R}^n, \mathscr{B}^n)$.

Now let us return again to the question of how we obtain a $P'$ on $(\Omega, \mathscr{F}'_t)$. The answer is that we need only have a *measurable* process $X_t$ with values in $(E, \mathscr{E})$. Necessary and sufficient condition for given $Y_t$ to have a measurable standard modification $X_t$ is well known; for example, either right- or left-continuity in probability suffices. (This requires the topology of $E$, but is much weaker than r.c.l.l. or l.c.r.l. paths.) Then we invoke representation theory by writing $P'(S') = P(X_{(\cdot)} \in S')$, $S' \in \mathscr{F}'_\infty$, thus obtaining our $(\Omega', \mathscr{F}'_t, P')$.

Here we already dropped the subscript $E$. This will not be harmful at all during the initial stages of our work, in view of the following result of Kuratowski. (For an elegant proof, see Dellacherie and Meyer (1975, Appendix to Chapter III)).

**Theorem 1.1** *All uncountable measurable Lusin spaces are isomorphic to* $([0, 1], \mathscr{B}[0, 1])$. *In other words, there exists a one-to-one, $\mathscr{E}/\mathscr{B}$-measurable mapping $\hat{\phi}$: $E$ onto $[0, 1]$, with $\mathscr{B}/\mathscr{E}$-measurable inverse.*

Because of this, as long as we do only measure-theoretic work we can and shall assume $([0, 1], \mathscr{B}[0, 1])$ as the state space. We let $(\Omega', \mathscr{F}'_t)$ (without subscripts) denote our canonical filtration for this case, and set

$$\mathscr{F}'_{t+} = \bigcap_{\varepsilon > 0} \mathscr{F}'_{t+\varepsilon}$$

($\bigvee_{\varepsilon > 0} \mathscr{F}'_{t-\varepsilon} = \mathscr{F}'_t$ here.) Also, note that our assertion $w_1 \equiv w_2$ iff $w_1 = w_2$ a.e. $t$ is quite obvious now since if $w_1 \neq w_2$ for positive $dt$-measure we cannot have $\int_0^t w_1(s) \, ds = \int_0^t w_2(s) \, ds$, $\forall t$. A remark is required on the uncountability assumption in the theorem. The case of finite or countable $E$ will be subsumed by regarding $E$ as a finite or countable subset of $[0, 1]$. Later it will be shown that in this case our prediction probabilities have probability 1 on this subset for all $t$, so nothing is lost. Another immediate advantage is that the necessary and sufficient condition for $Y_t$ to have a measurable standard modification, if $0 \leq Y_t \leq 1$ for all $t$, is simply that $Y_t$ be vector-valued-measurable (that is, a uniform limit of simple processes) as a process with values in $L^1(\Omega, \mathscr{F}, P)$ (Dellacherie and Meyer 1975, IV, Theorem 30). In particular, the $\sigma$-field $\mathscr{F}$ generated by the process $Y_t$ will be essentially countably generated.

Now we have seen how to obtain measures $P'$ on $\mathscr{F}'_\infty$, but a technical problem arises as to how to define a process $\tilde{X}$ to replace $X_t$. Note that the coordinates $w(t)$ on $\Omega'$ are, of course, not $\mathscr{F}'$-measurable ($\mathscr{F}' = \bigvee_t \mathscr{F}'_t$). Indeed, if we adjoined the coordinate $\sigma$-field, every atom of $\mathscr{F}'$ would become uncountably generated. It seems that we cannot give a meaning to $X_t$ for all $t$ in our set-up without further assumptions. Meanwhile, it is easy to give it a meaning for a.e. $t$. We introduce the following definition.

**Definition 1.2** Let $\Theta: \Omega' \to \Omega'$ be given by $\Theta_t w(s) = w(t+s)$, $0 \leq s, t$ (the translation operators). We have $\Theta_t \in \mathscr{F}'_s / \mathscr{F}'_{t+s}$ for all $0 \leq s, t$. Further, let

$$\tilde{\rho}(w) = \limsup_{n \to \infty} n \int_0^{1/n} w(s)\, ds,$$

and $\tilde{X}_t(w) = \tilde{\rho} \circ \Theta_t(w)$, $0 \leq t$. Similarly, let

$$\tilde{X}_t^-(w) = \limsup_{n \to \infty} n \int_{t-1/n}^t w(s)\, ds \qquad w(s) = 0, s < 0.$$

**Theorem 1.2**

(1) $\tilde{X}_s(w) \in \mathscr{B}[0, t) \times \mathscr{F}'_t$ on $[0, t) \times \Omega'$, and $\tilde{X}_s(w) \in \mathscr{B}[0, t] \times \mathscr{F}'_{t+}(w)$ on $[0, t] \times \Omega'$, for each $t$.
(2) For each $w$, $\tilde{X}_s(w) = w(s)$ for a.e. $s$. Thus for each $P'$,

$$\int_0^t \tilde{X}_s\, ds = \int_0^t w(s)\, ds$$

and $\sigma(\tilde{X}_s, s < t) \equiv \mathscr{F}'_t$.

**Proof**

(1) We have $n \int_0^{n^{-1}} w(s)\, ds \circ \Theta_t = n \int_t^{t+n^{-1}} w(s)\, ds$, continuous in $t$, and hence is in $\mathscr{B}[0, t] \times \mathscr{F}'_{t+1/n}$ on $[0, t] \times \Omega'$ (approximation by step functions of $t$). Then $\limsup_{n \to \infty} n \int_0^{n^{-1}} w(s)\, ds \circ \Theta_t \in \mathscr{B}[0, t] \times \mathscr{F}'_{t+1/n}$ for every $n$, which implies it is in $\mathscr{B}[0, t] \times \mathscr{F}'_t$ on $[0, t) \times \Omega'$. Finally, since

$$\limsup_{n \to \infty} \int_0^{n^{-1}} w(s)\, ds \circ \Theta_t \in \mathscr{F}'_{t+},$$

(1) follows easily.

(2) By a theorem of Lebesgue,

$$\tilde{X}_t = \frac{d}{dt} \int_0^t w(s)\, ds = w(t) \text{ for a.e. } t,$$

and so $\int_0^t w(s)\, ds = \int_0^t \tilde{X}s\, ds$. The final assertion follows directly from Theorem 2.8 of (Doob, 1953).

**Example 1** Let
$$w(s) = \begin{cases} 1 & 2^{-(2k+1)} < s \leq 2^{-2k} \\ 0 & 2^{-2k} < s \leq 2^{-2k+1}, \end{cases} \quad \text{all } k > 0.$$

Now
$$2^n \int_0^{2^{-n}} w(s)\, ds = \begin{cases} \tfrac{2}{3} & \text{if } n \text{ even} \\ \tfrac{1}{3} & \text{if } n \text{ odd}, \end{cases}$$

so there is no limit as $n \to \infty$. Even if we assume limits in probability it does not follow that lim sup can be replaced by lim. But if $w(t)$ is right-continuous, then $\tilde{X}_t(w) = w(t)$, $\forall t$.

**Remarks**

1. We could also have used $\tilde{X}_t^-$ in place of $\tilde{X}_t$ (setting $w(t) = 0$, $t < 0$). This would give even $(\mathscr{B}[0,t] \times \mathscr{F}'_t)$-measurability. Also, if $w(s)$ is left-continuous it would give $w(s)$. For brevity we choose to work mainly 'from the right'.

2. If we start with $(P, X_t)$ right-continuous in probability, we can obtain a replacement for $X_t$ on $(\Omega', \mathscr{F}', P')$ without extra assumptions as follows. For each $t$, let $n_k(t) \to \infty$ in such a way that $n_k(t) \int_t^{t+n_k^{-1}(t)} w(s)\, ds$ converges $P'$-a.s., and define $X'_t \in \mathscr{F}'_{t+}$ as (any) a.s. limit. Then clearly $X'_t$ has the same law for $P'$ as $X_t$ for $P$, and the same joint law with integrals $\int_0^t f(w_s)\, ds$ as (measurable) $X_t$ has with $\int_0^t f(X_s)\, ds$. Besides, this is again the same for any measurable standard modification of $X'_t$, say $X''_t$, if we use $\int_0^t f(X''_s)\, ds$. So, at least for fixed $P$, we obtain a valid replacement for the process $X_t$ on $(\Omega', \mathscr{F}'_{t+}, P')$. Unfortunately, it is not invariant under $\Theta_t$, which is a disadvantage in applications.

3. Finally, note that it is impossible to obtain a 'coordinate process' in general. Thus if
$$X_t = \begin{cases} 0 & \text{for all } t \neq t_0 \\ U & \text{for } t = t_0 \text{ (say, } U \text{ uniform on } (0,1)) \end{cases}$$

then clearly all paths of $X_t$ are equivalent to the identically 0 path. Thus the coordinate law cannot be obtained on $(\Omega', \mathscr{F}')$. Our view would be that such a process is usually unnatural since in reality an observation is not instantaneous in time.

The prediction process of a $(\Omega', \mathscr{F}'_t, P')$ will be a process whose values at time $t$ are probability measures on $\Theta_t^{-1}(\mathscr{F}'_\infty)$, which, of course, is isomorphic to $\mathscr{F}'_\infty$.

**Definition 1.3** Let $(M', \mathscr{M}')$ denote the space of all probability measures $z$ on $\mathscr{F}'_\infty$, with the $\sigma$-field generated by $z(S)$, $\forall S \in \mathscr{F}'_\infty$. We will write $P^z(S)$ for $z(S)$, and also $z(f)$ for $E^z f$, when convenient.

20 | Foundations of the prediction process

We note first that $\mathcal{M}'$ is generated by $z(S_n)$ where $S_n$ generate $\mathcal{F}'(=\mathcal{F}'_\infty)$. Hence $\mathcal{M}'$ is again countably generated.

We are going to provide $\Omega'$ and $M'$ with convenient topologies. This is largely an analytical device. A more natural topology will be introduced later on.

**Theorem 1.3** *Let* $d'(w_1, w_2) = \max_{t > 0} |\int_0^t e^{-s}(w_1(s) - w_2(s))\, ds|$. *Then* $d'$ *is a metric on (equivalence classes of)* $\Omega'$, *and with it* $\Omega'$ *is a compact metric space and* $\mathcal{F}'$ *are the Borel sets.*

**Proof** It is critical that the absolute values are *outside* the integral. For any $t$ we have

$$\left| \int_0^t e^{-s}(w_1(s) - w_2(s) + w_2(s) - w_3(s))\, ds \right|$$

$$\leq \left| \int_0^t e^{-s}(w_1(s) - w_2(s))\, ds \right| + \left| \int_0^t e^{-s}(w_2(s) - w_3(s))\, ds \right|,$$

from which we see that $d'$ satisfies the triangle inequality. Now if $d'(w_1, w_2) = 0$ then $\int_0^t e^{-s} w_1\, ds = \int_0^t e^{-s} w_2\, ds$, for all $t$, hence $w_1 = w_2$ a.e., so $d'$ is a metric on the equivalence classes (atoms) of $\mathcal{F}'$. Also $d' \leq 1$, so $\Omega'$ is bounded. Taking $w_1 \equiv 0$, the functions $\int_0^t e^{-s} w(s)\, ds$ are continuous in the metric $d'$ uniformly in $t$, and they generate $\mathcal{F}'$ since $\int_0^t \tilde{X}_s\, ds = \lim_{n \to \infty} \int_0^t e^s \{n \int_s^{s+n^{-1}} e^{-u} w(u)\, du\}\, ds$ by dominated a.e. convergence. On the other hand, balls in the metric $d'$ can be written as

$$\left\{ \sup_{t \in \mathbb{Q}} \left| \int_0^t e^{-s}(w(s) - w_0(s))\, ds \right| < \varepsilon \right\},$$

where $\mathbb{Q}$ denotes the (positive) rationals, and are clearly in $\mathcal{F}'$. It remains to show compactness. Now, by the Theorem of Ascoli and Arzela, any sequence $w_n$ has a subsequence (also denoted $w_n$) for which

$$\lim_{n \to \infty} \int_0^t e^{-s} w_n(s)\, ds = F(t)$$

exists for each $t$, and the convergence is uniform. Indeed since

$$0 \leq \int_{t_1}^{t_2} e^{-s} w_n(s)\, ds \leq e^{-t_1}(t_2 - t_1),$$

and the same is true for $F(t_2) - F(t_1)$. Hence $F$ is Lipshitz, and $0 \leq dF(t)/dt \leq e^{-t}$ for a.e. $t$. Then setting

$$w_\infty(t) = \begin{cases} e^t \dfrac{d}{dt} F(t) & \text{if } \exists\, \dfrac{d}{dt} \\ 0 & \text{elsewhere} \end{cases}$$

it is clear that $F(t) = \int_0^t e^{-s} w_\infty(s)\, ds$ and so $w_\infty$ defines the limit equivalence class. Indeed, convergence in $d'$ reduces to convergence at each *rational* $t$ of $\int_0^t e^{-s} w_n(s)\, ds$, and the tails are uniformly small.

**Corollary 1.4** *Let $M'$ be given the topology of weak-\* convergence w.r.t. $\Omega'$. Then $M'$ is a compact metric space, and $\mathcal{M}'$ are the Borel sets.*

**Proof** Let $f_n$ be dense in $\{f \in C(\Omega'),\ 0 \le f \le 1\}$. Convergence in $M'$ is metrized by $\sum_n |E^{z_1} f_n - E^{z_2} f_n| 2^{-n}$. If $z_k$ is a Cauchy sequence, then $\lim_{k \to \infty} E^{z_k} f_n$ defines a positive linear functional of norm less than or equal to 1 on $C(\Omega')$, hence by the Riesz representation theorem there exists a unique limit $z_\infty$. Clearly $z_\infty(1) = 1$, so it is a probability. By diagonal argument, any sequence $z_k$ has a Cauchy subsequence, so that $M'$ is compact.

Finally, since $E^z f_n$, $f_n \in C(\Omega')$ are continuous in $z$ and hence measurable, so are $E^z I_{S'} = P^z(S')$, $S' \in \mathcal{F}'$. So the metric Borel field is $\mathcal{M}'$.

The following result from general theory is not too well known, so we sketch the proof.

**Theorem 1.5** *Let $\mathcal{F}'_1$ and $\mathcal{F}'_2$ be σ-subfields of $\mathcal{F}'$. There exists a regular conditional probability $P^z(S|\mathcal{F}'_1)$, $S \in \mathcal{F}'_2$, which is $\mathcal{M}' \times \mathcal{F}'_1$-measurable in $(z, w)$, if and only if $\mathcal{F}'_1$ is countably generated.*

**Proof** The 'only if' is not needed here so we simply refer to Meyer and Yor (1976) for the proof. For the 'if', we can just as well replce $(\Omega', \mathcal{F}')$ by $([0, 1], \mathcal{B}[0, 1])$ by again using Theorem 1.1. Now let $B_1 \subset B_2 \subset \cdots$ be a sequence of *finite* subfields of $\mathcal{F}'_1$ ($\subset \mathcal{B}[0, 1]$), with $\mathcal{F}'_1 = V_n B_n$. Each $B_n$ is generated by a partition $B_{k,n}$ of $[0, 1]$. Set

$$P^z(S|B_n) = \begin{cases} P^z(S|B_{k,n}) & P^z(B_{k,n}) \ne 0,\ w \in B_{k,n} \\ z(S) & \text{otherwise.} \end{cases}$$

Clearly this is $(\mathcal{M}' \times B_n)$-measurable. By martingale convergence, $\lim_{n \to \infty} P^z(S|B_n)$ exists $P^z$—a.s. for each $z$, and defines a $P^z(S|\mathcal{F}'_1)$ for each $S$. Of course, the set of $(z, w)$ where the limit exists (fixed $S$) is in $\mathcal{M} \times \mathcal{F}'_1$. Now for rational $r \in [0, 1]$ set

$$P_0^z([0, r)|\mathcal{F}'_1) = \begin{cases} \lim_{n \to \infty} P^z([0, r)|B_n) & \text{if it exists for all } r \\ z([0, r]) & \text{elsewhere} \end{cases}$$

This is non-decreasing in $r$, and defines an $(\mathcal{M}' \times \mathcal{F}'_1)$-measurable conditional probability of $[0, r)$ as indicated. Now, for $0 \le x < 1$, set

$$P^z([0, x]|\mathcal{F}'_1) = \lim_{r \downarrow x} P_0^z([0, r)|\mathcal{F}'_1).$$

## 22 | Foundations of the prediction process

This has the desired properties for each $x$, and defines a distribution on $[0, 1]$ (using monotone convergence of conditional probabilities). Let $P^z(S|\mathscr{F}'_1)$ be the probability in $S \in \mathscr{B}[0, 1]$ having this distribution function. By analogous reasoning this satisfies the theorem for $\mathscr{F}'_2 = \mathscr{B}[0, 1]$, hence for any $\mathscr{F}'_2 \subset \mathscr{B}[0, 1]$ as asserted.

What we aim to do is to define $P^z(\Theta_t^{-1} S | \mathscr{F}'_{t+})$, $S \in \mathscr{F}'$, for all $(t, z, w)$ so as to have nice properties. But of course $\mathscr{F}'_{t+}$ is not separable, so it cannot be made $(\mathscr{M} \times \mathscr{F}'_{t+})$-measurable in $(z, w)$. We shall use supermartingale convergence.

**Lemma 1.6** *For* $z \in M'$ *and* $0 \leq h \in b(\mathscr{F}')$, *and* $\lambda > 0$,

$$e^{-\lambda t} E^z \left( \int_0^\infty e^{-\lambda s} h \circ \Theta_s \, ds \circ \Theta_t | \mathscr{F}'_t \right)$$

*is a $P^z$-supermartingale (for any choice of conditional probabilities).*

**Proof** We have only to write

$$E^z(e^{-\lambda(t_1+t_2)} \int_0^\infty e^{-\lambda s} f \circ \Theta_s \, ds \circ \Theta_{t_1+t_2} | \mathscr{F}'_{t_1})$$

$$= e^{-\lambda t_1} E^z \left( \int_0^\infty e^{-\lambda(s+t_2)} f \circ \Theta_{s+t_2} \, ds \circ \Theta_{t_1} | \mathscr{F}'_{t_1} \right)$$

$$\leq e^{-\lambda t_1} E^z \left( \int_0^\infty e^{-\lambda s} f \circ \Theta_s \, ds \circ \Theta_{t_1} | \mathscr{F}'_{t_1} \right),$$

which proves the assertion.

We now specialize our choice as follows.

**Notation 1.4** Let $0 \leq f_k \leq 1$ be uniformly dense in $\{f \in C(\Omega'), 0 \leq f \leq 1\}$, and let $W_r^z = z$ for $r < 0$, and, for $0 < r \in \mathbb{Q}$, let $W_r^z(S) = P^z(\Theta_r^{-1} S | \mathscr{F}'_r)$ be a regular $(\mathscr{M}' \times \mathscr{F}')$-measurable conditional probability over $S \in \mathscr{F}'$ (i.e. set $\mathscr{F}'_2 = \Theta_t^{-1} \mathscr{F}'$ in Theorem 1.5).

**Lemma 1.7** *For $z$ fixed, and $r_n \to t$, convergence of $W_{r_n}^z$ in the topology of $M'$ is equivalent to convergence of $W_{r_n}^z(\int_0^\infty e^{-\lambda s} f_k \circ \Theta_s \, ds)$ for all $k$ and $0 < \lambda \in \mathbb{Q}$.*

**Proof** Note first that for $\lambda_1 < \lambda_2$,

$$0 \leq \int_0^\infty e^{-\lambda_1 s} f_k \circ \Theta_s \, ds - \int_0^\infty e^{-\lambda_2 s} f_k \circ \Theta_s \, ds \leq \frac{1}{\lambda_1} - \frac{1}{\lambda_2},$$

so convergence for rational $\lambda > 0$ implies convergence for all $\lambda > 0$. Next, suppose that $f = \int_0^t w(s) \, ds (\leq t)$. Then $f \in C(\Omega')$ and $f \circ \Theta_s = \int_s^{s+t} w(u) \, du$ is uniformly continuous in $s$, uniformly in $w$. Now such $f$ separate $\Omega'$. Consider the algebra of functions of the form $g(f_1, f_2, \ldots, f_k)$, $1 \leq k$, where

The measure-theoretic prediction process | 23

each $f$ has the above form and $g(x_1, \ldots, x_k)$ is continuous on $\mathbb{R}^k$. Then $g(f_1, \ldots, f_k) \circ \Theta_s = g(f_1 \circ \Theta_s, \ldots, f_k \circ \Theta_s)$ is clearly uniformly continuous in $s$, uniformly on $\Omega'$, along with each $f \circ \Theta_s$ (since these are bounded). By the Stone–Weierstrass theorem these are uniformly dense in $C(\Omega')$, so the same continuity is true for every $f \in C(\Omega')$. For brevity, we now write $Pf$ for $E^P f$. Our hypothesis implies that $\int_0^\infty e^{-\lambda s} W_{r_n}^z f \circ \Theta_s \, ds$ converges for every $\lambda > 0$ and $f \in C(\Omega')$. By the continuity theorem for Laplace transforms, for each $f$ there is a limit $\mu_f(t)$ such that for a.e. $t$, $\int_0^t W_{r_n}^z f \circ \Theta_s \, ds \to \mu_f(t)$. The integrands on the left are uniformly equicontinuous in $s$. This implies $W_{r_n}^z f \circ \Theta_s \to d\mu_f(s)/ds$ uniformly in $s$, where $d\mu_f/ds$ has the same modulus of continuity. In particular (which is all we need) it implies that $\lim_{n \to \infty} W_{r_n}^z f$ exists for each $f \in C(\Omega')$.

**Remark** As pointed out by John Walsh, since $|\lambda \int_0^\infty e^{-\lambda s} f \circ \Theta_s \, ds - f| < \varepsilon$ for large $\lambda$, the continuity theorem can easily be avoided in the proof.

Conversely, since $\int_0^\infty e^{-\lambda s} f_k \circ \Theta_s \, ds \in C(\Omega')$ for each $\lambda$ and $f_k$, convergence of $W_{r_n}^z$ immediately implies that of the expectations $W_{r_n}^z$.

Now we are ready to construct, for given $P' = z$, a prediction process $Z_t^z$ of $z$. Note that we write $P^z$ for the probability $z$ itself ($P^z = z$). This redundancy seems intuitively helpful.

**Definition 1.5** Let

$$T_z = \sup\left\{t: \lim_{r \to s+, r \in \mathbb{Q}} W_r^z \text{ and } \lim_{r \to s-} W_r^z \text{ both exist for } 0 \leq s \leq t\right\}.$$

We define

$$Z_t^z = \begin{cases} \lim_{r \downarrow t} W_r^z & \text{for } t < T_z \\ z & \text{for } t \geq T_z. \end{cases}$$

**Theorem 1.8**

(1) $P^z\{T_z = \infty\} = 1$ and $\{T_z < t\} \in \mathcal{M} \times \mathcal{F}_t'$ for each $t$. In particular, for each $z$, $T_z$ is an $\mathcal{F}_{t+}'$-stopping time.

(2) $Z_t^z$ is right-continuous, with left limits $Z_{t-}^z$ except perhaps at $t = T_z < \infty$, and, for $\varepsilon > 0$, we have $Z_s^z(S) \in \mathcal{M}' \times B[0, t] \times \mathcal{F}_{t+\varepsilon}'$ on $M' \times [0, t] \times \Omega'$. In particular $Z_t^z \in \mathcal{F}_{t+}'$, and, since $\{Z_{t-}^z \text{ does not exist}\} \in \mathcal{F}_t'$, we have $Z_{t-}^z \in \mathcal{F}_t'$.

(3) We have for each $t$

$$P^z(\Theta_t^{-1} S | \mathcal{F}_{t+}') = Z_t^z(S), \quad S \in \mathcal{F}', \quad \text{for } t \geq 0,$$
$$P^z(\Theta_t^{-1} S | \mathcal{F}_t') = Z_{t-}^z(S), \quad S \in \mathcal{F}', \quad \text{for } t > 0.$$

## 24 | Foundations of the prediction process

**Remark** In the language of Dellacherie and Meyer (1975, IV, Section 6) $Z_t^z$ is $\mathscr{F}_{t+}^z$-optional, where $\mathscr{F}_{t+}^z$ is the usual $P^z$-augmentation of $\mathscr{F}_{t+}'$. As to $Z_{t-}^z$, it can be shown to be $\mathscr{F}_t'$-previsible (predictable) if we set $Z_{t-}^z = z$ when the left limit itself does not exist (Exercise 1.8). It is obviously $\mathscr{L}_t^z$-previsible, hence the average reader can go directly to the proof of Theorem 1.8.

**Exercise 1.8** (J. Walsh). Show that $Z_t^z$ is $\mathscr{F}_{t+}'$-optional, and that $Z_t^z$ is $\mathscr{F}_t'$ previsible.

For any $\varepsilon > 0$, let $T_0 = 0$ and inductively,

$$T_{n+1} = \inf\{t > T_n : |Z_t^z - Z_{T_n}^z| \geq \varepsilon\}.$$

Consider the process

$$Z_t^\varepsilon = \sum_n Z_{T_n \wedge T_z}^z I_{[T_n \wedge T_z, T_{n+1} \wedge T_z)}(t) + zI_{(t \geq T_z)}.$$

This is a sum of $\mathscr{F}_{t+}'$-optional processes, hence it is $\mathscr{F}_{t+}'$-optional. Since $\lim_{\varepsilon \to 0} \sup_t |Z_t^\varepsilon - Z_t^z| = 0$, $Z_t^z$ is also $\mathscr{F}_{t+}'$-optional. Now let $f_n$ be dense in $C(\Omega')$, and let $Y_n(t) = Z_t^z f_n$, which is $\mathscr{F}_{t+}'$-optional. Next set

$$\bar{Y}_n^m(t) = \sup_{t-(1/m) \leq r \in \mathbb{Q} < t} Y_n(r), \quad \text{and} \quad \underline{Y}_n^m(t) = \inf_{t-(1/m) \leq r < t} Y_n(r).$$

Now these are previsible: for example we can write

$$\{\bar{Y}_n^m > x\} = \bigcup_r \left(\{Y_n(r) > x\} \times \left\{r < t \leq r + \frac{1}{m}\right\}\right),$$

where the sets on the right are previsible.

Therefore the processes

$$\bar{Y}_n(t) = \lim_{m \to \infty} \bar{Y}_n^m(t) \quad \text{and} \quad \underline{Y}_n(t) = \lim_{m \to \infty} \underline{Y}_n^m(t).$$

are also previsible.

Now define

$$\hat{Y}_n(t) = \bar{Y}_n(t) I_{\cap_j (\bar{Y}_j(t) = \underline{Y}_j(t))} + E^z(f_n) I_{\cup_j (\bar{Y}_j(t) \neq \underline{Y}_j(t))}.$$

Note that these are all previsible, and moreover $\hat{Y}_n(t) = Y_n(t-)(=E^z(f_n))$ if $\not\exists Z_{t-}$) for all $t$. So it follows that $Z_{t-}^z$ is also previsible when we define $Z_{T_z-}^z = z$ on $\{\not\exists Z_{T_z-}^z\}$.

**Proof of Theorem 1.8** Since

$$E^z\left(\int_0^\infty e^{-\lambda s} f_n \circ \Theta_s \, ds \circ \Theta_r \Big| \mathscr{F}_r'\right) = W_r^z\left(\int_0^\infty e^{-\lambda s} f_n \circ \Theta_s\right) ds$$

for $r \in \mathbb{Q}$, and by Lemma 1.6 and Doob's supermartingale convergence theorem this has right and left limits at all $t$, $P^z$-a.s., Lemma 1.7 shows that $P^z\{T_z = \infty\} = 1$. Also the right side is in $\mathcal{M}' \times \mathcal{F}'_r$ for $r < t$, so the set where the number of upcrossings of some $(r_1, r_2)$ in $0 \le r < t$ is infinite is in $\mathcal{M}' \times \mathcal{F}'_t$, and so $\{T_z < t\} = \cup_{\varepsilon > 0}\{T_z < t - \varepsilon\} \in \mathcal{M} \times \mathcal{F}'_t$. Over $\{T_z \ge t\}$, the limit $Z^z_s$ is r.c.l.l. on $[0, t)$ in the topology of $\mathcal{M}'$, and this implies (since $Z^z_s \in \mathcal{M}' \times \mathcal{F}'_t$ for each $s < t$) that $Z^z_s \in \mathcal{M}' \times B[0, t) \times \mathcal{F}'_t$ on $M' \times [0, t) \times \Omega'_{t+\varepsilon}$. The rest of (2) follows easily.

Turning to (3), it follows by Hunt's lemma for conditional expectations (or, more easily, by uniform convergence) that

$$\lim_{\substack{r \downarrow t \\ (\text{resp. } r \uparrow t)}} W^z_r \int_0^\infty e^{-\lambda s} f_n \circ \Theta_s \, ds$$

$$= E^z\left(\int_0^\infty e^{-\lambda s} f_n \circ \Theta_s \, ds \circ \Theta_t \middle| \mathcal{F}'_{t+}\right) \quad (\text{resp. } |\mathcal{F}'_t)$$

$$= Z^z_t \int_0^\infty e^{-\lambda a} f_n \circ \Theta_s \, ds \quad (\text{resp. } Z^z_{t-})$$

$$= \int_0^\infty e^{-\lambda s}(Z^z_t f_n \circ \Theta_s) \, ds \quad (\text{resp. } Z^z_{t-}).$$

But since $E^z f_n \circ \Theta_s$ is uniformly continuous in $s$ (shown above) it follows that

$$E^z(f_n \circ \Theta_t | \mathcal{F}'_{t+}) = \lim_{\lambda \to \infty} E^z\left(\lambda \int_0^\infty e^{-\lambda s} f \circ \Theta_s \, ds \circ \Theta_t \middle| \mathcal{F}'_{t+}\right)$$

$$= \lim_{\lambda \to \infty} \lambda \int_0^\infty e^{-\lambda s} Z^z_t f_n \circ \Theta_s \, ds$$

$$= Z^z_t f_n \quad (\text{resp. } |\mathcal{F}'_t; Z^z_{t-}).$$

Now since $\{cf_n + d\}$ is dense in $C(\Omega')$, this extends easily to $I_S$, $S \in \mathcal{F}'$, as required.

We next extend the Markov properties of part (3) to corresponding strong Markov ones.

**Theorem 1.9**

(1) If $T$ is an $\mathcal{F}'_{t+}$-stopping time, $P^z(\Theta_T^{-1} S | \mathcal{F}'_{T+}) = Z^z_T(S)$ on $\{T < \infty\}$, $S \in \mathcal{F}'$.

(2) If $T^1 \le T_2 \le \cdots$ are $\mathcal{F}'_{t+}$-stopping times with $T_m < T = \lim_{n \to \infty} T_n$, $\forall m$, then $T$ is an $\mathcal{F}'_t$-stopping time and $P^z(\Theta_T^{-1} S | \mathcal{F}'_T) = Z^z_{T-}(S)$ on $\{T < \infty\}$, $S \in \mathcal{F}'$.

26 | Foundations of the prediction process

(3) *Setting, for convenience, $Z^z_{T-} = z$ when it does not exist as a limit, if $T > 0$ is any $\mathscr{F}'_t$-stopping time, then (as in (2)) $P^z(\Theta_T^{-1} S | \mathscr{F}'_T) = Z^z_{T-}(S)$, $S \in \mathscr{F}'$, on $\{T < \infty\}$.*

Indeed, if we also set $Z^z_{0-} = z$ and $\mathscr{F}'_0 = (\Phi, \Omega')$, then (3) holds even permitting $T = 0$.

(4) *If $T_1 \leq T_2 \leq \cdots$ as in (2), $T = \lim_{n \to \infty} T_n$, but only $P\{T_n < T\} = 1$, then $T$ is an $\mathscr{F}'_{t+}$-stopping time, and $P^z(\Theta_T^{-1} S | \mathscr{F}'_T) = Z^z_{T-}(S)$ on $\{T < \infty\}$, $S \in \mathscr{F}'$ (just as in (2)).*

**Proof** Since $\{T \leq t\} = \cap_m \{T_m < t\} \in \mathscr{F}'_t$, $T$ is an $\mathscr{F}'_t$-stopping time in (2). Now for (1) define the usual approximations $T_K = (k+1)2^{-K}$ on

$$\{k2^{-K} \leq T < (k+1)2^{-K}\},$$

so that $T < T_K \downarrow T$, and $T_K$ are $\mathscr{F}'_{t+}$-stopping times. Then, for $f \in C(\Omega')$, by Hunt's lemma,

$$E^z(f \circ \Theta_T^{-1} | \mathscr{F}'_{T+}) = \lim_{K \to \infty} E^z(f \circ \Theta_{T_K}^{-1} | \mathscr{F}'_{T_K+})$$

$$= \lim_{K \to \infty} Z^z_{T_K} f \quad \text{(since } T_K \text{ is countable valued)}$$

$$= Z^z_T f \quad \text{(since } Z^z_{T_K} \to Z^z_T \text{ in } M').$$

This now extends easily to $f = I_S$, proving (1). As to (2), Exercise 1.9 shows that $\mathscr{F}'_T = \bigvee_m \mathscr{F}'_{T_m+}$ for such $T$. Then, for $f \in C(\Omega')$,

$$E^z(f \circ \Theta_T | \mathscr{F}'_T) = \lim_{m \to \infty} E^z(f \circ \Theta_{T_m} | \mathscr{F}'_{T_m+}) = \lim_{m \to \infty} Z^z_{T_m} f = Z^z_{T-} f,$$

as needed.

**Exercise 1.9** Clearly, $\bigvee_m \mathscr{F}'_{T_m+} \subset \mathscr{F}'_T$. Conversely, we claim that

$$\mathscr{F}'_T \subset \left\{ S \in \mathscr{F}' \text{ such that } \int_0^s w_1(u)\, du = \int_0^s w_2(u)\, du,\ 0 < s \leq T(w_1), \right.$$

$$\left. \text{implies } w_1 \in S \Leftrightarrow w_2 \in S \text{ (i.e. } S \text{ is saturated for this equivalence)} \right\}$$

$$\subset \sigma\left\{ \int_0^{T \wedge s} w(u)\, du,\ 0 \leq s < \infty \right\}.$$

Indeed, for $S \in \mathscr{F}'_T$, we have easily $S \cap \{T = t\} \in \mathscr{F}'_t$, and each such set is clearly saturated for the asserted equivalence, hence so is $S$. Next, for any $w$ let $w'(u) = w(u)$, $u \leq T(w)$, and $w'(u) = 0$, $u > T(w)$. Then $w' \in \Omega'$ and $T(w') = T(w)$. It follows that, for $S \in \mathscr{F}'_T$, there is a Borel function $f(x_1, x_2, \ldots)$

and sequence $(t_n)$ for which

$$I_S = f\left(\int_0^{t_1} w(u)\,du, \ldots, \int_0^{t_n} w(u)\,du, \ldots\right)$$

$$= f\left(\int_0^{t_1} w'(u)\,du, \ldots, \int_0^{t_n} w'(u)\,du, \ldots\right)$$

$$= f\left(\int_0^{t_1 \wedge T} w(u)\,du, \ldots, \int_0^{t_n \wedge T} w(u)\,du, \ldots\right)$$

$$\in \sigma\left\{\int_0^{s \wedge T} w(u)\,du, 0 < s\right\}$$

as asserted. Now we have

$$\int_0^{s \wedge T} w(u)\,du = \lim_{m \to \infty} \int_0^{s \wedge T_m} w(u)\,du \in \bigvee_m \mathscr{F}'_{T_m},$$

so that $\mathscr{F}'_T \subset \bigvee_m \mathscr{F}'_{T_m} \subset \bigvee_m \mathscr{F}'_{T_m+}$ as required.

**Proof of Theorem 1.9** (continued) Since (3) is more involved than (2), and is really not essential, we will only give a sketch. Using Dellacherie and Meyer (1975, IV, Theorem 97) with $\int_0^t w(s)\,ds$ in the role of $w(t)$ (or more directly by Dellacherie and Meyer (1975), IV, 98 (d)) it follows that *every* $\mathscr{F}'_t$-stopping time is $\mathscr{F}'_t$-previsible. Therefore, by Dellacherie and Meyer (1975, IV, Theorem 56), $T$ is announcable with respect to the augmented filtration $\mathscr{F}^z_t$. Thus there are $T_1 \leq \cdots \leq T_n \leq \cdots < T$, $T_n \to T$, stopping times of $\mathscr{F}^z_{t+}$, and it follows as above that

$$P^z\left(\Theta_T^{-1} S \middle| \bigvee_n \mathscr{F}^z_{T_n}\right) = Z^z_{T-}(S).$$

Also, by Dellacherie and Meyer (1975, IV, Theorem 56) one has $\bigvee_n \mathscr{F}^z_{T_n} = \mathscr{F}^z_{T-}$. But an application of Blackwell's theorem, and Dellacherie and Meyer (1975, IV, 73 (c)) gives $\mathscr{F}'_{T-} = \mathscr{F}'_T$, so $\mathscr{F}^z_{T-} = \mathscr{F}^z_T$ and the result follows since $Z^z_{T-} \in \mathscr{F}'_T \subset \mathscr{F}^z_T$.

(4) Let $T'_n = T_n$ on $\{T_n < T\}$, and $T'_n = \infty$ elsewhere. Since $\{T_n < T\} \in \mathscr{F}_{T_n+}$, $T'_n$ is an $\mathscr{F}'_{t+}$-stopping time. Setting $T' = T$ on $\{T_n < T, \forall_n\}$, and $T' = \infty$ elsewhere, we have $T' = \lim T'_n$ and $T'_n < T'$ on $\{T' < \infty\}$. The proof of part (2) may now be applied to $T'_n$ and $T'$, which yields the result since $P\{T' = T\} = 1$.

Recalling the general definition of $(\Omega'_E, \mathscr{F}'_t)$, we can now state the following corollary.

**Corollary 1.10** Let $(M_E, \mathscr{M}_E)$ denote the space of probabilities on $(\Omega'_E, \mathscr{F}'_\infty)$ with $\mathscr{M}_E = \sigma(z(S'); S' \in \mathscr{F}')$ where $z$ ranges over $M_E$.

(1) *There is a $Z_t^z = Z_t^z(S, w')$ with values in $(M_E, \mathcal{M}_E)$ satisfying the same requirements as in Theorems 1.8 and 1.9 with subscripts E, except for the reference to a left-limit process $Z_{t-}^z$. Moreover, for each z these requirements determine $(Z_t^z, t \geq 0)$ uniquely up to a $P^z$-null set in $\mathcal{F}'_E$.*
(2) *There is a second process $Z_{t-}^z$ (not a left limit in any prescribed topology), such that $Z_{s-}^z \in \mathcal{M}'_E \in B[0, t] \times \mathcal{F}'_t$ on $M'_E \times [0, t] \times \Omega'$ for each t, and which satisfies the same requirements as in Theorems 1.8 and 1.9 concerning $Z_{t-}^z$. For each z, these requirements determine $(Z_{t-}^z, t > 0)$ uniquely up to a $P^z$-null set in $\mathcal{F}'_E$.*

**Proof** We consider first the case that $E$ is uncountable. By Theorem 1.1, we can then regard $(E, \mathcal{E})$ as $([0, 1], \mathcal{B}[0, 1])$ simply with a change of notation. This permits us to consider $z \in M'_E$ as a measure on $(\Omega', \mathcal{F}')$, and to construct the corresponding $Z_t^z$ and $Z_{t-}^z$, and finally to consider these as $M'_E$-valued processes. The requirements of Theorems 1.8 and 1.9 are immediately inherited. (Note that there even exists a topology on $E$, inherited from $[0, 1]$, so that $E$ becomes a compact metric space with Borel sets $\mathcal{E}$, and then we can view $Z_{t-}^z$ as a certain left limit except at $t = T_z < \infty$. But such a topology may be unrelated to a prescribed representation of $E$ as a Lusin space.)

The uniqueness assertions follow easily from the optional and previsible section theorems of Dellacherie and Meyer (1975, IV, 84, 85) and the corresponding parts of Theorem 1.9. Namely, as noted already, $Z_t^z$ is $\mathcal{F}'_{t+}$-optional and $Z_{t-}^z$ is $\mathcal{F}_t^z$-previsible (of course this simply means that the same is true of $f(Z_t^z)$ and $f(Z_{t-}^z)$ for every $f \in \mathcal{M}_E/\mathcal{B}$). Now Theorem 1.9 determines $Z_T^z$ (resp. $Z_{T-}^z$) at any optional $T \geq 0$ (resp. previsible $T > 0$), $P^z$-a.s., and by the section theorems this determines them up to the asserted $P^z$-null set. This argument involves, of course, capacity theory via the section theorems. We will later give another proof without any such prerequisite (Corollary 2.5).

It remains to consider $E$ countable (or finite). In this case, we map $E$ into a finite or countable subset of $[0, 1]$, say $\{x_n, 1 \leq n\}$. By hypothesis, then, we have, for all $t'$, $P'\{\int_0^{t'} f(w(s))\, ds = 0\} = 1$ provided that $f \in b(\mathcal{B}[0, 1])$ is such that $f(x_n) = 0$ for all $n$, where $P'$ is the measure induced on $(\Omega, \mathcal{F}')$ by our $\{x_n\}$-valued process. Then, by Theorem 1.9 we also have

$$Z_T^z\{\int_0^t f(w(s))\, ds = 0\} = 1,$$

$P^z$-a.s. for each optional $T < \infty$, and also $Z_{T-}^z\{\int_0^{t'} f(w(s))\, ds = 0\} = 1$, $P^z$-a.s. for each $\mathcal{F}_t^z$-previsible $0 < T < \infty$ (as usual, a.s. denotes 'with probability 1'). Now $Z_t^z\{\int_0^{t'} f(w(s))\, ds = 0\}$ is an optional process (resp. $Z_{t-}^z\{\int_0^{t'} f(w(s))\, ds = 0\}$ is $\mathcal{F}_t^z$-previsible), so by the section theorems these processes are 1 for all $t$, $P^z$-a.s. This means that $Z_t^z$ and $Z_{t-}^z$ can be interpreted, for all $t$, as measures on $\{w(s): w(s) \in \{x_n\}$ for all $s\}$. Indeed, we simply choose $f = 1 - I_{\{x_n\}}(x)$, and then $\int_0^\infty f(w(s))\, ds = 0$ means that $w(s) \in \{x_n\}$ for a.e. $s$, i.e. as an equivalence class $w(s) \in \Omega'_{\{x_n\}}$. Thus we can reinterpret $Z_t^z$ and $Z_{t-}^z$ as measures on $\{w(s): w(s) \in E$ for all $s\}$, as asserted by Corollary 1.10 in this case. (We may have to exclude a $P^z$-null set in $(\Omega'_E, \mathcal{F}'_E)$ depending on $z$, but then we can modify $Z_t^z$ after the stopping time $T = \inf\{t: Z_t^z \int_0^\infty f(w(s))\, ds \neq 0\}$, which, however, also depends on $z$. This type of difficulty disappears when we transfer to *prediction space* in Section 2.3.)

On the other hand, when dealing with a process that is known to have stronger properties than measurability, we can often reduce $\Omega'_E$ to a corresponding subset $S' \in \mathcal{F}'$ with $P(S') = 1$. This is illustrated by the following example.

**Example 2** Let $P^z$ be an ordinary Poisson process ($\lambda = 1$). We have easily $\{w(t) \in J$ (= integers), a.e. $t\}$ (= $S'$) $\in \mathcal{F}'$, with $P(S') = 1$. Now

$$\left\{ \int_{t_1}^{t_1+s} w(s')\, ds' \leq \int_{t_2}^{t_2+s} w(s')\, ds' < \infty, \forall t_1 < t_2, s > 0 \right\} \in \mathcal{F}'$$

since it is defined using only $t_1, t_2, s \in \mathbb{Q}$. Therefore $\{\tilde{X}_{t_1+s} \leq \tilde{X}_{t_2+s} < \infty$, a.e. $s\} \in \mathcal{F}'$, and on this set intersected with $S'$, we have

$$\tilde{X}_{t_1} = \lim_{n \to \infty} n \int_{t_1}^{t_1+1/n} \tilde{X}_s\, ds \leq \tilde{X}_{t_2}, \qquad \forall t_1 < t_2.$$

Further, $\{\tilde{X}_0 = 0\} \in \mathcal{F}'$, and on all these sets together, $\tilde{X}_t$ is r.c.l.l., nondecreasing, and integer valued. Finally, let

$$T_1 = \inf\{t: \tilde{X}_t > \tilde{X}_0\}, \qquad T_{n+1} = \inf\{t > T_n: \tilde{X}_t > \tilde{X}_{T_n}\}, \qquad \text{etc.}$$

Then

$$P\{\tilde{X}_{T_1} = 1, \tilde{X}_{T_2} = 2, \ldots, \tilde{X}_{T_n} = n, \ldots; \forall T_n < \infty\} = 1,$$

and defines the usual Poisson path space.

Now as to $T_1$, we have $T_1 \in \mathcal{F}'_{T_1+}$ and Theorem 1.14(a) applies. But

$$\{T_1 \leq t\} \notin \mathcal{F}'_t,$$

and so parts (b)–(d) are not applicable. Incidentally, in this particular case we have $\mathcal{F}'_{T_1-} \equiv \mathcal{F}'_{T_1+}$, because $T_1 \in \mathcal{F}'_{T_1-}$ holds by definition (Chung 1982, Section 1.3).

Although, generally speaking, we lose sight of the original topology of $E$ when we map $E$ onto $[0, 1]$ by $\hat{\varphi}$, a very important topological aspect is preserved by the times of discontinuity of $Z_t^z$. Namely, these are the times $(t, w)$ at which $Z_{t-}^z \neq Z_t^z$, and, by uniqueness of $Z_t^z$, this set is defined up to $P^z$-equivalence without dependence on $\hat{\varphi}$. Let us illustrate this by two more examples.

30 | Foundations of the prediction process

**Example 3** Let $B(t)$, $B(0) = 0$, be ordinary Brownian motion, that is, a Gaussian process with $EB(t) = 0$, variance $B(t) = t$, homogeneous, independent increments, and continuous paths. (These properties, of course, are somewhat redundant.)

We introduce the measurable process

$$X'_t = \begin{cases} B(t) & t \leq T(1) \\ B(t) + 1 & t > T(1), \end{cases}$$

where $T(1)$ is the first passage time of $B$ to $+1$. Of course, since $X'_t$ has r.c.l.l. paths, we may restrict $\Omega'$ to the corresponding subset. In this case,

$$T(1) = \lim_{n \to \infty} T(1 - 1/n)$$

is predictable, and $\mathscr{F}'_t = \sigma(\tilde{X}_s, s \leq t)$. Moreover, it is clear that

$$\mathscr{F}'_{T(1)} = \bigvee_n \mathscr{F}'_{T(1-1/n)},$$

whence we have

$$Z^z_{T(1)-}(\cdot) = E^z\left(\Theta^{-1}_{T(1)}(\cdot) \,\bigg|\, \bigvee_n \mathscr{F}'_{T(1-1/n)}\right)$$

$$= Z^z_{T(1)}(\cdot).$$

In other words, $Z^z_t$ is continuous, although $X'_t (\equiv \tilde{X}_t)$ is discontinuous at $t = T(1)$.

**Example 4** Let us return to the Poisson process of Example 2, and define $X'_t = \int_0^t P(s)\,ds$, where $P(s)$ denotes the process. Here, since $X'_t$ is continuous, we can restrict $\Omega'$ to all continuous paths and again have $X'_t \equiv \tilde{X}_t$ (in the sense of probability law on the continuous paths). Moreover $\{d^+\tilde{X}_t/dt^+$ exists for all $t$, and is r.c.l.l.$\}$ is defined in terms of the measurable process $\tilde{X}_t$ in such a way as to be $P^z$-analytic. Hence it is in $\mathscr{F}^z_t$, and we restrict further to this subset of probability 1. Then $T = \inf\{t : d^+\tilde{X}_t/dt^+ = 1\}$ is well defined, and has an exponential distribution with $\lambda = 1$. It is clear that $Z^z_t = z$ for $t < T$, and also that $Z^z_T \neq z$ since $d^+\tilde{X}_T/dt^+ = 1$. Hence $Z^z_t$ has a discontinuity at $T$, although $\tilde{X}_t$ is continuous.

Obviously, examples such as these exist in profusion and illustrate the point that an important aspect of continuity is expressed by $Z^z_t$ rather than by the original process. This will be developed further later in the book.

We turn now to studying the behaviour of the prediction processes $Z^z_t$ in measure-theoretic terms. For this we again may (and do) assume

$(\Omega'_E, \mathscr{F}'_E) = (\Omega', \mathscr{F}')$, i.e. $(E, \mathscr{E}) = ([0, 1], \mathscr{B}[0, 1])$. Let us introduce some notation.

**Definition 1.6** Let $q(t, z, A) = P^z\{Z^z_t \in A\}$, $t \geq 0$, $A \in \mathscr{M}$.

Note that for each $t$, $q(t, z, A)$ is $\mathscr{M}$-measurable in $z$. This follows by Theorem 1.8 (2), which implies $Z^z_t(\cdot) \in \mathscr{M} \times \mathscr{F}'_{t+\varepsilon}/\mathscr{M}$ for $t$ fixed. Then $I_A(Z^z_t) \in \mathscr{M} \times \mathscr{F}'/\mathscr{B}$, and so $q(t, z, A) = E^z(I_A(Z^z_t)) \in \mathscr{M}$. But since also $Z^z_t$ is right-continuous in $t$ (for a nice topology on $M$) it follows that $q(t, z, A) \in \mathscr{B} \times \mathscr{M}$ in $(t, z)$. Therefore, for each $t$, $q(t, z, A)$ is a (Borelian) Markov transition kernel on $(M, \mathscr{M})$, and it is measurable in $(t, z)$.

The above description of $q(t, z, A)$ in connection with $Z^z_t$ is very relevant. Indeed, the main theorem about the prediction processes is that, for given $(E, \mathscr{E})$, all the $Z^z_t$ are realizations of the *same* Markov process with transition function $q(t, z, A)$. Moreover, it is a strong Markov process with other regularity properties. This *intrinsic Markov property* in particular implies the identities

$$q(t_1 + t_2, z, A) = \int q(t_1, z, da)q(t_2, a, A) \tag{1.1}$$

of Chapman–Kolmogorov for $q$.

The proof of the strong Markov property requires a number of technicalities which are incidental to the main argument. Therefore, in order to break the ice, we shall start with a formal calculation of the simple Markov property which depends on the same idea but is free of technicalities. Letting $f \in b(\mathscr{F}'_{t_1})$, $g \in b(\mathscr{F}'_{t_2})$, and $A \in \mathscr{M}$, it is enough to show (subject to some details of measurability, and writing $w$ for $w'$) that we have

$$E^z(f(w)g(\Theta_{t_1}w)Z^{Z^z_{t_1}(w)}_{t_2}(\Theta_{t_1}w, A)) = E^z(f(w)g(\Theta_{t_1}w)Z^z_{t_1+t_2}(w, A)), \quad P^z\text{-a.s.} \tag{1.2}$$

Indeed, this implies that $Z^{Z^z_{t_1}}_{t_2}(w \circ \Theta_{t_1}) = Z^z_{t_1+t_2}(w)$, which gives the simple Markov property because we see that $Z^z_{t_1+t_2}$ depends on $\mathscr{F}'_{t_1}$ only through $Z^z_{t_1}$. Now the left side becomes

$$E^z(fE^z(g \circ \Theta_{t_1}Z^{Z^z_{t_1}(w)}_{t_2}(\Theta_{t_1}w, A)|\mathscr{F}'_{t_1})) = E^z(fE^{Z^z_{t_1}(w)}(gZ^{Z^z_{t_1}(w)}_{t_2}(A))), \tag{1.3}$$

by holding $Z^z_{t_1}$ fixed during the conditioning. The right side also becomes

$$E^z(fE^z(g \circ \Theta_{t_1}Z^z_{t_1+t_2}(A)|\mathscr{F}'_{t_1})) = E^z[fE^z(g \circ \Theta_{t_1}P^z(X_{t_1+t_2+\cdot} \in A|\mathscr{F}'_{t_1+t_2})|\mathscr{F}'_{t_1})]$$

$$= E^z(fE^z(g \circ \Theta_{t_1}I_{\{X_{t_1+t_2+\cdot} \in A\}}))$$

$$= E^z(fE^{Z^z_{t_1}}(gZ^{Z^z_{t_1}}_{t_2}(A))), \tag{1.4}$$

where the last equality follows by the definition of $Z^z_{t_1}$ and $Z^{Z^z_{t_1}}_{t_2}$ as conditional probabilities. Thus, in some sense, the Markov property of $Z^z_t$ reduces to an iteration of conditional expectations.

We turn now to the key theorem, due to Meyer and Yor (1976). It is elementary

32 | Foundations of the prediction process

and provides the simplest access to the theorem, due originally to Knight (1975).

**Theorem 1.11**

(1) *For $\mathscr{F}'_{t+}$-optional $T < \infty$, and any $z \in M$, $Z^z_{T+t}(\cdot, w) = Z^{Z^z_T(w)}_t(\cdot, \Theta_T w)$, $\forall t \geq 0$, $P^z$-a.s.*

(2) *For $\mathscr{F}'_t$-previsible $0 < T < \infty$ (we can as well assume $\exists T_n \uparrow T$, $T_n < T$, $P$-a.s. as in Theorem 1.9(4)) $Z^z_{T+t}(\cdot, w) = Z^{Z^z_{T-}(w)}_t(\cdot, \Theta_T w)$, $\forall t \geq 0$, $P$-a.s. (Here we detail the dependence on $w$, but use a dot for the dependence on $S$.)*

**Proof** Parts (1) and (2) are quite analogous so we prove (1) with parenthetical reference to the changes needed for (2). Certain preliminaries occupy our attention until Corollary 1.13. Then follows the main proof, where we incorporate a clarification due to John Walsh.

We first observe that it suffices to prove (1) and (2) for $t$ in any everywhere-dense set $\{s_k\}$ of strictly positive times, by right-continuity of $Z^z_t$ and $Z^{Z^z_T}_t$ in $t \geq 0$. Now it is easy to see from the construction of $Z^z_t$ (Definition 1.5) that $P^z\{Z^z_t = Z^z_{t-}$ except for countably many $t\} = 1$, and since $Z^z_{T+t}$ is $(\mathscr{B}^+ \times \mathscr{F}'/\mathscr{M}')$-measurable in $(t, w)$ (again by right-continuity) it follows by Fubini's theorem that there are dense $\{s_k > 0\}$ for which

$$P^z\{Z^z_{T+s_k-} = Z^z_{T+s_k}, \forall k\} = 1.$$

**Exercise 1.11** Show that, in fact, $P^z\{Z^z_{(T+t)-} = Z^z_{(T+t)}\} = 1$ except for countably many $t$.

For such $s_k$ we introduce, for $2^{-n} < s_k$,

$$T_n (= T_{n,k}) = \sup_{k'} \{k'2^{-n} : k'2^{-n} < T + s_k\}.$$

We claim that $T_n$ is an $\mathscr{F}'_t$-stopping time, and that $\mathscr{F}'_{(T+s_k)+} \equiv \bigvee_n \mathscr{F}'_{T_n}$, i.e. they differ only by $P^z$-null sets. Indeed, for each $k'$,

$$\{T_n = k'2^{-n} \leq t\} = \{k'2^{-n} < T + s_k \leq (k'+1)2^{-n}, k'2^{-n} \leq t\}$$
$$\in \mathscr{F}'_{((k'+1)2^{-n} - s_k)+} \subset \mathscr{F}'_t,$$

because $2^{-n} < s_k$ and $k'2^{-n} \leq t$. Also, of course, $T < T_n < T + s_k$ and

$$T_n \uparrow T + s_k$$

as $n \to \infty$, so that $\bigvee_n \mathscr{F}'_{T_n} \subset \mathscr{F}'_{(T+s_k)+}$. To show the reverse inclusion (up to

$P^z$-null sets) note that, by Theorem 1.9 and choice of $s_k$,

$$P^z(\Theta^{-1}_{T+s_k}S|\mathscr{F}'_{(T+s_k)+}) = P^z(\Theta^{-1}_{T+s_k}S|\bigvee_n \mathscr{F}'_{T_n})$$

for all $S \in \mathscr{F}'$.

Now for any $\lambda > 0$, we can write

$$E^z\left(\int_0^\infty e^{-\lambda s} w(s)\,ds\Big|\mathscr{F}'_{(T+s_k)+}\right) = \int_0^{t\wedge s_k} e^{-\lambda s} w(s)\,ds + E^z\left(e^{-\lambda(T+s_k)}\right.$$

$$\left.\int_0^\infty e^{-\lambda s} w(s)\,ds \circ \theta_{T+s_k}\Big|\mathscr{F}'_{(T+s_k)+}\right) = E^z\left(\int_0^\infty e^{-\lambda s} w(s)\,ds\Big|\bigvee_n \mathscr{F}'_{T_n}\right), \quad (1.5)$$

since $T \in \mathscr{F}'_{T_n}$. The same reasoning applies to any polynomial $f(y_1,\ldots,y_n)$ if $y_i = \int_0^\infty e^{-\lambda_i s} w(s)\,ds$, $1 \le i \le n$, by separating each integral into two parts and expanding the powers (all terms are bounded). Therefore, by the Stone–Weierstrass Theorem, (1.5) extends to any $g \in C(\Omega')$, and hence, by monotone class argument, to $b(\mathscr{F}')$. This implies that $\mathscr{F}'_{(T+s_k)+}$ and $\bigvee_n \mathscr{F}'_{T_n}$ differ only by $P^z$-null sets, since any $h \in b(\mathscr{F}'_{(T+s_k)+})$ satisfies

$$h = E^z(h|\bigvee_n \mathscr{F}'_{T_n}), \text{ $P$-a.s.}$$

Now the benefit of using $\mathscr{F}'_{T_n}$ in place of $\mathscr{F}'_{(T+s_k)+}$ is because of the following.

**Lemma 1.12** $\mathscr{F}'_{T_n} = \sigma\{\int_0^{T_n \wedge t} w(s)\,ds;\ 0 < t\}$.

(This avoids Galmarino's lemma used in the author's earlier proofs).

**Proof** Since $T_n$ is an $\mathscr{F}'_t$-stopping time it is trivial that

$$\{T_n = k2^{-n}\} \cap \left\{\int_0^{T_n \wedge t} w(s)\,ds \in B\right\} \in \mathscr{F}'_{k2^{-n}}$$

which implies

$$\sigma\left\{\int_0^{T_n \wedge t} w(s)\,ds,\ 0 < t\right\} \subset \mathscr{F}'_{T_n}.$$

Conversely, denoting $\mathscr{F}_n = \sigma\{\int_0^{T_n \wedge t} w(s)\,ds,\ 0 < t\}$, we first show by induction on $k \ge 1$ that $\{T_n = k2^{-n}\} \in \mathscr{F}_n$ for all $k$. Now if $k = 1$, since

$$\{T_n = 2^{-n}\} \in \mathscr{F}'_{2^{-n}},$$

there are rational $r_j < 2^{-n}$, and a Borel function $f_1(x_1, x_2, \ldots, x_j \ldots)$, such that

$$I_{\{T_n = 2^{-n}\}} = f_1\left(\int_0^{r_1} w(s)\,ds, \ldots, \int_0^{r_j} w(s)\,ds, \ldots\right)$$

(this is because such $\int_0^{r_j} w(s)\,ds$ generate $\mathscr{F}'_{2^{-n}}$). Since $T_n \geq 2^{-n}$, this is the same as $f_1(\int_0^{T_n \wedge r_1} w(s)\,ds, \ldots, \int_0^{T_n \wedge r_j} w(s)\,ds, \ldots)$, which is the indicator of a set in $\mathscr{F}_n$. Assuming that $\{T_n = j2^{-n}\} \in \mathscr{F}_n$, $1 \leq j \leq k$, consider

$$I_{\{T_n = (k+1)2^{-n}\}} = f_{k+1}\left(\int_0^{r_1} w(s)\,ds, \ldots, \int_0^{r_j} w(s)\,ds, \ldots\right)$$

where $r_j < (k+1)2^{-n}$ and $f_{k+1}(x_1, \ldots, x_j, \ldots)$ is Borel. Now we can also write

$$I_{\{T_n = (k+1)2^{-n}\}} = I_{\{T_n = (k+1)2^{-n}\}}\left(1 - \sum_{i=1}^{k} I_{\{T_n = i2^{-2}\}}\right)$$

where the last summands are already Borel functions of countably many $\int_0^{r_j} w(s)\,ds$, $r_j < i2^{-n}$. In this form, if we replace each $r_j$ by $T_n \wedge r_j$, the factors are all unchanged if $T_n \geq (k+1)2^{-n}$. On the other hand, if

$$T_n = i2^{-n} < (k+1)2^{-n},$$

then the factor on the right is $1 - I_{\{T_n = i2^{-n}\}} = 0$, which is unchanged if all $r_j$ are replaced by $T_n \wedge r_j$, so the entire expression equals 0 after the replacement, as it should. This completes the induction step.

Now since $T_n$ is $\mathscr{F}_n$-measurable, and we can write for $S \in \mathscr{F}'_{T_n}$,

$$S \cap \{T_n = k2^{-n}\} = f\left(\int_0^{r_1} w(s)\,ds, \ldots, \int_0^{r_j} w(s)\,ds, \ldots\right), \quad (r_j \leq k2^{-n})$$

$$= f\left(\int_0^{r_1 \wedge T_n} w(s)\,ds, \ldots, \int_0^{r_j \wedge T_n} w(s)\,ds, \ldots\right) I_{\{T_n = k2^{-n}\}}$$

we see that $S \cap \{T_n = k2^{-n}\} \in \mathscr{F}_n$ for each $k$, so $S \in \mathscr{F}_n$ as asserted.

The form in which Lemma 1.12 is to be applied is given in the following corollary.

### Corollary 1.13

(1) $\bigvee_n \mathscr{F}'_{T_n}$ is contained in the $\sigma$-field generated by all functions of the form $(g)(h \circ \Theta_T)$; $g \in b(\mathscr{F}'_{T+})$, $h \in b(\mathscr{F}'_{S_k})$.

(2) In case $0 < T < \infty$ is $\mathscr{F}'_t$-previsible, we may replace $g \in b(\mathscr{F}'_{T+})$ by $g \in b(\mathscr{F}'_{T-})$.

**Proof** By the lemma it suffices to show each $\sigma(\int_0^{T_n \wedge t} w(s)\,ds)$ is so contained. Writing

$$\int_0^{T_n \wedge t} w(s)\,ds = \int_0^{T \wedge t} w(s)\,ds + \int_0^{(T_n \wedge t) - (T \wedge t)} w(s) \circ \Theta_T\,ds,$$

the first term on the right is in $b(\mathscr{F}'_{T+})$ for (1) and in $b(\mathscr{F}'_{T-})$ for (2), so (take $h = 1$) we need only look at $\int_0^{(T_n \wedge t) - (T \wedge t)} w(s) \circ \Theta_T \, ds$. On $\{t < T\}$ it is 0, and $\{t < T\} \in \mathscr{F}'_{T+}$ (resp. $\{t < T\} \in \mathscr{F}'_{T-}$, case (2), since then of course $T$ is $\mathscr{F}'_{T-}$-measurable). Now we approximate $(T_n \wedge t) - (T \wedge t)$ by a sequence $\hat{T}_N = 0$ on $\{T < t\}$ and for $k \geq 0$, $\hat{T}_N = k 2^{-N}$ on

$$\{k 2^{-N} \leq (T_n \wedge t) - T < (k+1) 2^{-N}\}.$$

Then $\hat{T}_N < s_k$ and $\lim_{N \to \infty} \hat{T}_N = (T_n \wedge t) - (T \wedge t)$, and also, of course, $\{\hat{T}_N = k 2^{-N}\} \in \mathscr{F}'_{T+}$ (resp. $\mathscr{F}'_{T-}$, case (2)) so that

$$\int_0^{\hat{T}_N} w(s) \circ \Theta_T \, ds = \sum_k \left( \int_0^{k 2^{-N}} w(s) \, ds \circ \Theta_T \right) I_{\{\hat{T}_N = k 2^{-N}\}}$$

is in the required $\sigma$-field. To finish the proof we multiply by $I_{\{t < T\}}$ and let $N \to \infty$.

With these tools, we return to the proof of Theorem 1.11, recalling that it has become

$$Z^z_{T+s_k}(w) = Z^{Z^z_T}_{s_k}(\Theta_T w) \text{ for (1) (resp. } Z^z_{T-} \text{ for (2))}. \tag{1.6}$$

For brevity, we now write $s_k = t$, which is fixed, and, with $z$ also fixed, consider both sides of (1.6) as $(M, \mathscr{M})$-valued random variables. Let us check that both sides are $(\mathscr{F}'_{(T+t)+}/\mathscr{M})$-measurable. The left side is clearly so since $T + t$ is a stopping time. For the right side, by considering the usual discrete $T_n \downarrow T$ it follows easily that $\Phi_T$ is $(\mathscr{F}'_{T+t+\varepsilon}/\mathscr{F}'_{t+\varepsilon})$-measurable for $\varepsilon > 0$. Then, by Theorem 1.8(2) it follows that

$$Z^{(\cdot)}_t(\Theta_T(w)) \in \mathscr{M} \times \mathscr{F}'_{T+t+\varepsilon}.$$

Since $Z^z_T$ is $(\mathscr{F}'_{T+}/\mathscr{M})$-measurable ($z$ fixed), using a second variable $w' \in \Omega'$ it is clear that $Z^{Z^z_T(w')}_t \Theta_T(w)$ is $\mathscr{F}'_{T+} \times \mathscr{F}'_{T+t+\varepsilon}$-measurable in $(w', w)$. The mapping $w \to (w, w)$ being $(\mathscr{F}'_{T+t+\varepsilon}/(\mathscr{F}'_{T+} \times \mathscr{F}'_{T+t+\varepsilon}))$-measurable, it follows that the right side of (1.6) is also $\mathscr{F}'_{(T+t)+}$-measurable. In case (2) we need only replace $Z^z_T \in \mathscr{F}'_{T+}$ by $Z^z_{T-} \in \mathscr{F}'_{T-}$ in the last two steps.

Now for further brevity we write $z(f)$ for $E^z f$. Recalling that

$$\mathscr{F}'_{(T+t)+} \equiv \bigvee_n \mathscr{F}'_{T_n},$$

the proof of (1) will be complete if we show

$$0 = E^z(Y(Z^z_{T+t} f - Z^{Z^z_T}_t(f \circ \Theta_T))), \quad f \in C(\Omega'), \; Y \in b\left(\bigvee_n \mathscr{F}'_{T_n}\right), \tag{1.7}$$

for then we could take for $Y$ the factor after it to get a square (the same

## 36 | Foundations of the prediction process

holds for (2) using $Z^z_{T-}$). Then, by Corollary 1.13, it suffices to consider

$$Y = (g)(h \circ \Theta_T), \qquad g \in b(\mathcal{F}'_{T+}) \text{ (resp. } g \in b(\mathcal{F}'_{T-})\text{)}, \; h \in b(\mathcal{F}'_t),$$

since linear combinations of such form an algebra with monotone closure containing $b(\bigvee_n \mathcal{F}'_n)$, and (1.7) is linear in $Y$. For such a $Y$ we can write

$$\begin{aligned} E^z(Y Z^z_{T+t} f) &= E^z(g E^z(f \circ \Theta_{T+t} | \mathcal{F}'_{(T+t)+}) h \circ \Theta_T) \\ &= E^z(g E^z(h \circ \Theta_T \cdot f \circ \Theta_{T+t} | \mathcal{F}'_{T+})) \\ &= E^z(g Z^z_T(h \cdot f \circ \Theta_t)) \\ &= E^z(g Z^z_T(h Z^{Z^z_T}_t f)), \end{aligned} \quad (1.8)$$

where the last two $Z^z_T$'s are, of course, identical. The problem is to remove the first $Z^z_T$ in favour of a $\Theta_T$, keeping the second $Z^z_T$ in place. To see what the problem is here, let $w$ represent the argument of $Z^z_T$ and $w'$ that of $h$ and $Z_t f$, so that we can write $Z^{Z^z_T}_t f = Z^{Z^z_T(w)}_t(w', f)$. Then the last expression above becomes

$$\int g(w) \left[ \int Z^z_T(w, dw')(h(w') Z^{Z^z_T(w)}_t(w', f)) \right] z(dw). \quad (1.9)$$

What we need to show is that this equals

$$\int g(w) \left[ h \circ \Theta_T(w) \int Z^{Z^z_T(w)}_t(\Theta_T w, f) \right] z(dw).$$

This problem becomes easy when it is stated in more general terms. In place of the term $h(w') Z^{Z^z_T(w)}_t(w', f)$, let $K(w, w')$ be any bounded, $(\mathcal{F}'_{T+} \times \mathcal{F}')$-measurable function. We will show that

$$\int g(w) \int Z^z_T(w, dw') K(w, w') z(dw) = \int g(w) K(w, \Theta_T(w)) z(dw),$$

as needed. Indeed, for $K = K_1(w) K_2(w')$ this is just Theorem 1.9(1) and the definition of conditional expectation. However, both sides are linear and monotone in $K$, so by the monotone class theorem the proof is complete.

For the case of Theorem 1.11(2), the same argument applies with $Z^z_{T-}$ and only $g \in b(\mathcal{F}'_{T-})$, when we use Theorem 1.9(4).

We now have easily the main theorem.

**Theorem 1.14**

(1) (Strong Markov property) *For every $z$ and $\mathcal{F}'_{t+}$-stopping time $T < \infty$,*

$$P^z(Z^z_{T+t} \in A | \mathcal{F}'_{T+}) = q(t, Z^z_T, A), \qquad A \in \mathcal{M}, \; t \geq 0.$$

(2) (Moderate Markov property) *For $\mathscr{F}'_t$-stopping times $T > 0$ (automatically $\mathscr{F}'_t$-previsible, as in Theorem 1.9(3))*

$$P^z(Z^z_{T+t} \in A | \mathscr{F}'_T) = q(t, Z^z_{T-}, A), \quad A \in \mathscr{M}, t > 0.$$

**Remarks**

1. These become the usual Markov properties only after we identify $\mathscr{F}'_{t+} \equiv \sigma(Z^z_s, s \leq t)$ and $\mathscr{F}'_t \equiv \sigma(Z^z_s, s < t)$ below. Note further that for $t = 0$ and $T = c$ we get $P^z(Z^z_c \in A | \mathscr{F}'_c) = q(0, Z^z_{c-}, A)$. This is not just $I_A(Z^z_{c-})$. For example, a *discrete parameter process*, say

$$\tilde{X}_t = \begin{cases} 0 & t < 1 \\ \pm 2 & t \geq 1 \end{cases}$$

each with $p = \frac{1}{2}$, has $Z^z_{1-} \neq Z^z_1$ $P$-a.s.

2. To prove (2) we may as well assume $\exists T_n \uparrow T$, $T_n < T$, as in Theorem 1.9(4). The general case reduces to this after augmentation of $\mathscr{F}'_T$ to $\mathscr{F}^z_T$, and then modification of each $T_n$ on a $P^z$-null set. Again, the assumption $T > 0$ is only a convenience since, in general, one can replace 0 by $\infty$ on $\{T = 0\}$, then replace $T$ by $T \wedge N < \infty$.

**Proof** Now to prove (1), recalling Definition 1.6, it follows by Theorem 1.11(1), for $f \in b(\mathscr{M})$, that $E^z(f(Z^z_{T+t})|\mathscr{F}'_{T+}) = E^z(f(Z^{Z^z_T}_t) \circ \Theta_T | \mathscr{F}'_{T+})$ where $\Theta_T$ does not apply to the superscript $Z^z_T$. The same reasoning as for (1.9) above shows that this is $E^{Z^z_T}f(Z^{Z^z_T}_t) = \int_M f(m)q(t, Z^z_T, dm)$, as asserted. The proof of (2) goes the same way using Theorem 1.11(2).

**Remarks** It is important to appreciate the generality of Theorem 1.14. First, it was scarcely to be expected that the single transition function $q$ applies both to $Z^z_t$ and to $Z^z_{t-}$, or that it should be the same for every $z \in M$. Second, by application of Theorem 1.1 and Corollary 1.10, the entire theorem carries over immediately to any measurable Lusin space $(E, \mathscr{E})$. In particular, each has its 'universal' transition function $q_E(t, z, A)$ for both $Z^z_t$ and $Z^z_{t-}$ (as $(M_E, \mathscr{M}_E)$-valued prediction processes of $z$ on $\mathscr{F}'_E$). Of course, $q_E$ continues to be $(\mathscr{B} \times \mathscr{M}_E)$-measurable in $(t, z)$, and satisfies the Chapman–Kolmogorov identities (1.1). It is evident that countability of $E$ is no problem: we just restrict $q$ to the set of probabilities on $\Omega'_E = \{w(s) \in E, \text{a.e. } s\}$ as in Corollary 1.10, which is a Borel set of $\Omega'$ after mapping $E \leftrightarrow \{x_n\} \subset [0, 1]$.

We turn now to a result due to Meyer (1976b) whose proof uses the section theorem. A proof without section is in Corollary 2.53 below but it would require postponing the result which is better introduced now.

**Definition 1.7** The non-branch points $z \in M_E$ are $D = \{z: q(0, z, \{z\}) = 1\}$. If $q(0, z, \{z\}) \neq 1$, $z$ is called a branch point.

**Remark** It will be seen in a moment that $q(0, z, \{z\}) = 0$ or $1$ for all $z$, so that $z$ is a branch point iff $q(0, z, \{z\}) = 0$, but the following result is proved first.

**Theorem 1.15** $D \in \mathcal{M}$, and for every $z$, $P^z\{Z_t^z \in D \text{ for all } t \geq 0\} = 1$.

**Remark** This gives remarkable insight into how $Z_t^z(S)$ behaves for $S \in \mathcal{F}'_{0+}$ (which is not countably generated). Indeed, for $z \in D$ and $S \in \mathcal{F}'_{0+}$ we have $P^z(S|\mathcal{F}'_{0+}) = Z_0^z(S) = I_S$, $P^z$-a.s. while $Z_0^z = z$, $P^z$-a.s. by definition of $D$. Therefore $I_S = z(S)$, $P^z$-a.s., which implies that $z(S)$ is either $0$ or $1$. Now the theorem implies that

$$P^z\{Z_t^z(S) = 0 \text{ or } 1 \text{ for all } S \in \mathcal{F}'_{0+} \text{ and } t \geq 0\} = 1,$$

so that, in particular, the measures $Z_t^z$ on $\mathcal{F}'_{0+}$ are two-valued. They are also non-atomic except in very special cases. (In particular, if $z$ is derived from Brownian motion they are non-atomic.)

**Proof** It suffices, as usual, to take $E = [0, 1]$. Let $A_k^{(n)}$, $1 \leq k \leq n$, be a sequence of Borel partitions of $M'$ of maximum diameter $d_n \to 0$ (in some fixed metric for the compact space $M'$). Then

$$q(0, z, \{z\}) = \lim_{n \to \infty} \sum_{k=1}^n I_{\{z \in A_k^{(n)}\}} q(0, z, A_k^{(n)}),$$

so that $q(0, z, \{z\})$ is $\mathcal{M}$-measurable. It follows immediately that $q(0, Z_t^z, \{Z_t^z\})$ is $\mathcal{F}_{t+}^z$-optional. Now, for any $\mathcal{F}_{t+}^z$-optional $T < \infty$ we have, by the strong Markov property

$$P^z\{q(0, Z_T^z, \{Z_T^z\}) = 1\} = P^z\left\{\bigcap_n \left(\sum_k I_{\{Z_T^z \in A_k^{(n)}\}} q(0, Z_T^z, A_k^{(n)}) = 1\right)\right\}$$

$$= E^z\left(\sum_k I_{\{Z_T^z \in A_k^{(n)}\}} P^z(Z_{T+0}^z \in A_k^{(n)} | \mathcal{F}'_{T+})\right)$$

$$= 1.$$

Hence the optional section theorem implies the assertion (but the exceptional set need not be in $\mathcal{F}'$; it is in $\mathcal{F}^z$ for every $z$).

To prove the 0-1 law alluded to above, we must first connect $F'_{t+}$ and $\sigma\{Z_s^z, s \leq t\}$. At the same time we can improve on our previous coordinates $\tilde{X}_t$.

**Definition 1.8** Let $E = [0, 1]$. Then we set $E^z \tilde{X}_0 = \hat{\rho}(z)$.

**Theorem 1.16**

(1) We have $\hat{\rho}(z) \in \mathcal{M}/\mathcal{B}$, and for every $\mathcal{F}^z_{t+}$-optional $T < \infty$, we have $P^z\{\hat{\rho}(Z^z_T) = \tilde{X}_T\} = 1$, hence $P^z\{\hat{\rho}(Z^z_t) = \tilde{X}_t \text{ for a.e. } t\} = 1$.

(2) For each $z$, $\sigma\{Z^z_s, s \leq t\} \subset \mathcal{F}'_{t+}$ and they differ only by $P^z$-null sets.

**Remarks** It is important to note that (1) does *not* assert $P^z\{\hat{\rho}(Z^z_t) = \tilde{X}_t, 0 \leq t\} = 1$. Indeed, by (1) we see that $\hat{\rho}(Z^z_t)$ is $\mathcal{F}^z_{t+}$-optional, so this last would imply $\tilde{X}_t$ is also optional. We do not know if this is true. The same problem does not arise with the left coordinates $\tilde{X}_{t-}$, which are clearly previsible. Also, this problem with $\tilde{X}_t$ disappears on prediction space, to be introduced later. Note that neither $Z^z_t - Z^z_0$ nor $\hat{\rho}(Z^z_t) - \hat{\rho}(Z^z_0)$ is an additive functional on $\Omega$, i.e. $Z^z_{s+t} - Z^z_0 \neq (Z^z_t - Z^z_0) + (Z^z_s - Z^z_0) \circ \Theta_t$.

Assertion (2) completes the Markov property of Theorem 1.14(1) by identifying $\mathcal{F}'_{t+}$ with the natural filtration of $Z^z_t$ up to $P^z$-null sets. An anologous fact for 1.14(2) follows from this.

**Proof** Since $\tilde{X}_0 \in \mathcal{F}'_{0+} \subset \mathcal{F}'$, clearly $\hat{\rho}(z) \in \mathcal{M}/\mathcal{B}$. Therefore, since

$$Z^z_s \in \mathcal{M} \times \mathcal{B}[0, t] \times \mathcal{F}'_{t+\varepsilon}$$

on $M \times [0, t] \times \Omega'$, $\hat{\rho}(Z^z_s) \in \mathcal{M} \times \mathcal{B}[0, t] \times \mathcal{F}'_{t+\varepsilon}$. As to $\tilde{X}_T$, note that since $\Theta_T$ is $(\mathcal{F}'_t/\mathcal{F}'_{T+t})$-measurable it follows that $\tilde{X}_T (= \Theta_T \tilde{X}_0)$ is $\mathcal{F}'_{T+\varepsilon}$, hence $\mathcal{F}'_{T+}$-measurable. Therefore, we have, $P^z$-a.s.,

$$E^z(\tilde{X}(T)|\mathcal{F}'_{T+}) = \tilde{X}(T) = E^{Z^z_T}(\tilde{X}_0) = \hat{\rho}(Z^z_T).$$

Turning to (2), it follows by Fubini's Theorem that

$$P^z\left\{\int_0^t \hat{\rho}(Z^z_s) \, ds = \int_0^t \tilde{X}_s \, ds = \int_0^t w(s) \, ds, \forall t\right\} = 1.$$

Now for $h \in C(M')$, clearly $\int_0^t h(Z^z_s) \, ds \in \sigma(Z^z_s, s < t)$, and by the monotone class theorem this extends to $\hat{\rho} \in b(\mathcal{M}')$. Hence $\sigma(Z^z_s, s < t)$ contains $\mathcal{F}'_t$ up to $P^z$-null sets. As to $\mathcal{F}'_{t+}$, since clearly $\mathcal{F}'_{t+} \subset \cap_{\varepsilon>0} \sigma(Z^z_s, s \leq t + \varepsilon)$ up to $P^z$-null sets, it suffices to show that $\cap_{\varepsilon>0} \sigma(Z^z_s, s < t + \varepsilon)$ is contained in $\sigma(Z^z_s, s \leq t)$ up to $P^z$-null sets. Now this is a form of the Blumenthal 0-1 law for the strong Markov process $Z^z_t$. Indeed, the following is classical.

**Exercise 1.16** Show that $\cap_{\varepsilon>0} \sigma(Z^z_s, s \leq t + \varepsilon) \subset \sigma(Z^z_s, s \leq t)$ up to $P^z$-null sets. Hint: it suffices to show that for $f = \prod_{k=1}^n f_k(Z^z_{t_k})$, $f_k \in b(\mathcal{M})$, $t_1 < t_2 < \cdots < t_n$, we have $E^z(f|\mathcal{F}'_{t+}) = E^z(f|\sigma(Z^z_s, s \leq t))$.

**Corollary 1.17** $q(0, z, \{z\}) = 0$ or $1$ according as $z \in D^c$ or $z \in D$.

**Proof**

$$P^z\{Z_0^z = z\} = P^z\{Z_0^z = z \text{ and } Z_{0+0}^z = z\}$$
$$= P^z\{Z_{0+0}^z = z | Z_0^z = z\} P^z\{Z_0^z = z\}$$
$$= q(0, z, \{z\}) q(0, z, \{z\})$$

using Theorem 1.16(2). So $q(0, z, \{z\}) = q^2(0, z, \{z\}) = 0$ or $1$.

**Remarks** These results transfer immediately to $q_E(0, z, \{z\}) = 0$ or 1, for general $E$, as does also (2) of Theorem 1.16. We can also transfer (1) if we define $\hat{\rho}_E(z)$ via the image measure of $z$ under $E \leftrightarrow [0, 1]$, but we shall not do this because then $\hat{\rho}_E$ would lose relation to the prescribed topology of $E$. Note that on $[0, 1]$, if $w$ is r.c.l.l. ($P^z$-a.s.) then so is $\tilde{X}_t$, and hence so is $\hat{\rho}(Z_t)$. We want a similar result to hold on $E$, but this requires a more topological approach. However, for real-valued processes the present $\hat{\rho}$ suffices since $[0, 1]$ is homeomorphic to $[-\infty, \infty]$. Taking $\hat{\varphi} = (1/\pi)\tan^{-1} + \frac{1}{2}$, we can use $\tan(\pi(\hat{\rho} - \frac{1}{2}))$ as our coordinate.

# 2

# Topological considerations and prediction space: application to Markov processes

Mit allen augen sieht die Kreatur das Offene. Nur unsre Augen sind wie umgekehrt und ganz um sie gestellt

R. M. Rilke, 1922. Eighth Duino Elegy

## 2.1  The prediction topology

In the case of a general Lusin state space $E$, it has been emphasized that the processes $Z_t^z$ and $Z_{t-}^z$ are not related to each other through the prescribed Lusin topology of $E$. In fact, since our Kuratowski map $\hat{\varphi}: E \leftrightarrow [0,1]$ is by no means unique, it is far from obvious that the definitions of $Z_{t\pm}^z$ do not depend on the choice of the map. It was only by use of section theorems for Corollary 1.10 that we obtained uniqueness for all $t$ up to a $P^z$-null set, so in fact we know that $Z_{t\pm}^z$ are essentially independent of the map. But to understand this better, and also to free it from section theorems, we wish to introduce a topology for $Z_{t\pm}^z$ which is dictated by the prescribed topology of $E$, as a Borel subset of a compact metric space $\bar{E}$. In this topology we will have $Z_{t-}^z = \lim_{s\uparrow t} Z_s^z$, except at $t = T_z$ with $P^z\{T_z = \infty\} = 1$ (just as in the case $E = [0,1]$, except that even in this case we will have a different topology on $(M, \mathcal{M})$, and so a slightly different $T_z$ as well).

The following results transfer immediately to general $E$. (It does not involve a topology for $E$.)

**Theorem 2.1**  *For every $f \in b(\mathcal{F}')$ and $\lambda > 0$, $E^{Z_t^z} \int_0^\infty e^{-\lambda s} f \circ \Theta_s \, ds$ is r.c.l.l., $P^z$-a.s.*

**Remarks**  We know from the proof of Theorem 1.8 that this is true for $f \in C(\Omega')$. Here, however, the exceptional $P^z$-null set depends both on $z$ and on $f \in b(\mathcal{F}')$. It cannot be reduced to depend only on $z$, as in the case $f \in C(\Omega')$. Note further that we can write

$$E^z \int_0^\infty e^{-\lambda s} f \circ \Theta_s \, ds = E^z \int_0^\infty e^{-\lambda s} E^z(f \circ \Theta_s | \mathscr{F}'_{s+}) \, ds$$

$$= E^z \int_0^\infty e^{-\lambda s} E^{Z^z_s} f \, ds$$

$$= \int_0^\infty e^{-\lambda s} \int_M q(s, z, dm) E^m f \, ds.$$

Introducing the notation

$$R^z_\lambda h(z) = \int_0^\infty e^{-\lambda t} \int_M q(t, z, dm) h(m) \, dt, \qquad \lambda > 0, \tag{2.1}$$

for the *resolvent* of $Z^z_t$, for $h \in b(\mathcal{M})$, Theorem 2.1 says that, for $h = E^z f$, $R^z_\lambda h(Z^z_t)$ is r.c.l.l., $P^z$-a.s. This is a special case of a general result about *right processes*; see, for example, Getoor (1975, Theorem (9.4)). The general result implies that $g(Z^z_t)$ is r.c.l.l. for any excessive (or $\lambda$-excessive) function $g$ (for $q$). This is proved using capacity theory, but again an elementary direct proof can also be given. We return later (Corollary 2.20) to this point.

A key fact (not essential for Theorem 2.1, but needed later for Corollary 2.27 and Corollary 2.28) is

**Lemma 2.2** (Meyer 1966) *Let $0 \leq x_1(t) \leq \cdots \leq x_n(t) \leq \cdots$ be a uniformly bounded sequence of supermartingales on a $(\Omega, \mathscr{F}_t, P)$. If each $x_n(t)$ is r.c.l.l., then so is the limit $x(t) = \lim_{n \to \infty} x_n(t)$.*

**Proof** This result is due to P.-A. Meyer 1966, VI, T16 (using capacity theory) but we give the elementary proof from Doob (1984). Since right-continuous $x_n(t)$ is also a supermartingale with respect to $\mathscr{F}_{t+}$, we may assume without loss of generality that $\mathscr{F}_t = \mathscr{F}_{t+}$. Now

$$E(x(t_2) | \mathscr{F}_{t_1}) = \lim_{n \to \infty} E(x_n(t_n) | \mathscr{F}_{t_1}) \leq \lim_{n \to \infty} x_n(t_1) = x(t_1),$$

so $x(t)$ is indeed a supermartingale. It remains to show that it is r.c.l.l. This is not difficult once we show: for each $\varepsilon > 0$ and for each finite stopping time $T$ there is a stopping time $T'$ with $P\{T' > T\} = 1$ such that $|x(t) - x(T)| \leq \varepsilon/2$ for $T \leq t < T'$. To show this, by applying the reasoning to the supermartingale $(x(T + t), \mathscr{F}_{T+t})$, it suffices to assume $T = 0$. (Here we observe that if $S$ is an $\mathscr{F}_{T+t}$-stopping time then $T' = T + S$ will be an $\mathscr{F}_t$-stopping time.)

Now, with $T = 0$, let $S = \limsup_{n\to\infty} S_n$, where

$$S_n = \inf\{t: |x_n(t) - x(0)| > \varepsilon/2\}.$$

It is easy to see that $S_n$, and hence $S$, are stopping times, and $|x_n(S_n) - x(0)| \geq \varepsilon/2$ on $\{S_n < \infty\}$, while $|x(t) - x(0)| \leq \varepsilon/2$ for $0 \leq t < S$. We need to show that $P\{S > 0\} = 1$. Otherwise, set $A = \{S = 0\}$, so that $P(A) > 0$ and $A \in \mathscr{F}_0$. Then, for $w \in A$, $S_n < \infty$ holds for large $n$ and $|x_n(S_n) - x(0)| \geq \varepsilon/2$, so either $x_n(S_n) - x(0) \geq \varepsilon/2$ for all large $n$, or there is $n_k \to \infty$ with $x_{n_k}(S_{n_k}) - x(0) \leq -\varepsilon/2$ for all $k$. In the latter case $x_{n_k}(S_{n_k}) - x(0) \leq -\varepsilon/2$ for $n_k \geq n$, and letting $k \to \infty$ we get $x_n(0) - x(0) \leq -\varepsilon/2$ by right-continuity of $x_n$. But letting $n \to \infty$ gives a contradiction, so only the former case remains. But then since $x(t)$ is a supermartingale,

$$E(x(0)I_A) \geq \lim_{n\to\infty} E(x(S_n)I_A)$$

$$\geq E\left(\lim_{n\to\infty} x(S_n)I_A\right)$$

$$\geq E\left(\lim_{n\to\infty} x_n(S_n)I_A\right)$$

$$\geq E((x(0) + \varepsilon/2)I_A),$$

(by Fatou's lemma) which is a contradiction unless $P(A) = 0$.

Now let $\mathcal{O}(t) = \lim_{\delta\to 0} \sup_{t < s < t+\delta} |x(s) - x(t)|$, which is in $\mathscr{F}'_{t+}$ because $x(t)$ is a measurable process, so that $\mathcal{O}(t)$ is the *right oscillation* of $x(t)$. Then $\mathcal{O}(t) < \varepsilon$ for $T \leq t < T'$, with $T, T'$ as above. Also let $R(\varepsilon) = \sup\{t: \mathcal{O}(s) < \varepsilon$ for $s < t\}$. To complete the proof we need only show $P\{R(\varepsilon) = \infty\} = 1$. Suppose $T_1, T_2, \ldots$ are any stopping times with $P\{T_n \leq R(\varepsilon)\} = 1$. Then $\sup_n T_n$ is a stopping time with the same property. For fixed $K$, it follows that there exist $T_1 \leq T_2 \leq \cdots$ having this property and such that

$$\lim_n E(T_n \wedge K) = \sup_T \{E(T \wedge K)\}$$

where the sup is over all such $T$. Then with $T = \lim_n T_n$ we obtain $E(T \wedge K) = \sup_T\{E(T \wedge K)\}$. But then $E(T \wedge K) = K$, for if not we could find $T' > T$ as above and $E(T' \wedge K) > E(T \wedge K)$. Thus $E(R(\varepsilon) \wedge K) = K$, and letting $K \to \infty$ finishes the proof.

**Exercise 2.1** Extend Lemma 2.2 to unbounded $x_n(t)$. Hint: consider $x_n(t) \wedge K$.

Returning to Theorem 2.1, it is clearly enough to prove it in the case $E = [0, 1]$, and transfer the result by Theorem 1.1. In this case we already

## 44 | Foundations of the prediction process

know that $E^{Z\bar{t}} \int_0^\infty e^{-\lambda s} f \circ \Theta_s \, ds$ is r.c.l.l. for $f \in \mathscr{C}(\Omega')$. Now the class of $f \in b(\mathscr{F}')$ for which it is r.c.l.l. (meaning, of course, r.c.l.l. $P^z$-a.s.) is a vector space. By Lemma 2.2 it is closed under monotone increasing limits of positive elements, but by adding constants the positivity restriction is lifted. The monotone class theorem now completes the proof.

Continuing with general $(E, \mathscr{E})$, it will be convenient to specify a metric compatible with the $E$-topology. None of our results not explicitly referring to the metric depends on the choice, but for convenience we specify it as follows. Let $0 \leq f_n \leq 1$ be uniformly dense in the continuous functions on $\bar{E} \to [0,1]$, where $\bar{E}$ is compact containing $E$ as Borel subset.

**Definition 2.1** $\quad d_E(e_1, e_2) = \sum_{n=1}^\infty 2^{-n} |f_n(e_1) - f_n(e_2)|.$

We note that $(f_n(e)): \bar{E} \to \times_1^\infty [0,1]$ defines a homeomorphism of $E$ onto a Borel subset of $\times_1^\infty [0,1]$.

We will introduce, now, a topology for $\Omega'_E$ considered as a space of pseudo-paths, i.e. of equivalence classes modulo Lebesgue null sets.

**Definition 2.2** The pseudo-path topology of $(\Omega'_E, \mathscr{F}'_E)$ is the topology of convergence in measure of the pseudo-paths $w(s)$ with respect to scaled Lebesgue measure $e^{-s} \, ds$.

It is known that this does not depend on the choice of metric $d_{\bar{E}}$, and that it is metrizable in such a way that $\Omega'_E$ becomes complete (but not compact). To see that $\mathscr{F}'_E$ are still the Borel sets, one can use the following equivalent (but apparently weaker) description of the topology.

**Definition 2.3** The sojourn measures of $w(s)$ are $\mu_t(A, w) = \int_0^t I_A(w_s) \, ds$; $A \in \mathscr{E}, 0 < t$.

We note the obvious properties

$$\mu_t(E) = t \quad \text{and} \quad \int_E \mu_t(dx, w) f(x) = \int_0^t f(w_s) \, ds, \quad f \in b(\mathscr{E}).$$

**Theorem 2.3** *The pseudo-path topology is the same as the topology of weak-\* convergence of sojourn measures for each $t$. In other words, $w_n \to w$ if* and only if, for every rational $r$ and $f \in \mathscr{C}(\bar{E})$,

$$\int \mu_r(w_n, \mathrm{d}x) f(x) \to \int \mu_r(w, \mathrm{d}x) f(x).$$

**Remark** Integration by parts and the continuity theorem of Laplace transforms show that this is equivalent to

$$\int_0^\infty e^{-\lambda t} \int \mu_t(w_n, \mathrm{d}x) f(x) \, \mathrm{d}t \to \int_0^\infty e^{-\lambda t} \int \mu_t(w, \mathrm{d}x) f(x) \, \mathrm{d}t, \qquad \forall \lambda > 0.$$

It follows in the same way that, for every $t \geq 0$,

$$\int_t^\infty e^{-\lambda s} f(w_n(s)) \, \mathrm{d}s \to \int_t^\infty e^{-\lambda s} f(w(s)) \, \mathrm{d}s.$$

**Proof** If $w_n \to w$ in $e^{-s} \, \mathrm{d}s$-measure on $[0, t]$, then, since $f$ is uniformly continuous,

$$\int_0^t f(w_n(s)) \, \mathrm{d}s \to \int_0^t f(w(s)) \, \mathrm{d}s$$

and we have $\mu_t(w_n) \to \mu_t(w)$ weakly.

Conversely, if this holds then, for each $f_k$ as in Definition 2.1 and each $g(t)$ continuous with compact support,

$$\int_0^t g(s) f_k(w_n(s)) \, \mathrm{d}s \to \int_0^t g(s) f_k(w(s)) \, \mathrm{d}s,$$

as follows by approximating $g(s)$ by $\sum_{k=1}^n g(kt/n) I_{((k-1)t'n, kt/n]}(s)$. Then, for any bounded continuous $g(s)$,

$$\int_0^\infty g(s) f_k(w_n(s)) \, e^{-s} \, \mathrm{d}s \to \int_0^\infty g(s) f_k(w(s)) \, e^{-s} \, \mathrm{d}s,$$

and since these $g$ are dense in $L^2(e^{-2} \, \mathrm{d}s)$ this implies that $f_k(w_n(s)) \to f(w(s))$ weakly in $L^2(e^{-s} \, \mathrm{d}s)$.

On the other hand, we also have

$$\int_0^\infty f_k^2(w_n(s)) \, e^{-s} \, \mathrm{d}s \to \int_0^\infty f_k^2(w(s)) \, e^{-s} \, \mathrm{d}s,$$

46 | Foundations of the prediction process

since $f_k^2 \in C(\bar{E})$, hence $\|f_k w_n\|_2 \to \|f_k w\|_2$ for $L^2(e^{-s} ds)$, and hence by a known criterion $\|f_k w_n - f_k w\|_2 \to 0$. In particular, then $f_k w_n$ converges in measure to $f_k w$ for each $k$. Referring to the metric $d_E$, the $e^{-s} ds$-measure of $\{t: d(w_n, w) > \varepsilon\}$ is at most

$$\sum_{k=1}^{N} (\text{measure of } \{t: |f_k w_n - f_k w| > \varepsilon/2\}),$$

provided that $2^{-(N+1)} < \varepsilon/2$. Hence, choosing $N$ large and letting $n \to \infty$ we obtain convergence in measure with respect to $d_{\bar{E}}$ (which does not depend, for such $w_n$, on the choice of either $\bar{E}$ or $d_{\bar{E}}$).

Analogous to our earlier procedure, we introduce the following definition.

**Definition 2.4** The *prediction topology* of $M_E$ is the topology of weak-* convergence of measures with respect to the pseudo-path topology of $(\Omega'_E, \mathscr{F}'_E)$.

To give rigorous meaning to this, it must first be shown that $(\Omega'_E, \mathscr{F}'_E)$ is a topological Lusin space (it is not compact, as will be seen). It is easiest to see this by observing that, by Theorem 2.3, the topology is generated by the functions $\int_0^r f_k(w(s)) ds$, $0 < r \in \mathbb{Q}$. Therefore, our mapping $e \to (f_k(e))$ induces on $\Omega'_E$ a bicontinuous mapping $w(s) \to (f_k(w(s)))$ of $\Omega'_E \to \times_1^\infty \Omega'_{[0,1]}$, which is one-to-one. We see without difficulty that $\mathscr{F}'_E$ is the inverse of $\times_1^\infty \mathscr{F}'_{[0,1]}$, since this makes each $\int_0^r f_k(w(s)) ds$ measurable, hence $\int_0^r f(w(s)) ds$ measurable for $f \in b(\mathscr{E})$. (Define the measurable coordinates $\tilde{X}_k$ as before.) Now we also see that the pseudo-path topology is metrized by $\sum_{k=1}^{\infty} 2^{-k} d'(f_k(w_1), f_k(w_2))$ (with $d'$ as in Theorem 1.3). Thus it need only be seen that the range is an element of $\times_1^\infty \mathscr{F}'_{[0,1]}$. Let $\rho$ denote the mapping $e \xrightarrow{\rho} (f_k(e))$, and $\rho(E)$ its range ($\rho(E) \in \times_1^\infty \mathscr{B}[0,1]$ by Lusin's theorem). Then the range of $w \to (f_k(w))$ is characterized by

$$\int_0^r I_{\rho(E)}(f_k(w(s))) ds = r = \int_0^r I_{\rho(E)}(\tilde{X}_k(w(s))) ds, \qquad \forall r \in \mathbb{Q}^+.$$

This defines a set in $\times_1^\infty \mathscr{F}'_{[0,1]}$ as needed.

To see that it is not compact, even if $E = [0,1]$, consider the sequence $w_n(t) = \frac{1}{2}(1 + \sin nt)$. Then no subsequence converges in measure, although $\int_0^t w_n(s) e^{-s} ds \to (1 - e^{-t})/2$. Thus $w_n(s) \to w \equiv \frac{1}{2}$ in our Chapter 1 topology, but has no limit in the pseudo-path topology. Nevertheless, weak-* convergence on $M_E$ is uniquely defined as follows.

**Definition 2.5** $\lim_{n\to\infty} z_n = z$ (in the weak-* sense) if and only if $\lim_{n\to\infty} E^{z_n}f = E^z f$ for every bounded continuous function on $\Omega'_E$.

Now for any compact metric $\bar{\Omega}_E \supset \Omega'_E$ as a Borel subset, and any such $f$, there are $f_n \uparrow f$, $f_n \in C(\bar{\Omega}_E)$, and $g_n \in C(\bar{\Omega}_E)$ with $g_n \downarrow f$ on $\Omega'_E$. For the details of this, see, for example, Getoor (1975, p. 101, (14.8)). In particular, we may consider $\bar{\Omega}_E$ to be the compactification of $\Omega'_E$ induced by the map $\Omega'_E \to \times_1^\infty \Omega'_{[0,1]}$ defined above, since $\times_1^\infty \Omega'_{[0,1]}$ is a compact metric space. It will often be convenient to use this particular compactification. Therefore, if $z_n \to z$ as measures on $(\bar{\Omega}_E, \bar{\mathscr{F}}_E)$ then they also converge in the weak-* sense as measures on $(\Omega_E, \mathscr{F}'_E)$. The converse being immediate, we see that weak-* convergence does not depend on the choice of $\bar{\Omega}_E$. Note that the prediction topology is metrizable, via any particular choice of $\bar{\Omega}_E$, just as in Corollary 1.4, and $M_E$ becomes a Borel subset of compact $\bar{M}_E$. Moreover, the Borel field of $M_E$ is again $\mathscr{M}_E$, just as before. It is easy to see that $M_E$ is compact if and only if $\Omega'_E$ is compact.

**Exercise 2.2** Show that, even if $E$ consists of just two points, $\Omega'_E$ is not compact. Conclude that $\Omega'_E$ is never compact unless $E$ is a singleton (or void?). Hint: take $E = \{0, 1\}$ and write $w(t) = I_S(t)$, $S \in \mathscr{B}^+$. Then consider $S_n = \bigcup_{0 \le k} (2k/2^n, (2k+1)/2^n]$.

**Theorem 2.4** *For* $z \in M_E$,

$$P^z\{Z^z_t \text{ is right-continuous in the prediction topology, with left limits } Z^z_{t-} \text{ for all } t > 0\} = 1$$

(where $Z^z_t$ is the same as in Corollary 1.10).

**Proof** Let $\bar{\Omega}_E \supset \Omega'_E$ be a compact metric space as before, and $g_n$ a uniformly dense sequence $0 \le g_n \le 1$. It suffices to show that

$$P^z\{Z^z_t g_n \text{ is right continuous with left limit } Z^z_{t-} g_n\} = 1$$

(recall $z(g) = E^z g$). More precisely, we can choose $g_n$ of the form

$$g_n(w) = g'_{m,k}\left(\int_0^{t_1} f_1(w(s))\,ds, \ldots, \int_0^{t_m} f_m(w(s))\,ds\right)$$

for $f_1, \ldots, f_m \in C(\bar{E})$, and $0 < t_1 \le \cdots \le t_m$, $1 \le k$, $1 \le m$, and

$$g'_{m,k}(x_1, \ldots, x_m) \in C(R^m).$$

Indeed, as in Lemma 1.7, these are an algebra separating points. It follows as before that $g_n \circ \Theta_t$ is uniformly continuous in $t$, uniformly on $\Omega'_E$. (One can also see this directly from the metric following Definition 2.4, which makes $\Theta_t$ uniformly continuous.) Writing $z(f)$ for $E^z f$, we have by Theorem 2.1 that, for $\lambda > 0$, $P^z$-a.s., $Z_t^z(\int_0^\infty \lambda e^{-\lambda s} g_n \circ \Theta_s \, ds)$ is r.c.l.l. in $t$. Since, however, $Z_t^z g_n \circ \Theta_s$ is uniformly continuous in $s$, uniformly in $t$, we have

$$\lim_{\lambda \to \infty} \int_0^\infty \lambda e^{-\lambda s} Z_t^z(g_n \circ \Theta_s) \, ds = Z_t^z g_n,$$

uniformly in $t$. So, applying Fubini's theorem, we obtain that the right side is r.c.l.l. It remains to see that the left limits are still given by $Z_{t-}^z$. This is trivial if we use the previsible section theorem, because, if $T_k \uparrow T$, $T_k < T$, are stopping times, it follows by Hunt's lemma and the moderate Markov property that

$$\lim_{k \to \infty} Z_{T_k}^z g_n = E^z(g_n \circ \Theta_T \mid V_k \mathscr{F}'_{T_k+})$$
$$= Z_{T-}^z g_n, \qquad P^z\text{-a.s. on } \{T < \infty\}.$$

Thus the set where the left limits are unequal (an $\mathscr{F}'_t$-previsible set when we define the left-limits to be $\mathscr{F}'_t$-previsible, as in Exercise 1.8) does not intersect the graph of any previsible $T$ with positive probability.

On the other hand, we can prove directly that

$$P^z\left\{Z_{t-}^z\left(\int_0^\infty e^{-\lambda s} f \circ \Theta_s \, ds\right) = \lim_{u \uparrow t} Z_u^z\left(\int_0^\infty e^{-\lambda s} f \circ \Theta_s \, ds\right), \forall t > 0\right\} = 1$$

for any $f \in b(\mathscr{F}')$ and $z \in M_{[0,1]}$, which certainly suffices for our result. This is known for $f \in C(\Omega'_{[0,1]})$ and the class of $f$ for which it holds is a vector space. As in Theorem 2.1, it suffices to show that if $0 \leq f_n \uparrow f \in b(\mathscr{F}')$, where $f_n$ is in the class, then $f$ is also. But by the maximal inequality for positive r.c.l.l. supermartingales, for $\lambda > 0$,

$$\lim_{n \to \infty} P^z \sup_t \left\{e^{-\lambda t} Z_t^z\left(\int_0^\infty e^{-\lambda s}(f - f_n) \circ \Theta_s \, ds\right) > \lambda\right\}$$

$$\leq \lambda^{-1}\left(\lim_{n \to \infty} E^z \int_0^\infty e^{-\lambda s}(f - f_n) \circ \Theta_s \, ds\right) = 0$$

by monotone convergence. Thus the left limit processes converge uniformly in $t$ when $n = n_k \to \infty$ sufficiently fast. Since

$$\lim_{n \to \infty} Z_{t-}^z \left( \int_0^\infty e^{-\lambda s} f_n \circ \Theta_s \, ds \right) = Z_{t-}^z \left( \int_0^\infty e^{-\lambda s} f \circ \Theta_s \, ds \right)$$

for every $t$ and every path, this shows that $f$ is in the class, as required.

**Corollary 2.5** *The prediction process $Z_t^z$ on $(\Omega_E', \mathscr{F}_E')$, for each $z \in M_E'$, is a.s. independent of the isomorphism $\hat{\varphi}$ of Theorem 1.1. Indeed, it is a.s. uniquely defined by the requirements*

(1) $Z_r^z(S) = P^z(\Theta_r^{-1} S \mid \mathscr{F}_{r+}')$, $r \in \mathbb{Q}^+$, $S \in \mathscr{F}_E'$; and

(2) $Z_t^z$ is r.c.l.l. in the prediction topology of $M_E'$.

**Proof** Now obvious. Note, however, that we have now avoided the section theorems.

The prediction topology also plays an important role in applications to Markov processes. To carry out such a step we must return to the coordinate problem. The coordinates $\tilde{X}_t$ of Theorem 1.2 do not transfer to $\Omega_E'$ in such a way as to preserve $w(t)$ which are already r.c.l.l. (i.e. $\tilde{X}_t(\hat{\varphi}w)$ is not r.c.l.l. when $w$ is r.c.l.l., at least for general $\hat{\varphi}$), and the same holds true for $\hat{\rho}(Z_t^{z\hat{\varphi}^{-1}}(\hat{\varphi}w))$ of Theorem 1.16, even when $Z_t^z\{\text{r.c.l.l. } w\} = 1$ for all $t$. Nevertheless, the same ideas can be made to apply by use of the homeomorphism

$$\rho: \rho(e) = (f_k(e)) \text{ of } \bar{E} \xrightarrow{\text{into}} \times_1^\infty [0, 1].$$

**Definition 2.6** Let **e** be a fixed but arbitrary element of $E$, and let

$$\tilde{\rho}_k(w) = \limsup_{n \to \infty} n \int_0^{1/n} f_k(w(s)) \, ds.$$

We define

$$\tilde{\rho}_E(w) = \begin{cases} e & \text{if } \exists e \in E \text{ such that } e = \rho^{-1}(\tilde{\rho}_k(w)) \\ \mathbf{e} & \text{otherwise,} \end{cases}$$

and

$$\tilde{X}_E(t) = \tilde{\rho}_E(w \circ \Theta_t), \quad 0 \leq t.$$

From now on we drop the subscript and just write $\tilde{X}_t$. (This does not coincide with $\tilde{X}_t$ of Definition 1.2 when $E = [0, 1]$, but no confusion arises.)

**Lemma 2.6** *Theorem 1.2 holds without change for the new $\tilde{X}_t$.*

**Proof** The measurability assertions are treated below. To see that, for each $w$, $\tilde{X}_t(w) = w(t)$ for a.e. $t$, it suffices to observe that Theorem 1.2 applies to each path $f_k(w(s))$, hence $\tilde{\rho}(f_k(w) \circ \Theta_s) = f_k(w(s))$ for all $k$, a.e. $s$. Of course, since $(\tilde{\rho}_k(w))$ is $(\mathscr{F}'_{0+} / \times_1^\infty \mathscr{B}[0,1])$-measurable, the set $\{w: (\tilde{\rho}_k(w)) \in \rho(E)\}$ is in $\mathscr{F}'_{0+}$, and so $\rho^{-1}(\tilde{\rho}_k(w))$ is $(\mathscr{F}'_{0+}/\mathscr{E})$-measurable on this set. Hence $\tilde{\rho}_E$ is measurable, and the joint measurability follows as before.

**Remarks** To see what $\tilde{\rho}_E$ means one can write

$$n \int_0^{1/n} f_k(w(s)) \, ds = n \int_E \mu_{1/n}(dx, w) f_k(x)$$

(using Definition 2.3). Thus the limit as $n \to \infty$ exists for all $k$ if and only if the sojourn measures are *weakly differentiable* at $t = 0$. This is well known to be the case for $w(s) \circ \Theta_t$, for a.e. $t$ (as a right density of $w(s)$ at time $t$), hence the lim sup is actually a limit for a.e. $t$. Note in particular that if $w(t)$ is right-continuous (i.e. equal a.e. to a right-continuous $\tilde{w}(t)$) then $\tilde{X}_t = \tilde{w}(t)$ for all $t$. Needless to say, one could also define *left coordinates* $\tilde{X}_t^-$ by using $n \int_{t-1/n}^t$, with $w(t) = w(0)$ for $t < 0$, and such $\tilde{X}_t^-$ would be $\mathscr{F}'_t$-previsible along with $n \int_{t-1/n}^t f_k(w(s)) \, ds$. Another improvement would be to replace $n$ by a subsequence, such as $2^n$, which would accord better with the right (or left) limits in probability when they exist. However, it does not seem to be possible to obtain a definition which accords $P^z$-a.s. with the right (or left) limit in probability for *every* $z$ such that such a limit exists except by invoking the continuum hypothesis (Mokobodski ultrafilters, etc.). In view of the independence thereof (Cohen's theorem) this seems to be too radical a step.

We also want a topological extension of $\hat{\rho}(z)$ from Definition 1.8 and Theorem 1.16.

**Definition 2.7** Let $\hat{\rho}_E(z) = e$ or $\mathbf{e}$ according as there exists an $e \in E$ for which $\rho(e) = (E^z \tilde{\rho}_k)$, $\mathbf{e}$ being fixed as before.

**Theorem 2.7** *Theorem 1.16(1) holds without change for $\hat{\rho}_E$. In particular, $\hat{\rho}_E \in \mathscr{M}/\mathscr{E}$ and*

$$P^z \left\{ \int_0^t f_k(\hat{\rho}_E(Z_s^z)) \, ds = \int_0^t f_k(\tilde{X}_s) \, ds = \int_0^t f_k(w(s)) \, ds \text{ for all } k \text{ and } t \right\} = 1,$$

*so that the $\mathscr{F}_{t+}^z$-optional process $Z_t^z$ generates $\mathscr{F}_t'$ up to $P^z$-null sets. Furthermore, since $P^z\{\hat{\rho}_E(Z_T^z) = \tilde{X}_T\} = 1$ for $\mathscr{F}'_{t+}$-optional $T < \infty$, $\hat{\rho}_E(Z_t^z)$ defines coordinates for the same processes as $\tilde{X}_t$ does.*

**Remarks** $\hat{\rho}_E(Z_t^z)$ depends on $z$, unlike $\tilde{X}_t$. This drawback disappears when we introduce prediction space. From now on we drop the subscript $E$, and just write $\hat{\rho}$. The proof of Theorem 2.7 is the same as that of Theorem 1.16(1), and need not be repeated. We note further that if $P^z\{\text{r.c.l.l.}\} = 1$, i.e. we start with a process having r.c.l.l. paths to generate $P^z$ (where {r.c.l.l.} is universally measurable over $\mathscr{F}'$ by Dellacherie and Meyer (1975, IV, 34), and if $E$ is Polish it is in $\mathscr{F}'$ by application of Lusin's theorem) then since $\tilde{X}_t$ is r.c.l.l. on {r.c.l.l.}, it is also $\mathscr{F}^z_{t+}$-optional ($\mathscr{F}^z_E$ being, as before, the $P^z$-augmentation of $\mathscr{F}'_E$). Therefore, using Meyer's 0-1 law (Theorem 1.15) the optional section theorem implies that

$$P^z\{\tilde{X}_t = \hat{\rho}_E(Z^z_t) \text{ for all } t \geq 0\} = 1$$

in this special situation. This is handy for certain applications (to r.c.l.l. processes).

**Exercise 2.7** Show that if $(E, \mathscr{E})$ is metrized by $d_E$ (Definition 2.1), and if $X_t$ is an $(E, \mathscr{E})$-valued process right-continuous in probability, with $P\{d_E(X_{t+s}, X_t) > \delta\} = O(s^{1+\varepsilon})$, $\varepsilon = \varepsilon(t, \delta) > 0$, as $s \to 0+$, then the law of $X_t$ is inherited by $\tilde{X}_t$ when $P$ is used (along with a measurable version of $X_t$) to define $z$ on $\mathscr{F}'_E$. If we replace $n$ by $2^n$ in defining $\tilde{X}_t$, then $O(s^\varepsilon)$ suffices. Hint: note that $d_E(e_1, e_2) \geq 2^{-k}|f_k(e_1) - f_k(e_2)|$, and let $(X_k(t))$ denote a limit in probability of $n \int_0^{1/n} X_k(t + s) \, ds$ as $n \to \infty$. Show that in fact the convergence is $P$-a.s., so that $(X_k(t)) = \rho(\tilde{X}_t)$, $P$-a.s.

## 2.2 Application to Markov processes

The question of whether $\tilde{X}_t$ suffices to express the law of a *Markov* process can be answered (at least in general terms) without any *ad hoc* hypotheses such as in the above exercise. We turn to an application of this nature. Some familiarity with Markov processes will be assumed.

As our starting point we take a

**Definition 2.8** $p(t, x, B)$, where $0 < t$, $x \in E$, $B \in \mathscr{E}$, is a Markov transition function if it is a probability kernel $\mathscr{E} \to \{\text{prob. on } \mathscr{E}\}$ for each $t$, and satisfies the Chapman–Kolmogorov equations (1.1).

We will assume throughout the following hypothesis.

**Hypothesis** Let $p(t, x, B)$ be a Markov transition function such that:

(1) it is $(\mathscr{B}^+ \times \mathscr{E})$-measurable in $(t, x)$ for $B \in \mathscr{E}$;

(2) the collection $p(\cdot, x, \cdot)$ $(0 < t, B \in \mathscr{E})$ separates points of $E$.

**Discussion** The measurability restriction is not serious, and some such condition is necessary to even define a corresponding resolvent. We set $T_0 f = f$, and $T_t f(x) = \int p(t, x, dy) f(y)$, $0 < t$, $f \in b(\mathscr{E})$. Then $T_t$ is the *semigroup* of $p$: we have $T_{t_1+t_2}f = T_{t_1}T_{t_2}f$ by Chapman–Kolmogorov (without assuming (1) and (2)), so $T_t$ is a semigroup on $b(\mathscr{E})$. Incidentally, it follows from Hille and Phillips (1957, Theorems 3.5.5 and 10.2.3) that if $\int T_t f(x)\mu(dx)$ is measurable in $t$ for finitely additive measures $\mu$ and $f \in b(\mathscr{E})$, and if for each $f$ the range of $T_t f$, $0 \leq t$, is almost separably valued (i.e. separable apart from a $dt$-null set) then $T_t f$ is continuous in the uniform norm for $t > 0$. In this case it is easy to see that (1) is satisfied. But the trivial example $T_t f(x) = f(x+t)$, $f \in b(R, \mathscr{B})$, shows that (1) may hold yet $T_t f$ is not almost separably valued (take $f = I_{[0,\infty]}$). A matter of greater importance concerns the resolvent defined as

$$R_\lambda f(x) = \int_0^\infty e^{-\lambda t} T_t f(x)\, dt, \qquad \lambda > 0.$$

It is included below as an appendix to this chapter that the family $\{p(\cdot, x, \cdot)\}$ generates the same $\sigma$-field as the family $\{R_\lambda f(x); 0 < \lambda, f \in b(\mathscr{E})\}$. Therefore, Hypothesis (2) is the same as assuming that the resolvent separates points in $E$. If this fails to hold then it is known that one may consider $p$ as a transition function on the generated $\sigma$-field instead of on $\mathscr{E}$, and obviously (1) still holds. However, it does *not* follow that this $\sigma$-field gives a Lusin-measurable structure to $\mathscr{E}$ (unless one assumes, for instance, that the range $(\lambda R_\lambda f_n): E \to \times_1^\infty [0,1]$, with $f_n$ as in Definition 2.1, is a (Borel) set in $\times_1^\infty \mathscr{B}[0,1]$). As we need (2) at various places below, we take it as a starting point. We now have a crucial lemma as follows.

**Lemma 2.8** *For each $x$, there is a* measurable *process $X_t'$ (for the coordinate $\sigma$-field $\mathscr{F}^\circ = \times_{t\geq 0}(\mathscr{E}_t)$) with law*

$$P^x\left\{\bigcap_{k=1}^n (X_{t_k} \in B_k)\right\} = \int_{B_{n-1}} \cdots \int_{B_1} p(t_1, x, dx_1)$$

$$\cdots p(t_{n-1} - t_{n-2}, x_{n-2}, dx_{n-1}) p(t_n - t_{n-1}, x_{n-1}, B_n).$$

**Proof** Since $E$ is Borel in $\bar{E}$ one knows from Kolmogorov extension that there exists a $P^x$ with this law on $(\times_{t\geq 0} E_t, \mathscr{F}^\circ)$. We need to show that it has a measurable standard modification. Now we consider the familiar $(\mathscr{F}_t^\circ, P^x)$-supermartingales $e^{-\lambda t} R_\lambda f_n(X_t)$ (Chung 1982, p. 46) and let $M_{\lambda,n}(t)$ denote a right limit along rational $t$ when this exists, and set $M_{\lambda,n}(t) = 0$ elsewhere. The exceptional set is contained in a single $P^x$-null set for all $t$, $\lambda > 0$, and $n$ at once. Letting $\mathscr{F}_{t\pm}^\circ$ denote the usual right and left coordinate $\sigma$-fields, we have, for each $t$, $P^x$-a.s., by dominated convergence,

$$E^x(M_{\lambda,n}(t) \mid \mathscr{F}_t^\circ) \leq e^{-\lambda t} R_\lambda f_n(X_t),$$
$$E^x(e^{-\lambda t} R_\lambda f_n(X_t) \mid \mathscr{F}_{t-}^\circ) \leq M_{\lambda,n}(t-)$$

where $M_{\lambda,n}(t-)$ denotes the left limit, existing $P^x$-a.s. Thus, if $t$ has the property that $P^x\{M_{\lambda,n}(t-) = M_{\lambda,n}(t)\} = 1$, then $E^x(M_{\lambda,n}(t) \mid \mathscr{F}_t^\circ) = M_{\lambda,n}(t-)$, and so $M_{\lambda,n}(t-) \leq e^{-\lambda t} R_\lambda f_n(X_t)$, which implies (by taking expectations) that

$$e^{-\lambda t} R_\lambda f_n(X_t) = M_{\lambda,n}(t-) = M_{\lambda,n}(t), \qquad P^x\text{-a.s.}$$

On the other hand, since $M_{\lambda,n}(t)$ is r.c.l.l. ($P^x$-a.s.), we have

$$P^x\{M_{\lambda,n}(t) = M_{\lambda,n}(t-)\} = 1$$

except for $t$ in some countable set. (As in Exercise 1.11, consider

$$\{t: P^x\{|M_{\lambda,n}(t) - M_{\lambda,n}(t-)| > \varepsilon\} \neq 0\} \quad .)$$

The same is then simultaneously in $\lambda$ and $n$ where $\lambda$ may be taken rational. Thus except for a countable set of $t$ we have

$$P^x\{e^{-\lambda t} R_\lambda f_n(X_t) = M_{\lambda,n}(t)\} = 1.$$

We will modify $X_t$ by changing it only for those $t$ such that

$$P^x\{e^{-\lambda t} R_\lambda f_n(X_t) = M_{\lambda,n}(t)\} = 1.$$

Then if $(\lambda e^{\lambda t} M_{\lambda,n}(t))$ is not in the range of the mapping $(\lambda R_\lambda f_n): E \to \times_1^\infty [0,1]$, we set $X_t' = \mathbf{e}$ for some fixed element $\mathbf{e} \in E$, and elsewhere we define $X_t'$ as the inverse of $\lambda e^{\lambda t} M_{\lambda,n}(t)$ under this map. To see that we have thus obtained a measurable standard modification, note that the values on the fixed countable set of $t$ do not impair measurability. Furthermore, the above change is clearly on a $P^x$-null set for each $t$ if the set $\{(\lambda e^{\lambda t} M_{\lambda,n}(t))$ not in the range$\}$ is measurable. But $\lambda R_\lambda f_n: E \to \times_1^\infty [0,1]$, is one-to-one by Hypothesis (1), and it is measurable, so by Lusin's theorem the range is in $\times_1^\infty \mathscr{B}[0,1]$. Thus the above set is in $\mathscr{F}^\circ$ for each $t$. Finally, since $\lambda e^{\lambda t} M_{\lambda,n}(t)$ is a measurable process, it remains measurable if we replace it by $\mathbf{e}$ on the above set, and elsewhere, for all but countably many $t$, by its inverse in $E$ under the map $\lambda R_\lambda f_n: E \to \times_1^\infty [0,1]$. This defines the required standard modification $X_t'$.

Next we recall that the definition $P^z(S) = P^x\{X_\cdot' \in S\}$, $S \in \mathscr{F}_E'$, does not depend on which measurable standard modification we choose. It is uniquely determined, therefore, by $p(t, x, B)$. We will base our arguments on this map, so we introduce the following notation.

**Notation 2.9** For fixed $p$ (satisfying the above hypothesis) let $\varphi(x): E \to M_E$ be the above map:

$$P^{\varphi(x)}(S') = P^x\{X_{(\cdot)}' \in S'\}, \qquad S' \in \mathscr{F}_E'.$$

**Lemma 2.9** $\varphi(x)$ is $\mathscr{E}/\mathscr{M}_E$-measurable.

**Proof** For any isomorphism $\hat{\varphi}: E \leftrightarrow [0,1]$, (Theorem 1.1), the induced map $M_E \leftrightarrow M_{[0,1]}$ is clearly an isomorphism $\mathcal{M}_E \leftrightarrow \mathcal{M}_{[0,1]}$, so in proving the lemma we can assume that $E = [0,1]$. That is, we replace $p(t,x,A)$ by $p(t, \hat{\varphi}(x), \hat{\varphi}(A))$. It suffices then to show that for any continuous $g(x_1, \ldots, x_n)$ and $t_1, \ldots, t_n$, $1 \leq n$, $E^{\varphi(x)}g(\int_0^{t_1} w(s)\,ds, \ldots, \int_0^{t_n} w(s)\,ds)$ is $\mathscr{E}/\mathscr{B}$-measurable. Then by the Stone–Weierstrass theorem it is enough to prove this for polynomials $g$. If $g$ is linear this reduces to the measurability in $x$ of $\int_0^{t_1} \int_0^1 y p(s, x, dy)\,ds$, valid by $(s, x)$-measurability of $p$. Of the remaining cases, we will write only the case $(\int_0^{t_1} w(s)\,ds \int_0^{t_2} w(s)\,ds)$. Readers can convince themselves that the other cases can be handled analogously. This reduces to calculating (we can assume $t_1 \leq t_2$)

$$E^{\varphi(x)}\left[\int_0^{t_1}\int_0^{t_1} \tilde{X}(s_1)\tilde{X}(s_2)\,ds_1\,ds_2 + \int_0^{t_1}\int_{t_1}^{t_2} \tilde{X}(s_1)\tilde{X}(s_2)\,ds_1\,ds_2\right].$$

Now since, for each $x$, $\tilde{X}(s)$ has the joint law determined by $p$ except for a Lebesgue null set of $s$, this becomes

$$2\int_0^{t_1}\int_0^{s_2}\int_0^1\int_0^1 y_1 y_2 p(s_1, x, dy_1) p(s_2 - s_1, y_1, dy_2)\,ds_1\,ds_2$$

$$+ \int_0^{t_1}\int_{t_1}^{t_2}\int_0^1\int_0^1 y_1 y_2 p(s_1, x, dy_1) p(s_2 - s_1, y_1, dy_2)\,ds_1\,ds_2,$$

which is also measurable in $x$, as required.

**Remark** In fact, these formulas follow immediately from the definition of $\varphi(x)$ in terms of $X_t'$, without introducing $\tilde{X}$.

For the range of $\varphi$ we write $\varphi(E)$. Note that, since $\varphi$ is one-to-one (by Hypothesis (2) above and the appendix) and Borel measurable (Lemma 2.9) it follows by Lusin's theorem that $\varphi(E) \in \mathcal{M}_E$.

**Theorem 2.10** *For each $x$, we have*

(1) $P^{\varphi(x)}\{\varphi(\tilde{X}_t) = Z_t^{\varphi(x)} \text{ for a.e. } t\} = 1$; *and*

(2) *For a.e. $t$,* $P^{\varphi(x)}\{\varphi(\tilde{X}_t) = Z_t^{\varphi(x)}\} = 1$.

**Proof** Since $\varphi(\tilde{X}_t)$ and $Z_t^{\varphi(x)}$ are both measurable processes, (1) follows from (2) by Fubini's theorem. It remains to prove (2). We return to our measurable standard modification of the coordinate process with law $P^x$, denoted by $X_t'$ (Lemma 2.8). The Markov property implies that $P^x(\Theta_t^{-1}S^\circ \mid \mathscr{F}_t^\circ) = P^{X_t'}(S^\circ)$, $S^\circ \in \mathscr{F}^\circ$. When we try to translate this fact on to the space $(\Omega', \mathscr{F}', P^{\varphi(x)} = z)$ two difficulties appear. First, we cannot replace $X_t'$ on the right by $\tilde{X}_t$ because nothing ensures that $\tilde{X}_t$ has the same law as $X_t'$, and, second, we cannot replace $\mathcal{F}_t^0$ by $\mathcal{F}_{t+}'$ on the left because, even for measurable $X_t'$, only for $s \leq t$ are the integrals $\int_0^s f(X_u') \, du$ $\mathcal{F}_t^0$-measurable so the translation only gives us $\mathcal{F}_t'$ (and some coordinate information in $\mathcal{F}_t^0$ may be lost since $X_s'$ has no real counterpart on $\mathcal{F}_t'$).

Suppose, however, that $t$ has the following two properties:

(1) $P^x \left\{ \limsup_{n \to \infty} n \int_t^{t+1/n} f_k(X_s') \, ds = f_k(X_t') \right\} = 1$

for all the $f_k$ used to construct $\tilde{X}_t$ (in Definition 2.6); and

(2) $P^{\varphi(x)} \{ Z_{t-}^{\varphi(x)} = Z_t^{\varphi(x)} \} = 1$.

Then property (1) implies that, when we translate to $(\Omega', \mathcal{F}')$, $\tilde{X}_t$ has the same joint law with the variables generating $\mathcal{F}'$ as $X_t'$ has with their counterparts on coordinate space. Moreover, if we look closely at the definition of $X_t'$ from $X_t$, we observe that it depends on $x$ (in $P^x$) only at a countable set of $t$. Elsewhere it depends only on the coordinates $X_r$, $r \in \mathbb{Q}^+$, in such a way that, for $S' \in \mathcal{F}'$, $\{X_{(\cdot)}' \in \Theta_t^{-1}(S')\} \in \Theta_t^{-1} \mathcal{F}^0$ on coordinate space, and does not depend on $x$. Furthermore, for each $x$ we could have defined $X_s'$ using $X_r$, $r + t \in \mathbb{Q}$ instead of $r \in \mathbb{Q}^+$ and the result would be the same, $P^x$-a.s. for all $s$ (simultaneously). Consequently, $\{X_{(\cdot)}' \in \Theta_t^{-1}(S')\}$ and $\Theta_t^{-1}\{X_{(\cdot)}' \in S'\}$ are equivalent. Putting these two observations together, it follows from the simple Markov property of $X_t'$ and (1) that, for $t$ satisfying (1) and (2),

$$P^{\varphi(x)}(\Theta_t^{-1} S' \mid \mathcal{F}_t' \vee \sigma(\tilde{X}_t)) = P^{\varphi(\tilde{X}_t)}(S'), \qquad S' \in \mathcal{F}'.$$

On the other hand, (2) implies (Theorem 1.16(2)) that $\mathcal{F}_t' \equiv \mathcal{F}_{t+}'$ for $P^{\varphi(x)}$, so we can replace $\mathcal{F}_t'$ by $\mathcal{F}_{t+}'$ on the left. Then it follows by definition of $Z_t^z$ that $P^{\varphi(x)} \{ Z_t^{\varphi(x)} = \varphi(\tilde{X}_t) \} = 1$. Therefore, to complete the proof, we need only observe that (1) holds for a.e. $t$ by (Lebesgue) differentiations and Fubini's theorem, and (2) holds for a.e. $t$ (in fact, for all but countably many $t$) because $Z_t^{\varphi(x)}$ is r.c.l.l.

Theorem 2.10 provides the link between the transition function $p(t, x, B)$ and the prediction process $Z_t^{\varphi(x)}$. The coordinates $\tilde{X}_t$ suffice for sufficiently many values of $t$ to permit them to be used to 'regularize' both the semigroup $T_t$ and the coordinate process so as to obtain the features of $Z_t$, such as the strong Markov property, r.c.l.l. paths, etc. This theory has a long history, going back at least to Knight (1961, pp. 591–613), which is now entirely superseded.

We consider for our new state space

$$M_p = \left\{ z \in M_E : \int_0^\infty \lambda e^{-\lambda t} q(t, z, \varphi(E)) \, dt = 1 \text{ for } \lambda > 0 \right\}.$$

Noting that $\lambda$ can be taken countable, we see that $M_p \in \mathcal{M}_E$, and by Theorem 2.10 we have $\varphi(E) \subset M_p$.

**Lemma 2.11** *For* $z \in M_p$,

$$P^z\{Z_t^z \in M_p \text{ for } t \geq 0\} = 1 \quad \text{and} \quad P^z\{Z_{t-}^z \in M_p \text{ for } t > 0\} = 1.$$

**Proof** We rely on the section theorems for brevity (but another proof could be based on showing (by Theorem 2.25) that $\int_0^\infty e^{-\lambda t} q(t, Z_s^z, \varphi(E))\, dt$ is r.c.l.l. along the paths of $Z_s^z$ for any $z$). Since $Z_t^z$ is optional and $Z_{t-}^z$ is previsible, it is enough to show that $Z_T^z \in M_p$ for optional $T < \infty$ and $Z_{T-}^z \in M_p$ for previsible $0 < T < \infty$, $P^z$-a.s. But this is easy: we have, by the strong Markov property,

$$1 = E^z \left( \int_0^\infty \lambda e^{-\lambda t} I_{\varphi(E)}(Z_t^z)\, dt \right)$$

$$= E^z \left( \int_0^T \lambda e^{-\lambda t} I_{\varphi(E)}(Z_t^z)\, dt \right) + e^{-\lambda T} Z_T^z \left( \int_0^\infty \lambda e^{-\lambda t} I_{\varphi(E)}(Z_t^z)\, dt \right)$$

$$\leq E^z \left( \int_0^T \lambda e^{-\lambda T}\, dt + e^{-\lambda T} Z_T^z \left( \int_0^\infty \lambda e^{-\lambda t} I_{\varphi(E)}(Z_t^z)\, dt \right) \right)$$

$$= E^z \left( 1 - e^{-\lambda T} + e^{-\lambda T} Z_T^z \left( \int_0^\infty \lambda e^{-\lambda t} I_{\varphi(E)}(Z_t^z)\, dt \right) \right)$$

$$\leq 1.$$

The equality implies, as required, that

$$1 = Z_T^z \left( \int_0^\infty \lambda e^{-\lambda t} I_{\varphi(E)}(Z_t^z)\, dt \right),$$

$P^z$-a.s. The argument for previsible $T$ is analogous, using the moderate Markov property.

**Remark** We called such a stochastically closed set for both $Z_t^z$ and $Z_{t-}^z$ a *complete Borel packet* in Knight (1981a, Essay I, Definition 2.1). Here we will propose the term *demesne* (Definition 2.14).

It is clear that, for $x \in E$, $Z_t^{\varphi(x)}$ is a process with state space $M_p$ which can be regarded as a regularized version of $\tilde{X}_t$ for the same $P^{\varphi(x)}$. More concretely, let us replace $M_p$ by $E_p = E \cup (M_p - \varphi(E))$, with $\mathscr{E}_p$ the Borel field $\sigma(\mathscr{E}) \vee \sigma(\mathscr{M}|M_p - \varphi(E))$ generated by both components (when we identify $x \leftrightarrow \varphi(x)$ this is $(\mathscr{M}|M_p)$ with a change of notation, where we use the bar to denote the trace on the set $M_p$). We also replace $Z_t^{\varphi(x)}$ by the following process.

## Definition 2.10

$$X_t^x = \begin{cases} \varphi^{-1}(Z_t^{\varphi(x)}) & \text{if } Z_t^{\varphi(x)} \in \varphi(E) \\ Z_t^{\varphi(x)} & \text{otherwise.} \end{cases}$$

Then $X_t^x$ is a process in $(E_p, \mathscr{E}_p)$, and $P^{\varphi(x)}\{X_t^x = \tilde{X}_t \text{ for a.e. } t\} = 1$ by Theorem 2.10. In this form it is clear that the original topology of $E$ need not be the best for the regularization. Instead, we should consider $E$ with the topology induced by $\varphi(x)$ from the prediction topology of $M_E$. For the given transition function $p$ generating $\varphi$ we could call this the $\varphi$-topology of $E$. It is clear that $(E, \mathscr{E})$ is again a Lusin space for the $\varphi$-topology. Indeed, we can extend the topology to $E_p$ in the obvious way so that $(E_p, \mathscr{E}_p)$ is also a topological Lusin space. Nor is the definition of $X_t^x$ limited to $x \in E$. For any $e \in E_p$ we can define $X_t^e$ by the same recipe (with $e$ in place of $\varphi(x)$) and it is clear from the Markov property and the definition of $M_p$ that the following holds.

## Theorem 2.12

(1) *For each $e \in E_p$, $X_t^e$ is r.c.l.l. and strong Markov on $E_p$ with transition function inherited from $q(t, z, A)$ by replacing $z$ by $\varphi^{-1}(z)$ for $z \in \varphi(E)$, and $A$ by $\varphi^{-1}(A \cap \varphi(E)) \cup (A - \varphi(E))$.*

(2) *For $e \in E_p$, $P^z\{X_t^e = \tilde{X}_t \text{ for a.e. } t\} = 1$ and moreover, $P^z\{X_t^e \in E \text{ for a.e. } t\} = 1$, where $z = \varphi(e)$ if $e \in E$, and $z = e$ otherwise.*

Of course, $X_{t-}^e$ exists and is moderately Markov for the same transition function, and also satisfies (1) and (2), (although the left coordinates $\tilde{X}_t^-$ might appear to be more appropriate for (2)). It is quite worthwhile to examine the connection of this new transition function with the given $p(t, x, E)$ for $x \in E$. For some compact metric $\bar{M}_E \supset M_E$ let $\overline{\varphi(E)}$ denote the closure of $\varphi(E)$ in $\bar{M}_E$ (not compact). We can define in this way an extension $\bar{\bar{E}}$ of $E$ to a closed set as follows.

**Definition 2.11** $\bar{\bar{E}}$ is obtained by completing $E$ using the metric induced by $\varphi$, and retaining only the elements which correspond to elements in $\overline{\varphi(E)}$.

Now if $N_x$, $x \in E$, denotes $\{t: P^x\{X_t^x \neq \tilde{X}_t\} > 0\}$, it is clear that for $g \in C(\bar{M}_E)$, $\lim_{s \to t+, s \notin N_x} T_s g \circ \varphi(x) = E^{\varphi(x)} g \circ \varphi(X_t^x)$ for all $t \geq 0$ and $n$, and defines a positive linear functional on $C(\bar{M}_E)$ with corresponding measure concentrated on $\overline{\varphi(E)}$ (since each $p(s, x, \varphi^{-1}(dy))$ is concentrated on $\varphi(E)$). In this way we obtain a semigroup $T_{t+}$ for $X_t^e$ consisting of measures concentrated on $\bar{\bar{E}} \cap E_p$ by defining, for $f = g \circ \varphi$,

**Definition 2.12** $T_{t+} f(e) = E^e g \cdot \varphi(X_t^e)$, $g \in b(\bar{\mathscr{M}}_E)$, $e \in E_p$, $t > 0$.

Then $T_{t+}$ is right-continuous in $t$ for the weak-* topology of measures on $\bar{E} \cap E_p$, since the paths of $X_t^e$ are right-continuous. In analogous fashion we can define a left-continuous semigroup $T_{t-}$ for $X_{t-}^e$, and it follows from Theorem 2.10 that we have, for $x \in E$ and for a.e. $t$,

$$T_t f(x) = T_t f(x) = T_{t+} f(x), \quad \text{for all } f \in b(\mathscr{E}_p).$$

Indeed, we have only to combine the a.e. $t$ of Theorem 2.10 with the a.e. $t$ at which $X_{t-}^x = X_t^x (= X_{t+}^x)$, $P^x$-a.s. (we write $X_t^x$ for $X_t^e$ if $e \in E$).

A little more thought will show, in fact, that if we reinstate $\Omega'_E$ as the space of $\mathscr{B}^+/\mathscr{E}$-measurable paths (rather than pseudo-paths), we can make a direct comparison of $X_t^x$ with the $P^x$-coordinate process as follows.

**Theorem 2.13** *For each $x \in E$, let $P^{\varphi(x)}$ be extended to $\mathscr{F}' \vee F^\circ$, where $\mathscr{F}^\circ = \sigma\{w(t), 0 \leq t\}$, by giving $\mathscr{F}^\circ$ the $P^x$-law of $X_t'$. Then setting $X_t' = w(t)$ on $\Omega'$ we have $P^{\varphi(x)}\{X_t' = X_t^x\} = 1$ except for countably many $t$. In particular, for each $x$, $T_{t-}f(x) = T_t f(x) = T_{t+}f(x)$, $\forall f \in b(\mathscr{E}_p)$, except for a countable set of $t$ (depending on $x$).*

**Proof** We first observe that, since $X_t'$ has measurable paths, and by definition $P^{\varphi(x)}(S') = P^x\{X'_{(\cdot)} \in S'\}$, $S' \in \mathscr{F}'$, there is an extension of $P^{\varphi(x)}$ to $\mathscr{F}' \vee \mathscr{F}^\circ$ which preserves the law of $X'$ on $\mathscr{F}^\circ$, namely that given by the above expression for $S' \in \mathscr{F}' \vee \mathscr{F}^\circ$. Next, we introduce the r.c.l.l. modifications $M_{\lambda,n}(t)$ of $e^{-\lambda t} R_\lambda f_n(w(t))$ on $\Omega'$, just as in the construction of $X_t'$ from $X_t$ (Lemma 2.8). (Of course, these are equivalent if we substitute $X_t'$ for $X_t$, $t \in \mathbb{Q}$, in defining them.) We claim that, for all $n$ and $\lambda > 0$,

$$P^{\varphi(x)}\{M_{\lambda,n}(t) = e^{-\lambda t} R_\lambda f_n(X_t^x) \text{ for all } t\} = 1,$$

when we extend $R_\lambda f_n$ to $e \in E_p - E$ by setting

$$R_\lambda f_n(e) = E^e \int_0^\infty e^{-\lambda t} f_n(w(t)) \, dt.$$

Indeed, we have

$$E^z \int_0^\infty e^{-\lambda t} f_n(w(t)) \, dt = \lambda E^z \int_0^\infty e^{-\lambda t} \left( \int_0^t f_n(w(s)) \, ds \right) dt,$$

which is continuous on $M$ since $\int_0^t f_n(w(s)) \, ds$ is continuous on $\Omega'$, uniformly in $t$. Since $X_t^x$ is r.c.l.l. in the topology of $E_p$ induced by $\varphi$ from $M$, we see that $R_\lambda f_n(X_t^x)$ is r.c.l.l. along with $e^{\lambda t} M_{\lambda,n}(t)$, so it suffices to show that they are equal $P^{\varphi(x)}$-a.s. on a dense set of $t$. But such a set is given by

$$\{t : X_t^x = \tilde{X}_t = w(t) \text{ and } M_{\lambda,n}(t) \equiv e^{-\lambda t} R_\lambda f_n(w(t))\}.$$

Now we had $P^{\varphi(x)}\{M_{\lambda,n}(t) = M_{\lambda,n}(t-)\} = 1$ except for countably many $t$, and for such $t$ we also had $P^{\varphi(x)}\{M_{\lambda,n}(t) = e^{-\lambda t} R_\lambda f_n(w(t))\} = 1$. Consequently, we have

$$P^{\varphi(x)}\{R_\lambda f_n(w(t)) = R_\lambda f_n(X_t^x)\} = 1$$

except for countably many $t$. Finally, since by transform inversion the functions

$$E^z \int_0^\infty e^{-\lambda t}\left(\int_0^t f_n(w(s))\,ds\right)dt, \qquad 1 \le n, \lambda > 0,$$

separate points of $M$, the $R_\lambda f_n$ separate points of $E$ (and of $E_p$ when extended), so for $t$ in the above set, $P^{\varphi(x)}\{w(t) = X_t^x\} = 1$, as required.

**Remarks and examples**

0. As a first type of example, let $X_t$, $X_0 = 0$, be a real-valued process with stationary, independent increments (often called a Lévy process). A good deal of the theory of stochastic processes was developed using such $X_t$ as a prototype (see for example Ito (1942)). Using our method it is easy to prove the existence of an r.c.l.l. version of $X_t$ without using any separability argument. (The case of non-stationary increments can also be approached, but it is, of course, more complicated, and needs extra hypotheses.)

To start with, we need a measurable version of $X_t$. It is easy to see that $X_t$ must have an infinitely divisible law for each $t$, and then it is an exercise (Breiman 1968, Chapter 14, Problem 3) to show that either $X_t$ is continuous in probability with log characteristic function linear in $t$, or else it differs from such a process by a fixed, non-measurable solution $\beta$ of the equation $\beta(t + s) = \beta(t) + \beta(s)$. We assume the former case, which is necessary and sufficient for the existence of a measurable standard modification. Then the homogeneous transition function $p(t, x, A) = P\{X_t + x \in A\}$, $A \in \mathscr{B}$, is measurable in $(t, x)$ and satisfies our Hypothesis. For example, the Chapman–Kolmogorov equation becomes the convolution identity

$$P\{X_{s+t} + x \in A\} = \int P\{X_t + x \in dy\}P\{X_{s+t} - X_t + y \in A\}.$$

Thus we can apply our method, and in particular Theorem 2.13.

For fixed $x$, let $\{t_j(x)\}$ be the (at most countable) set of $t$ at which $P^{\varphi(x)}\{X_t' \ne X_t^x\} > 0$. By homogeneity it is clear that

$$P\{Z_t^{\varphi(x)} \in \varphi(A)\} = P\{Z_t^{\varphi(x+y)} \in \varphi(A + y)\},$$

so $P^{\varphi(x)}\{Z_t^{\varphi(x)} \in \varphi(R)\}$ does not depend on $x$. By the same reasoning, $P^{\varphi(x)}\{X_t' \ne X_t^x \text{ and } X_t^x \in R\}$ does not depend on $x$, so $\{t_j(x)\} = \{t_j\}$ is also free of $x$. Now the Markov property of the pair $(X_t', X_t^x)$ with respect to the

## 60 | Foundations of the prediction process

extended $P^{\varphi(x)}$ shows that if $t \notin \{t_j\}$ then

$$P^x\{X'_{t+s} \neq X^x_{t+s}\} = E^x P^{\varphi(X_t)}\{X'_s \neq X^{\varphi(X_t)}_s\},$$

where the inner probability on the right does not depend on $\varphi(X'_t)$. Thus if $t \notin \{t_j\}$ then $t + s \in \{t_j\}$ if and only if $s \in \{t_j\}$, hence $\{t_j\}$ is invariant under all but countably many translations. This implies $\{t_j\}$ is empty, so $X'_t$ has a $P^{\varphi(x)}$-standard modification $X^x_t$ which is r.c.l.l. in the topology induced by $\varphi: R \to E$.

It remains to see that this induced topology is the natural topology of $R$. But $x_n \to x$ in $R$ if and only if, for every $w' \in \Omega'_R$,

$$e^{-s}(w'(s) + x_n) \to e^{-s}(w'(s) + x)$$

in $ds$-measure, and then for $f$ bounded, continuous on $\Omega'_R$

$$\lim_{n \to \infty} E^{\varphi(x_n)} f = \lim_{n \to \infty} E^{\varphi(0)} f(w' + x_n)$$
$$= E^{\varphi(0)} f(w' + x)$$
$$= E^{\varphi(x)} f,$$

proving that the topology of $R$ is finer. Conversely, consider

$$f_\lambda(w') = \int_0^1 e^{i\lambda w'(s)} \, ds, \quad -\infty < \lambda < \infty,$$

which are bounded and continuous on $\Omega'_R$ (separating the real and imaginary parts). Then since $E^{\varphi(x_n)} f_\lambda = e^{i\lambda x_n} E^{\varphi(0)} f_\lambda$, if $x_n$ converges in the topology induced by $\varphi$ then either $E^{\varphi(0)} f_\lambda = 0$ or $e^{i\lambda x_n}$ also converges. But $E^{\varphi(0)} f_\lambda$ is a characteristic function (of a variable on $(0, 1) \times \Omega'_R$) and therefore is continuous at $\lambda = 0$. It follows that $x_n \to x$ for a unique $x \in R$, for if $|x_n| \to \infty$ while $e^{i\lambda x_n} \to g(\lambda)$ for $|\lambda| < \varepsilon$ small, then by dominated convergence

$$\lim_{n \to \infty} \int_0^y e^{i\lambda x_n} \, d\lambda = \int_0^y g(\lambda) \, d\lambda = 0,$$

proving that $g(\lambda) = 0$ for a.e. $|\lambda| < \varepsilon$, and contradicting the fact that $|g(\lambda)| = 1$. Therefore, the r.c.l.l. version $X^x_t$ is the required process with homogeneous independent increments.

1. To understand the necessity of extending beyond $E$ in defining the regularization $X^x_t$ (Definition 2.10) we take a simple example: that of the reflecting Brownian motion kernel

$$p(t, x, y) = (2\pi t)^{-1/2} \left[ \exp\left\{ -\frac{(y - x)^2}{2t} \right\} + \exp\left\{ -\frac{(y + x)^2}{2t} \right\} \right].$$

It is well known that if $E = [0, \infty)$ then there is a well-behaved process with continuous paths and transition density $p$, and that this process reaches 0 from any $x$. But, considered in the abstract, suppose we did not know this and had chosen $E = (0, \infty)$. It is clear that our Hypothesis is equally satisfied by $p$ on $(0, \infty)$; indeed, we could omit any Borel set of Lebesgue measure 0. It is equally clear that no regular Markov process with transition density $p$ (or even $p$ except for countably many $t$) exists on $E = (0, \infty)$. This is simply the wrong state space, and to get a sensible process we must extend to $[0, \infty) = E_p \cap \overline{E}$. We note that $E_p$ alone contains a process for every initial distribution $\mu$ on $[0, \infty)$, so it has many superfluous branch points.

2. We know that $T_t f(x)$ is $(t, x)$-measurable, and so are $T_{t+} f(x)$ and $T_{t-} f(x)$ since they have one-sided weak-* continuity in $t$. Therefore, we have $\{(t, x): T_{t-} = T_t = T_{t+} \text{ at } x\} \in \mathscr{B}^+ \times \mathscr{E}$, and its complement has countable section in $t$ at each $x$. The following example shows that the countable exceptional set must be allowed to depend on $x$. Let $E = [0, \infty)$ and

$$T_t f(x) = \begin{cases} f(x+t) & \text{if } x+t < 1 \\ \frac{1}{2}f(1) + \frac{1}{4}(f(2) + f(3)) & \text{if } x+t = 1 \\ \frac{1}{2}(f(x+t+1) + f(x+t+2)) & \text{if } x+t > 1. \end{cases}$$

Thus, we have a process of uniform motion to the right except when it approaches 1, where it jumps to 2 or 3 with probability $\frac{1}{2}$, but at this time it is either right- or left-continuous, also with probability $\frac{1}{2}$ and independently of whether it jumps to 2 or 3. Thus $T_{(1-x)+} f(x) = \frac{1}{2}(f(2) + f(3))$ and $T_{(1-x)-} f(x) = f(1)$, $0 < x < 1$. We see that for $0 < x < 1$, $t = 1 - x$ is in the exceptional set, which therefore cannot be reduced to a countable set not depending on $x$. Returning to the state space of $E_p$, one notes that the exceptional $(t, x)$ are contained in the set of times $t$ such that $P^{\varphi(x)}\{Z^{\varphi(x)}_{t-} \neq Z^{\varphi(x)}_t\} > 0$, i.e. the fixed discontinuity times of $Z^{\varphi(x)}_t$.

3. Let us clarify the role of $\overline{E}$. There are, in general, many 'superfluous' elements in $E_p$. Any branching point in $E_p$ will never be visited for $t \geq 0$ by any r.c.l.l. process with semigroup $T_{t+}$ (but they may be visited by the l.c.r.l. process with $T_{t-}$ as semigroup). In 1. above, all initial distributions except those with $\mu = \delta_x$, $x \in \overline{E}$, can be discarded as far as representing the processes is concerned, because they will never be visited (even as left limits, unless by definition at $t = 0$). In general, we really only need $E_p \cap \overline{E}$, which equals $\overline{E}$ in this example. This is because any initial value in $E_p$ leads immediately (i.e. for arbitrarily small $t$) into $E$, and any process starting in $E$ remains thereafter in $\overline{E}$, along with its left-limit process.

## 62 | Foundations of the prediction process

**Exercise 2.13** Let $P\{\hat{e} > t\} = e^{-t}, t \geq 0$, and consider a particle in $\mathbb{R}^3$ which starts at $(x, 0, 0)$, $x \geq 0$, and translates at uniform velocity 1 along the $x$-axis until time $t = \hat{e}$. For $t \geq \hat{e}$ the $x$-coordinate is fixed but the particle translates either parallel to the $y$-axis or $z$-axis, each with probability $\frac{1}{2}$. Show that $Z_t^{\varphi(x)}$ can be viewed as a process with values in $\mathbb{R}_x^+ \cup \mathbb{Q}_{x,y}^+ \cup \mathbb{Q}_{x,z}^+$ where $\mathbb{Q}_{x,y}^+$ and $\mathbb{Q}_{x,z}^+$ are closed, disjoint quadrants. Thus identify $E_p \cap \overline{\overline{E}}$ in this example (noting that the process is a simple Markov process) and describe $\{t: X_t^x \neq \widetilde{X}_t\}$ (noting that $\widetilde{X}_t$ is r.c.l.l. in the obvious topology). Solution: $\{t: X_t^x \neq \widetilde{X}_t\} = \{\hat{e}\}$.

## 2.3 Prediction space: Ray processes and right processes

Besides applying $Z_t^z$ to the study of Markov processes, it is also useful to consider applying the theory of Markov processes to $Z_t^z$. In the first place, this suggests that we eliminate the dependence on $z$ by introducing a canonical space of paths for all of the processes $Z_t^z$, for fixed $E$ and $M$. In other words, we treat $q(t, z, A)$ of Definition 1.6 in the role of $p(t, x, B)$. But this raises the question of whether $q$ needs any regularization such as we carried out for general $p$. It turns out that it does not. To see what is involved, recall (Theorem 2.13) that regularization of $T_t$ is needed only for $t$ with $P^{\varphi(x)}\{e^{-\lambda t} R_\lambda f_n(X_t') = M'_{\lambda,n}(t)\} \neq 1$ for some $0 < \lambda \in \mathbb{Q}$ and $n \geq 1$. Now for $Z_t^z$ we have

$$e^{-\lambda t} R_\lambda^Z f_n(Z_t^z) = e^{-\lambda t} E^{Z_t^z} \int_0^\infty e^{-\lambda s} f_n(Z_s^{Z_t^z}) \, ds,$$

and no regularization is needed provided that these are $P^z$-a.s. right-continuous along the paths of $Z_t^z$. In the particular cases $f_n(z) = E^z g_n$, $g_n \in b(\mathcal{F}')$, this becomes

$$e^{-\lambda t} E^{Z_t^z} \int_0^\infty e^{-\lambda s} E^{Z_t^z}(g_n \circ \Theta_s \mid \mathcal{F}'_{s+}) \, ds = e^{-\lambda t} E^{Z_t^z} \int_0^\infty e^{-\lambda s}(g_n \circ \Theta_s) \, ds,$$

which is right-continuous in $t$ by Theorem 2.1. In particular, if we take $g_n \in C(\overline{\Omega}')$, these $f_n$ are continuous on $\overline{M}$ as required by our definitions (Definition 2.1). But of course the difficulty is that such $f_n$ are not uniformly dense in $C(\overline{M})$, hence they do not suffice to avoid the regularization.

Nevertheless, postponing this problem, we introduce a basic definition.

**Definition 2.13** The canonical prediction space of $E$ is $(\Omega_Z, \mathcal{L}^\circ_{t+}, \Theta_t^Z, P^z)$, where

$\Omega_Z = \{$r.c.l.l. paths $w_z(t)$ with values in $M_E \cap D$ (prediction topology for the non-branch points) and left limits in $M_E\}$,

$$\mathscr{L}^\circ_{t+} = \bigcap_{\varepsilon > 0} \sigma\{w_Z(s) \in A; A \in \mathscr{M}_E, s \leq t + \varepsilon\}, \qquad \Theta^z_t w_Z(s) = w_Z(t + s),$$

and

$$P^z(S) = P^z\{Z^z_{(\cdot)} \in S\}, \qquad S \in \mathscr{L}^\circ\left(= \bigvee_t \mathscr{L}^\circ_{t+}\right), z \in M_E.$$

Now we can summarize the implications of Theorem 1.14 as follows.

**Theorem 2.14** *The coordinate process $w_Z(t)$ on $(\Omega_Z, \mathscr{L}^\circ_{t+}, \Theta^Z_t, P^z)$ is an r.c.l.l., strong Markov process with transition function $q$ on $M_E$. Moreover, its left-limit process $w_Z(t-)$, $t > 0$, is moderately Markov with the same $q$.*

**Remarks**

1. We also, of course, define $P^\mu = \int P^z \mu(dz)$ for any initial distribution $\mu$ on $(M, \mathscr{M})$, and we may augment the filtrations $\mathscr{L}^\circ_{t+}$ by inclusion of all sets which are $P^z$-null in the $P^z$-completion of $\mathscr{L}^\circ$ for all $z$, in the usual manner of probabilistic potential theory. Indeed, all of the theory of Blumenthal and Getoor (1968) not requiring quasi-left-continuity applies to the prediction process. From now on we let $Z_t$ denote $w_z(t)$ and we continue to drop the subscript $E$. Note that the space $\Omega_Z$ is in some sense much larger than the range of any $Z^z_t$. In particular, we now have, of course, $Z_{t+s} = Z_s \circ \Theta^Z_t$, whereas no such relation was possible on $(\Omega', \mathscr{F}')$.

2. The space $\Omega_Z$ is not particularly 'nice' from an analytical viewpoint. The best that we can say, using Dellacherie and Meyer (1975, IV, 19) is that it is a 'complement of an analytic set'. This implies that it is universally measurable, but it is not a Lusin or even a Suslin space. This should not be of much concern; we needed a Lusin space in order to define $Z^z_t$, but once this is done there is no more need for special conditional distributions.

3. Let us trace how the space can be applied to study a process $X_t$ with values in $(E, \mathscr{E})$, i.e. a given $P$ for such $X_t$. We first must have a measurable standard modification $X'_t$, thus defining a $P^z$ on $(\Omega', \mathscr{F}')$ by $P^z(S') = P\{X'_{(\cdot)} \in S'\}$. Then we introduce the coordinates $\hat{\rho}(Z^z_t)$, and recall that $P^z(S') = P^z\{\hat{\rho}(Z^z_{(\cdot)}) \in S'\}$ because $\hat{\rho}(Z^z_t) = \tilde{X}_t$ for a.e. $t$. Now the coordinates $\hat{\rho}(Z^z_t)$ become just $\hat{\rho}(Z_t)$ on the canonical space, so we have finally

$$P\{X'_{(\cdot)} \in S'\} = P^z\{\hat{\rho}(Z_{(\cdot)}) \in S'\}, \qquad S' \in \mathscr{F}'.$$

Now we are ready to use the theory of Markov processes on $\Omega_Z$ to help in studying probabilities of this type.

## 64 | Foundations of the prediction process

As we saw in our first application, it is important to be able to narrow down both $M$ and $\Omega_z$ to fit particular purposes.

**Definition 2.14**

(1) A Borel prediction demesne is a set $U \in \mathcal{M}$ such that
$$P^z\{Z_t^z \in U \text{ for all } t \geq 0\} = 1 \quad \text{for all } z \in U.$$

(2) We say that $U$ is *complete* if, in addition,
$$P^z\{Z_{t-}^z \in U \text{ for all } t > 0\} = 1 \quad \text{for } z \in U.$$

**Remarks** In our application we saw (Lemma 2.11, ff.) that for a Markov transition function $p$ satisfying the Hypothesis of Section 2.2 is associated a complete Borel demesne $M_p \cap \overline{\varphi(E)}$. Of course if we do not ask for completeness we can replace any demesne $U$ by $U \cap D$, also a demesne (by Theorem 1.15) and it suffices for the same class of processes $P^z$ as $U$ does. We can then introduce the canonical subspace $\Omega_U$ (or $\Omega_{U \cap D}$) consisting of all r.c.l.l. paths with values in $U$ (or in $U \cap D$) and if $U$ is complete then it may be restricted further to have left limits in $U$. Finally, in some instances one does not have $U \in \mathcal{M}$, but if the defining properties hold with respect to the augmented $\mathcal{F}^z$ and if $U$ is universally measurable in $(M, \mathcal{M})$, one still refers to $U$ as a demesne (or complete demesne), without the 'Borel'. For construction of demesnes, see also Knight (1981a, Essay I, pp. 21–4), where the reference to outer measure is superfluous.

Now let us return to the $p(t, x, B)$ of the Hypothesis of Section 2.2 and ask the question: Under what further conditions do we have, for each $x \in E$, $P^x\{\tilde{X}_t = X_t^x, \forall t \geq 0\} = 1$, and also $T_t f(x) = T_{t+} f(x), \forall t \geq 0, x \in E, f \in b(\mathcal{E})$? In other words, when is the modification of Definition 2.10 redundant on $E$ (so that in particular $\varphi(E)$ is a Borel demesne)? If we go back to the basic argument of Theorem 2.10 (and Theorem 2.13) we see that the key is justifying the equation $Z_t^{\varphi(x)} = \varphi(\tilde{X}_t)$, which was true for 'non-exceptional $t$' with $P^x$-probability 1. But if it can be shown simultaneously in $t \geq 0$, then we do not need modification. Now an obvious sufficient condition is the following.

**Criterion**

(1) For each $x$, there exists a right-continuous choice of the process $X_t'$ on $E$ for $P^x$, and we have the Markov property relative to $\mathcal{F}_{t+}^\circ$ (not just $\mathcal{F}_t^\circ$).

(2) With probability 1 for $P^{\varphi(x)}$, $\varphi(\tilde{X}_t)$ is right-continuous in $t$ for the topology of $M$.

Indeed, (1) implies that (1) in the proof of Theorem 2.10 is met for all $t$, and then the Markov property relative to $\mathcal{F}_{t+}^\circ$ implies the Markov property with $\mathscr{F}'_{t+}$ in place of $\mathscr{F}'_t$, namely for each $t$, $P^x\{Z_t^{\varphi(\tilde{X}_t)} = \varphi(\tilde{X}_t)\} = 1$. But then (2) implies that both sides of this equality are right-continuous in $t$, so it extends simultaneously over $t \geq 0$. Finally, we note that, by Hunt's lemma as $\tau \downarrow t$, $\tau \in \mathbb{Q}$, the $M_{\lambda,n}(t)$ of Lemma 2.8 satisfy

$$M_{\lambda,n}(t) = e^{-\lambda t} E^x\left(\int_0^\infty e^{-\lambda s} f_n(X_{t+s})\, ds \mid \mathscr{F}^o_{t+}\right),$$

whereas

$$e^{-\lambda t} R_\lambda f_n(X'_t) = e^{-\lambda t} E^x\left(\int_0^\infty e^{-\lambda s} f_n(X_{t+s})\, ds \mid \mathscr{F}^o_t\right)$$

by the Markov property. But since $X'_t$ is Markov for $\mathscr{F}^o_{t+}$ these are equal, and it follows as in Theorem 2.13 that $T_t = T_{t+}$ on $E$.

To connect the Criterion with the literature of Markov processes, we will show that it holds for two basic types, called respectively Borel *right processes*, and *Ray processes*. It also emerges that the Markov property relative to $\mathscr{F}^o_{t+}$ follows from (2) and the first part of (1), so this was redundant.

Recalling the dense sequence $(f_k)$ on $\overline{E}(\supset E)$ of Definition 2.1, the following is a key analytical result.

**Lemma 2.15** *Under the Hypothesis in Section 2.2, $\varphi(x)$ is continuous, $E \to M$, if and only if $R_\lambda f(x)$ is continuous on $E$ for bounded $f \in C(E)$.*

**Remark** The result does not depend on the choice of $\overline{E}$. However, following Definition 2.5 it suffices to show continuity of $E^{\varphi(x)} g$ for $g \in C(\overline{\Omega})$, where we can take $\overline{\Omega}$ to be the closure of $\Omega'_E$ under the map given there.

**Proof** First of all, we have for bounded $f \in C(E)$,

$$\int_0^\infty e^{-\lambda t} f(\tilde{X}_t)\, dt = \lambda \int_0^\infty e^{-\lambda t}\left(\int_0^t f(\tilde{X}_s)\, ds\right) dt$$

where $\int_0^t f(\tilde{X}_s)\, ds$ is continuous on $\Omega$, uniformly in bounded $t$ (say $t \leq N$), while the tails $\int_N^\infty (\ )\, dt$ are uniformly small (since $\int_0^t f(\tilde{X}_s)\, ds \leq t$). Therefore $E^z \int_0^\infty e^{-\lambda t} f(\tilde{X}_t)\, dt$ is continuous on $M$, and so the condition of Lemma 2.15 is necessary.

Conversely, since $\{\int_0^t f_k(\tilde{X}_s)\, ds\}$ separates points of $\overline{\Omega}$ and are continuous in $t$, it follows by inversion in $\lambda$ that $\{\int_0^\infty e^{-\lambda s} f_k(\tilde{X}_s)\, ds\}$ also separates points in $\overline{\Omega}$. Then by the Stone–Weierstrass theorem, linear combinations of $\prod_{k=1}^m \int_0^\infty e^{-\lambda_k t} f_{n_k}(\tilde{X}_t)\, dt$, $1 \leq m$, are uniformly dense in $C(\overline{\Omega})$. So it is enough to show each $E^{\varphi(x)} \prod_{k=1}^m \int_0^\infty e^{-\lambda_k t} f_{n_k}(\tilde{X}_t)\, dt$ is continuous, where the case

$m = 1$ is our hypothesis. Writing the general case as

$$E^{\varphi(x)} \int_0^\infty \cdots \int_0^\infty \exp\left(-\sum_{k=1}^m \lambda_k s_k\right) \Pi_{k=1}^m f_{n_k}(\tilde{X}_{s_k}) \, ds_1 \cdots ds_m,$$

we separate it into $m!$ possible orderings of $s_1, \ldots, s_m$. It will suffice to write the case $s_1 < \cdots < s_m$. By the Markov property of Theorem 2.12 this becomes

$$E^{\varphi(x)} \int_0^\infty e^{-\lambda_1 s_1} f_{n_1}(\tilde{X}_{s_1}) \int_{s_1}^\infty e^{-\lambda_2 s_2} f_{n_2}(\tilde{X}_{s_2}) \int_{s_2}^\infty \cdots \int_{s_{m-1}}^\infty e^{-\lambda_m s_m} f_{n_m}(\tilde{X}_{s_m}) \cdots ds_1$$

$$= E^{\varphi(x)} \int_0^\infty e^{-\lambda_1 s_1} f_{n_1}(\tilde{X}(s_1)) E^{\varphi(\tilde{X}_{s_1})} \Bigg[\int_0^\infty e^{-\lambda_2(s_1+s_2)} f_2(\tilde{X}_{s_2})$$

$$\times \int_{s_2}^\infty \cdots \int_{s_{m-1}}^\infty e^{-\lambda_m(s_1+s_m)} f_{n_m}(\tilde{X}_{s_m}) \cdots ds_2 \Bigg] ds_1$$

$$= E^{\varphi(x)} \int_0^\infty \exp\left\{-\left(\sum_{k=1}^m \lambda_k\right) s_1\right\} f_{n_1}(\tilde{X}_{s_1}) E^{\varphi(\tilde{X}_{s_1})}(T_{m-1}) \, ds_1,$$

where

$$T_{m-1} = \int_0^\infty e^{-\lambda_2 s_2} f_{n_2}(\tilde{X}_{s_2}) \int_{s_2}^\infty e^{-\lambda_3 s_3} f_{n_3}(\tilde{X}_{s_3}) \cdots \int_{s_{m-1}}^\infty e^{-\lambda_m s_m} f_{n_m}(\tilde{X}_{s_m}) \cdots ds_2.$$

Now $T_{m-1}$ is an expression of the same type as each of the $m!$ terms in $\Pi_{k=1}^m (\int_0^\infty e^{-\lambda_k t} f_n(\tilde{X}(t)) \, dt)$, but with $m-1$ in place of $m$, so by the induction hypothesis $E^{\varphi(x)} T_{m-1}$ may be assumed continuous in $x$. But then the integrand $f_{n_1}(x) E^{\varphi(x)} T_{m-1}$ is also bounded and continuous in $x$, so by the case $m = 1$ the first line above is continuous in $x$. This completes the induction, and the proof.

**Remark** A further observation which will be useful is that according to the above proof,

$$E^{\varphi(x)}\left(\Pi_{k=1}^m \int_0^\infty e^{-\lambda_k t} f_{n_k}(\tilde{X}_t) \, dt\right) = R_{\sum_{k=1}^m \lambda_k} g(x)$$

where $g(x) = \sum_\sigma f_{\sigma(1)}(x) E^{\varphi(x)} T_{m-1}(\sigma)$ and $\sigma$ ranges over permutations of $n_1, \ldots, n_m$. This fact did not require any continuity assumption on $R_\lambda f$.

**Corollary 2.16** $\varphi(x)$ is continuous, $\bar{E} \to \bar{M}$, if and only if $R_\lambda f$ maps $C(\bar{E}) \to C(\bar{E})$, when extended from $E$ by continuity.

**Proof** Same as above.

We turn now to the case where $T_t$ is a Borel right semigroup in the Hypothesis of Section 2.2. The following definition is based on the definition of Walsh and Meyer (1971, pp. 143–66) or Getoor (1975, p. 60).

**Definition 2.15** $T_t$ is a Borel right semigroup if

(1) for each probability $\mu$ on $(E, \mathscr{E})$ there is a process $X_t$ with coordinate law $P^\mu$, with right-continuous paths, and $P^\mu(X_0 \in A) = \mu(A)$ for $A \in \mathscr{E}$ (note that as $T_t$ is Borel, it suffices to use $\mathscr{E}$ instead of the universal extension $\mathscr{E}^*$ of Getoor (1975)); and

(2) for the above processes (then called *Borel right processes*) if $f$ is $\lambda$-excessive then $f(X_t)$ is right-continuous $P^\mu$-a.s.

Now we can easily obtain the following theorem.

**Theorem 2.17** *For a Borel right semigroup, $P^{\varphi(x)}\{\tilde{X}_t = X_t^x, \forall t \geq 0\} = 1$. In other words, the coordinate process $\tilde{X}_t$ satisfies $P^{\varphi(x)}\{\varphi(\tilde{X}_t) = Z_t^{\varphi(x)}, \forall t \geq 0\} = 1$. Also $T_t = T_{t+}$ on E. Therefore, $\tilde{X}_t$ is a strong Markov realization of the process with semigroup $T_t$, and its left limits in $E_p \cap \bar{E}$ are moderately Markov for $T_t$ on $E_p \cap \bar{E}$, and $\tilde{X}_t$ is r.c.l.l. ($P^{\varphi(x)}$-a.s.).*

**Proof** Let us first prove Criterion (2), p. 64 above. According to the Remark before Corollary 2.16, it suffices that $R_{\sum_1^m \lambda_k} g(\tilde{X}_t)$ be right-continuous in $t$ with probability 1 for each $P^\mu$. Now since the process $X_t$ can be chosen right-continuous by (1), it follows that $\tilde{X}_t$ has the same coordinate law as $X_t$ for $P^\mu$. In fact, $P^\mu$ is concentrated on the set $\{r.c.\} = \{\text{right-continuous paths}\}$, which is universally measurable (Dellacherie and Meyer 1975, IV, Theorem 34), and on this set $\tilde{X}_t$ is right-continuous. Therefore, on this set $\tilde{X}_t$ is also a right-continuous realization of the process with semigroup $T_t$ and initial measure $\mu$. Now $R_{\sum_1^m \lambda_k} g$ is $(\sum_1^m \lambda_k)$-excessive for $T_t$, so it follows from Criterion (2) that $R_{\sum_1^m \lambda_k} g(\tilde{X}_t)$ is $P^\mu$-a.s. right-continuous, where $P^\mu$ is restricted to $\{r.c.\}$. But $P^\mu\{r.c.\} = 1$ since $\{r.c.\}$ is universally measurable, so we have shown (2). Now according to the proof following the Criterion this already implies $P^x\{\tilde{X}_t = X_t^x, \forall t \geq 0\} = 1$. It remains to show that $T_t = T_{t+}$, or, equivalently, that $X_t$ is Markov relative to $\mathscr{F}_{t+}^o$. But when we represent this statement on $(\Omega', \mathscr{F}')$ with $\tilde{X}_t$ in place of $X_t$, which is possible as shown above, it suffices that $\tilde{X}_t$ be Markov relative to $\mathscr{F}_{t+}'$ (since $\sigma(\tilde{X}_s, s \leq t) \equiv \mathscr{F}_t'$). Since $\varphi(\tilde{X}_t) = Z_t^{\varphi(x)}$, and $\varphi$ is one-to-one on $E$, this is just another way of saying that $Z_t^{\varphi(x)}$ is Markov relative to $\mathscr{F}_{t+}'$, which holds by Theorem 1.14(1).

**Remarks** This argument shows that, as mentioned before, the hypothesis '$X_t'$ is Markov relative to $\mathscr{F}_{t+}^o$', in Criterion (1) is redundant. Moreover, Definition 2.15(1) is largely superfluous for Theorem 2.17. To see this, recall that the definition of $\lambda$-excessive function for $T_t$ does not involve any topology on $E$. It depends only on $\mathcal{E}$ and $T_t$. Now (2) implies as we have seen that $\varphi(X_t)$ can be made right-continuous in $t$. (We simply reformulate (2) to state that for *some* process $X_t$ with coordinate law $P^\mu$, $f(X_t)$ is a.s. right-continuous for $\lambda$-excessive $f$.) So it follows that $X_t$ can be chosen right-continuous in the topology on $E$ induced by $\varphi$ from that of $M$. Since $\varphi$ is one-to-one and Borel, $\mathcal{E}$ is again the Borel field of $E$ for this topology, and $E$ is a Lusin space. Therefore we have (1) in this new topology, and (2) still holds so the proof is unchanged.

We insert here a special case which is of fundamental importance in probabilistic potential theory.

**Definition 2.16** A Hunt process is a Borel right process which has left limits and is *quasi-left-continuous*: equivalently, if $T_1 < T_2 < \cdots < T_n < \cdots$ are $\mathcal{F}_t$ ($\equiv \mathcal{F}^o_{t+}$)-stopping times with $\lim_n T_n = T$, then

$$P^x\left\{\lim_{n\to\infty} X_{T_n} = X_T \text{ on } T < \infty\right\} = P^x\{T < \infty\}.$$

Now we have quite easily

**Theorem 2.18** *If $T_t$ is a Borel right semigroup, then the process $X_t$ is a Hunt process in the $\varphi$-induced (alternatively in the Ray) topology if and only if the demesne $\varphi(E)$ is complete.*

**Remark** The Ray topology (Definition 2.18 below) is analogous. It follows as in Dellacherie and Meyer (1987, XVII, Section 11) that any Hunt process is also a Hunt process in the $\varphi$-induced and Ray topologies. This result depends on the previsible section theorem.

**Proof** We must have $\varphi(E) \cap B = \Phi$ by right-continuity at $t = 0$ for any Borel right semigroup. Moreover, completeness of $\varphi(E)$ is obviously necessary for left limits. Conversely, let $T_1 < T_2 < \cdots$ be as above. After translating to $(\Omega', \mathcal{F}')$ and changing on a $P^{\varphi(x)}$-null set, we can take $T_n$ to be $\mathcal{F}'_{t+}$-stopping times. Then we have

$$P^{\varphi(x)}(\Theta_T^{-1} S' \mid \mathcal{F}'_{T+}) = P^{Z_T}(S'), \quad Z_T = \varphi(X_T),$$

and

$$P^{\varphi(x)}(\Theta_T^{-1} S' \mid \mathcal{F}'_{T-}) = P^{Z_{T-}}(S'), \quad Z_{T-} = \varphi(X_{T-}).$$

So

$$\{X_T \neq X_{T-}\} \stackrel{\text{a.s.}}{=} \{Z_T \neq Z_T\} \stackrel{\text{a.s.}}{=} \{Z_{T-} \in B\}.$$

(Take $S'_k = \{Z_0 \in A_k\}$ where $A_k$ generate $\mathcal{E}$.) Therefore, $P^x\{X_T \neq X_{T-}\} = 0$.

Assumption (2) of a right semigroup, however, looks artificial and unverifiable, at least to the non-potential-theorist. The following definition contains a purely analytical assumption which has far-reaching consequences.

**Definition 2.17** $R_\lambda$ is a Ray resolvent if

(1) $E$ is compact and $R_\lambda: C(E) \to C(E)$;

(2) the $\alpha$-supermedian functions separate points. ($0 \le f$ is $\alpha$-*supermedian*, $\alpha > 0$, if $\lambda R_{\alpha+\lambda} f \le f$ for $\lambda \ge 0$.)

Actually under the Hypothesis in Section 2.2, (2) holds automatically because $\{R_\lambda f, 0 \le f \le 1\}$ separates points and such $R_\lambda f$ are $\lambda$-supermedian. It should be noted further that since

$$R_\lambda f(x) = \int_0^\infty e^{-\lambda t} T_t f(x)\, dt = \int_0^\infty e^{-\lambda t} T_{t+} f(x)\, dt, \quad x \in E,$$

the resolvent $R_\lambda$ serves equally for $T_t$ and for $T_{t+}$ on $E$ (that is, for $f \in b(\mathscr{E}_p)$, $x \in E$, and $T_{t+}$ from Definition 2.12).

**Theorem 2.19**

(1) *If $R_\lambda$ is a Ray resolvent on $E$, then $\varphi(E)$ is a complete Borel demesne, and for $z \in \varphi(E)$, $P^z\{r.c.l.l.\} = 1$ (i.e. $P^z\{\hat\rho(Z_{(\cdot)})$ is r.c.l.l.$\} = 1$). Then for $x \in E$,*

$$P^{\varphi(x)}\{\tilde X_t = X_t^x, \forall t \ge 0, \tilde X_{t-} = X_{t-}^x, \forall t > 0\} = 1.$$

(2) *If $E$ is not compact, but $R_\lambda: C(\bar E) \to C(\bar E)$ as in Corollary 2.16, then we can extend $T_{t+}$ to a Markov semigroup on the compactification $E^\varphi$ of $E$ induced by $\varphi$ from $\overline{\varphi(E)}$ (close in $M$, as following Definition 2.5) in such a way that the situation of (1) prevails on $E^\varphi$.*

**Proof** (1) By Corollary 2.16, $\varphi(x)$ is continuous on (compact) $E$, hence its range $\varphi(E)$ is compact in $M$. Since $P^{\varphi(x)}\{Z_t^{\varphi(x)} \in \varphi(E)$ for a.e. $t\} = 1$, by Theorem 2.12(2), and $Z_t^{\varphi(x)}$ is r.c.l.l., it follows that

$$P^{\varphi(x)}\{Z_t^{\varphi(x)} \in \varphi(E), \forall t \ge 0\} = 1.$$

This implies $P^{\varphi(x)}\{X_t^x$ is r.c.l.l. on $E\} = 1$ because $X_t^x = \varphi^{-1}(Z_t^{\varphi(x)})$ and $\varphi^{-1}$ is continuous. Now $X_t^x$ is Markov with transition semigroup $T_{t+}$, so $T_{t+} I_E(x) = 1$ for $x \in E$. Therefore, since $\varphi(x)$ is the same for $T_{t+}$ as it is for $T_t$, $P^{\varphi(x)}\{r.c.l.l.\} = 1$, or in other words $P^{\varphi(x)}\{\tilde X_t$ is r.c.l.l.$\} = 1$. But then, from Theorem 2.12(2) again, it follows that

$$1 = P^{\varphi(x)}\{\tilde X_t = X_t^x, \forall t \ge 0 \text{ and } \tilde X_{t-} = X_{t-}^x, \forall t > 0\},$$

as asserted. We remark that when we need the analogue of $\{r.c.l.l.\}$ on $\Omega_Z$, we have to note that the map $\hat\rho(Z_{(\cdot)}): (\Omega_Z, \mathscr{L}_\infty^o) \to (\Omega_E', \mathscr{F}_E')$ is $(\mathscr{L}_\infty^o/\mathscr{F}_E')$-measurable, because $\hat\rho(Z_t)$ is a measurable process. We have seen (Dellacherie and Meyer 1975, IV, Theorem 34) that $\{r.c.l.l.\}$ is universally $\mathscr{F}_E'$-measurable, so $\{\hat\rho(Z_{(\cdot)}) \in r.c.l.l.\}$ is also universally $\mathscr{L}_\infty^o$-measurable.

70 | Foundations of the prediction process

We shall not insist on the details of the proof of (2), which is not entirely self-contained. By Corollary 2.16, $\varphi$ has a unique continuous extension from $E$ to $\bar{E}$. Since the range $\varphi(\bar{E})$ is compact, it must contain the complete Borel demesne $\overline{(\varphi(E)\cap M_p)}$. Now when we compactify $E$ to $E^\varphi$ in the metric induced by $\varphi$ from $\bar{M}$, we obtain a one-to-one continuous extension of $\varphi$ to $E^\varphi$, whose range is also $\varphi(\bar{E})$. Moreover, we can extend $R_\lambda$ to map $C(E^\varphi) \to C(E^\varphi)$. Indeed, $E^\varphi$ is the quotient space of $\bar{E}$ by the map $R_\lambda$, since $R_\lambda$ separates two points of $\bar{E}$ if and only if $\varphi$ does so (see the Remark before Corollary 2.16). In particular, we have $(E_p \cap \bar{E}) \subset E^\varphi$ as a Borel subset, since $E_p \cap \bar{E} = \varphi^{-1}(\overline{\varphi(E) \cap M_p})$ and Lusin's theorem applies. Thus we can view $T_{t+}$ as a semigroup on a Borel subset of $E^\varphi$.

The one point at which our argument is not self-contained is the assertion that $T_{t+}$ can be extended to a semigroup on $E^\varphi$, whose resolvent is the extended $R_\lambda$. For this step we appeal to Ray's original theorem concerning existence of semigroups for *Ray resolvents*. When we do not have a semigroup to start with, as here, we define a Ray resolvent by the resolvent identity

$$R_\lambda - R_\mu = (\mu - \lambda) R_\lambda R_\mu, \qquad 0 < \lambda, \mu, \tag{2.2}$$

together with the usual $0 \leq \lambda R_\lambda f$ for $0 \leq f \in b(E^\varphi, \mathscr{E}^\varphi)$, $\lambda R_\lambda 1 = 1$, separation of points, and preservation of $C(E^\varphi)$. Now all of these hold for our extended $R_\lambda : C(E^\varphi) \to C(E^\varphi)$. Only the first requires comment, but since $R_\lambda f \in C(E^\varphi)$ for $f \in C(E^\varphi)$ we see that for $x \in E^\varphi$, $R_\lambda R_\mu f(x) = \lim_{y \to x, y \in E} R_\lambda R_\mu f(y)$, so (2.2) is inherited from $E$. Now Ray's theorem asserts the existence of a unique semigroup $T_{t+}$ on $C(E^\varphi)$ which is right-continuous in $t$ at each $x$, and since $\varphi$ is continuous it is seen by Definition 2.12 that this $T_{t+}$ equals the former for $x \in E_p \cap \bar{E}$, completing the proof.

**Remark** Various proofs of Ray's theorem are known. The most elementary one seems to be that of Meyer (1966, X, Theorem 19). Moreover, this proof has an interesting corollary not stated there, but for completeness we include it here. We recall that, for compact $E$, a *Feller semigroup* $T_t$ is a Markov semigroup (satisfying the Hypothesis in Section 2.2) such that $T_t$ maps $C(E) \to C(E)$, and, for $f \in C(E)$, $T_t f \to f$ uniformly as $t \to 0+$ (strong continuity).

**Corollary 2.20** *If a Ray semigroup $T_{t+}$ has no branch points (i.e. $\varphi(E)$ has none) then it is a Feller semigroup.*

**Proof** In this case $T_{0+} = I$, and hence by Meyer (1966, (19.2)) the continuous $\lambda$-supermedian are $\lambda$-excessive. Letting $\mathscr{S}_\lambda$ denote the continuous, $\lambda$-excessive functions, it is known (Meyer 1966) that $\mathscr{S}_\lambda - \mathscr{S}_\lambda$ is uniformly dense in $C(E)$, and (by the Hille–Yosida theorem) $T_{t+}$ is strongly continuous on the closure of $R_\lambda(C(E))$. Thus it suffices to show that $\mathscr{S}_\lambda - \mathscr{S}_\lambda \subset \overline{R_\lambda(C(E))}$. But, for $f \in \mathscr{S}_\lambda$, $\lim_{\mu \to \infty} \mu R_{\mu+\lambda} f = f$ and the limit is monotone increasing. Hence by

Dini's theorem it is uniform. Since $R_\lambda(C(E)) = R_\mu(C(E))$ the assertion follows.

We have thus shown that for Borel right processes, and for Ray processes, the modification of Theorem 2.12 is superfluous. For such processes $\varphi(E)$ is already a Borel demesne and, for any initial distribution, the process may be identified with its own prediction process by just applying $\varphi$ simultaneously in $t \geq 0$ (with probability 1). But this leaves open what is perhaps the main question, namely that of $Z_t$ itself. As pointed out prior to Definition 2.13, it is not yet clear whether $Z_t$ can be identified with *its* prediction process for all $t$, and in particular we have not yet shown that $Z_t$, restricted to $D$, is a Borel right process. We turn now to studying this question (an affirmative answer follows directly from Getoor (1975, (9.4) (i), p. 53) but this invokes capacity theory and perhaps misses some insight into the underlying reasons). Recalling our Criterion, (1) is immediately clear for $Z_t$, i.e. it is right-continuous and Markov relative to $\mathcal{L}_{t+}^\circ = \mathcal{L}_t^\circ$ up to $P^z$-null sets for each $z$. But it remains to prove (2), which is the same as showing $P^z\{\varphi(Z_t)$ is right-continuous$\} = 1$, where $\varphi: M \to M_M$ is derived from the prediction process of $Z_t$ itself.

**Lemma 2.21**

(1) *In the case of $Z_t$, the topology induced by $\varphi$ is finer than the prediction topology of $M$.*

(2) $P^z\{\varphi(Z_t) \text{ is r.c., with left-limits } \varphi(Z_{t-}), t > 0\} = 1$.

**Proof** (1) As noted at the start of the proof of Lemma 2.15, the topology induced by $\varphi$ is finer than that induced by $R_\lambda^z f(z)$, $f \in C(\bar{M})$, where $\bar{M}$ is compact metric, $\bar{M} \supset M$. In particular, suppose $f(z) = E^z h$ where $h \in C(\bar{\Omega})$. Then

$$R_\lambda^z f(z) = E^z \int_0^\infty e^{-\lambda t} E^{Z_t^z} h \, dt$$

$$= E^z \int_0^\infty e^{-\lambda t} E^z(h \circ \Theta_t \mid \mathcal{F}_{t+}') \, dt$$

$$= E^z \int_0^\infty e^{-\lambda t} h \circ \Theta_t \, dt.$$

As in the proof of Theorem 2.4, $h \circ \Theta_t$ is uniformly continuous in $t$, uniformly on $\bar{\Omega}$, and so it follows that $E^z h$ ($= \lim_{\lambda \to \infty} \lambda E^z \int_0^\infty e^{-\lambda t} h \circ \Theta_t \, dt$) must be continuous in the topology induced by $\varphi$. But $E^z h$, $h \in C(\bar{\Omega})$, generate the prediction topology of $M$, so the proof of (1) is complete.

Turning to (2), which is really a crucial step in understanding the nature of $Z_t$, let us recall what it means for $Z_t^z$ to be r.c.l.l. Again as for Theorem 2.4, this means that $E^{Z_t^z} g(\int_0^{t_1} f_1(w(s))\, ds, \ldots, \int_0^{t_n} f_n(w(s))\, ds)$ is r.c.l.l. for any $f_1, \ldots, f_n \in C(\bar{E})$, $0 < t_1, \ldots, t_n$, and $g(x_1, \ldots, x_n) \in C(\mathbb{R}^n)$, $1 \le n$. Translated to the canonical space $(\Omega_Z, \mathscr{L}_t^\circ, \Theta_t^Z, P^z)$ this means that $E^{Z_t} g(\int_0^{t_1} f_1(w(s))\, ds, \ldots, \int_0^{t_n} f_n(w(s))\, ds)$ is r.c.l.l., $P^z$-a.s. for each $z$.

On the other hand, let $H_t^z$ denote the $P^z$-prediction process of $Z_t$, so that the state space of $H_t^z$ consists of probability measures on the space of measurable, $M$-valued paths. We know from Theorem 2.12(2) that $P^z\{\varphi^{-1}(H_t^z) = \tilde{Z}_t$ for a.e. $t\} = 1$, where $\tilde{Z}_t$ is the analogue of $\tilde{X}_t$ on the space of $H_t^z$. Since $Z_t$ is r.c.l.l., we can restrict to the (universally measurable) set where $\tilde{Z}_t = Z_t$. Moreover, by part (1) we know that $E^{\varphi^{-1}(h)} g(\int_0^{t_1} f_1(w(s))\, ds, \ldots, \int_0^{t_n} f_n(w(s))\, ds)$ is continuous $\varphi(M) \to M$. Therefore, since $H_t^z$ is r.c.l.l. in its (stronger) topology, we can take limits to obtain

$$P^z\{\varphi^{-1}(H_t^z) = Z_t, \forall t \ge 0, \text{ and } \varphi^{-1}(H_{t-}^z) = Z_{t-}, \forall t > 0\} = 1.$$

This is equivalent to (2).

The same reasoning immediately proves the following theorem.

**Theorem 2.22** $P^z\{Z_{(\cdot)}$ equals its regularization of Definition 2.10$\} = 1$, $z \in M$.

An interesting and important observation concerning this result and its proof is that they apply without change to the restriction of $Z_t$ to any Borel demesne $U$. Formally, the proof of Lemma 2.21(2) is slightly different since now $H_t^z$ has as state space the probabilities on measurable $U$-valued paths, so that the definition of $\varphi$ is changed. However, the restriction on $M$-valued paths that $t = \int_0^t I_U(w(s))\, ds$ for all $t$ defines a Borel subset, so we can regard $H_t^z$ as a (random) measure on $M$-valued paths giving probability 1 to $U$-valued paths. Then also $\varphi(z)$, $z \in U$, is the same as before. Let us record formally the following corollary.

**Corollary 2.23** For any Borel prediction demesne $U$, $P^z\{Z_t$ is r.c. when $U$ is given the $\varphi$-induced topology$\} = 1$, $z \in U$.

We can also prove the following more general result.

**Theorem 2.24** For any Borel demesne $U \subset D$, $Z_t$ on $U$ is a Borel right process.

**Proof** It remains to show that for any $\lambda$-excessive function $h(z)$, $h(Z_t)$ is right-continuous, $P^z$-a.s. for $z \in U$. We observe now that it suffices to show this for $h = R_\lambda^Z f$ with $f \in b^+(\mathscr{M})$. Indeed, as is well known from probabilistic potential theory, any such $h$ is an increasing limit of such potentials.

(This is Proposition (2.9) of Getoor (1975), which is proved by elementary but intricate arguments which we omit.) Then it follows from Meyer's theorem (Lemma 2.2 above) that $h(Z_t)$ is also r.c.l.l.

Next we see from the start of the proof of Lemma 2.15 that $R^Z_\lambda f(z)$ is continuous in the topology induced by $\phi$ when $f$ is bounded and continuous on $M$. Thus, if we know that $Z_t$ is r.c.l.l. in the topology induced by $\varphi$, then for such $f$ it follows that $R^Z_\lambda f(Z_t)$ is also r.c.l.l. But this was proved in Corollary 2.23. Finally, the class of $f \in b^+(\mathcal{M})$ for which $R^Z_\lambda f(Z_t)$ is r.c.l.l. is a monotone class by Lemma 2.2. Therefore, it equals $b^+(\mathcal{M})$, and the proof is complete.

We have not used the existence of left limits here, since they need not be in $U$.

On the other hand, we should also remark that for any Borel demesne $U$ there is a natural extension to a complete Borel demesne, namely (using section theorems).

**Observation** *For a Borel demesne $U$, let $U^* = \{z; \lambda R^Z_\lambda I_U(z) = 1\} \cap \bar{U}$, where $\bar{U}$ is the closure in the prediction topology. Then $U^*$ is a complete Borel demesne, $U \subset U^*$, and for $z \in U^*$,*

$$P^z\{Z_t \text{ is r.c.l.l. in the } \varphi\text{-induced topology of } U^*\} = 1.$$

**Proof** The first assertion follows as in Lemma 2.11 and Remark 3. following Theorem 2.13. In fact, for $z \in U^*$, $P^z\{Z_t \in U \text{ for all } t > 0\} = 1$, as is clear since $P^z\{Z_\varepsilon \in U\} = 1$ for some arbitrarily small $\varepsilon > 0$ and the Markov property applies. The last assertion follows by the same reasoning as Corollary 2.23.

**Remark** By Theorem 1.15, if $U$ is a Borel demesne then $U \cap D$ is also a Borel demesne (containing no branch points). Then the Observation above applies to $U \cap D$.

According to Lemma 2.21, extended to $U$, we can consider a whole sequence of topologies on $U$ as follows. First, we give $U$ the prediction topology—call this the $\varphi^{(0)}$ topology. Then, the topology induced by $\varphi$ on $U$ (which depends on the topology of $E$ through the $\varphi^{(0)}$ topology)—call this the $\varphi^{(1)}$ topology. By Lemma 2.21 the $\varphi^{(1)}$ topology is finer than the $\varphi^{(0)}$ topology. Since $(U, \mathcal{U} (= \mathcal{M}|U))$ is again a Lusin space in the $\varphi^{(1)}$ topology, we can introduce again the $\varphi$ topology relative to this $\varphi^{(1)}$ topology—call it the $\varphi^{(2)}$ topology, and so forth to the $\varphi^{(n)}$ topology for every $n$. Now we observe, as a general fact, that if $(E, \mathcal{E})$ has two Lusin topologies, one finer than the other, then the induced prediction topologies of $(M, \mathcal{M})$ are also ordered in the same direction. Indeed, the induced order

# 74 | Foundations of the prediction process

of the two topologies of $(\Omega', \mathscr{F}')$ gives the same order to the weak-* topologies of $(M, \mathscr{M})$. Therefore, we have $\varphi^{(n)} < \varphi^{(n+1)}$ in the sense of increasing fineness, for every $n$. Now we can introduce an important definition.

**Definition 2.18** The coarsest topology on $U$ finer than every $\varphi^{(n)}$-topology is the Ray topology of $U$.

**Remarks** This is slightly different from the Ray topology of Getoor (1975). The precise relationship will be considered following Lemma 2.34. We do not need to introduce any compactifications of $U$ for this Ray topology to make sense. We simply interpret weak-* convergence in terms of all bounded continuous functions on the corresponding $(\Omega', \mathscr{F}')$ (i.e. in the $n$th topology of $\Omega'$). The general question of compactifications of $U$ will be considered subsequently. Meanwhile, we have the following theorem.

**Theorem 2.25**

(1) $(U, \mathscr{M}|U)$ *is again a Lusin space in the Ray topology.*

(2) *In the Ray topology, $R_\lambda^Z$ maps bounded continuous functions into bounded continuous functions.*

(3) $P^z\{Z_t$ *is right-continuous, with left limits* $Z_{t-}$ *(in $M$) in the Ray topology*$\} = 1$.

**Proof** We return again to the proof of Lemma 2.15 to observe the following.

**Lemma 2.26** *The topology on $E$ generated by $\varphi$ is generated by a countable collection*

$$\{R_{\lambda_i} h_i;\ h_i \in b(\mathscr{E}),\ 0 \le h_i \le 1\} \supset \{R_{\lambda_i} f_k,\ f_k \in C(\bar{E})\},$$

*with $f_k$ (from Definition 2.1) uniformly dense.*

Indeed, we showed first that $E^z \int_0^\infty e^{-\lambda t} f(\tilde{X}_t)\,dt$ is continuous on $\bar{M}$ for $f \in C(\bar{E})$. Since $E^{\varphi(x)} \int_0^\infty e^{-\lambda t} f(\tilde{X}_t)\,dt = R_\lambda f(x)$ in view of Theorem 2.12(2), the topology induced by $\varphi(x)$ makes $R_\lambda f(x)$ continuous. On the other hand we can restrict to $\lambda \in \mathbb{Q}$ and $f \in \{f_k\}$, so $\{R_\lambda f\}$ generates the same topology as a countable set. Now, by the induction step of Lemma 2.15, the topology generated by $\varphi$ is generated by a collection of the form

$$\left\{ \sum_{\sigma(s_1,\ldots,s_m)} R_{\sum_1^m \lambda_k}(f_n(x) E^{\varphi(x)} T_{m-1}),\ 1 \le m \right\},$$

where (if we restrict to $\lambda_k \in \mathbb{Q}^+$) it is easy to see that the set of possible $T_{m-1}$ is countable for each $m$, and of course the number of permutations $\sigma(s_1,\ldots,s_m)$ is $m!$. Now we claim that for each $\sigma(s_1,\ldots,s_m)$ the corresponding summand is also continuous. It suffices, again, to consider the case of

$$E^{\varphi(x)} \int \cdots \int_{\{s_1 < \cdots < s_m\}} \exp(-\sum \lambda_k s_k) \Pi_k f_{n_k}(\tilde{X}(s_k)) \, ds_1 \cdots ds_m,$$

and we proceed by induction on $m$, the case $m = 1$ being true by hypothesis. By definition of the prediction topology we are reduced to showing that

$$\int_0^\infty e^{-\lambda_1 s_1} f_{n_1}(\tilde{X}(s_1)) \int_{s_1}^\infty e^{-\lambda_2 s_2} f_{n_2}(\tilde{X}(s_2)) \cdots \int_{s_{m-1}}^\infty e^{-\lambda_m s_m} f_{n_m}(\tilde{X}_{s_m}) \, ds_m \cdots ds_2 \, ds_1$$

is continuous on $\Omega'_E$. We know by the Remark to Theorem 2.3 that $\int_{s_{m-1}}^\infty e^{-\lambda_m s_m} f_{n_m}(\tilde{X}_{s_m}) \, ds$ is continuous for fixed $s_{m-1}$, and since this is uniformly continuous in $s_{m-1}$, a uniform step function approximation extends the continuity to the double integral, hence by induction to the integral over $\{0 < s_1 < \cdots < s_m\}$ as required.

Returning to the proof of Theorem 2.25, we see that the $\varphi$-topology is metrizable by, say,

$$d_2(x_1, x_2) = \sum_{\lambda \in \mathbb{Q}^+, n} \sum \lambda c_{\lambda, n} |R_\lambda h_n(x_1) - R_\lambda h_n(x_2)|$$

where $\sum \sum c_{\lambda, n} = 1$, $0 < c_{\lambda, n}$. Now if we replace $E$ by $U$ and $R_\lambda$ by $R_\lambda^Z$, this applies equally to $U$. But in this case we know by Lemma 2.21(1) that the topology generated by $\varphi$ is finer than that of $U$. We can use $d_2(z_1, z_2)$ to compactify $U$ in the $\phi$-topology, say to $\bar{U}_2$, then repeat the argument using $d_3(z_1, z_2)$ to obtain $\bar{U}_3$, and so forth. Then it is clear that the Ray topology is that of the metric $d_\infty(z_1, z_2) = \sum_{n=2}^\infty 2^{-n} d_n(z_1, z_2)$, so that we can compactify $U$ to a compact metric space $\bar{U}_\infty$ under $d_\infty$, which will be considered in detail later. In any case, it is clear by Lusin's theorem that $U$ is Borel in $\bar{U}_\infty$, and hence $(U, \mathcal{M}|U)$ is a Lusin space in the Ray topology.

As to (2), if $f \in C(\bar{U}_\infty)$ then $f$ must be continuous in each metric $d_n$, $n > 1$. Conversely, if $f$ is uniformly continuous in each $d_n$ then it extends continuously to $\bar{U}_\infty$, so uniform continuity in each $d_n$ on $U$ is necessary and sufficient for extendability to an element of $C(\bar{U}_\infty)$. But such an $f$ is continuously extendable to $\bar{U}_{n+1}$ for every $n \geq 2$, i.e. $R_\lambda^Z f$ is uniformly continuous on $U$ for $d_n$, $n \geq 3$, and we also have seen that $d_{n+1}$ generates a finer topology than $d_n$ on $U$. In fact, it is easy to see by the proof of Lemma 2.21(1) that the metric $d_{n+1}$ is actually stronger than $d_n$, $n \geq 1$ (where $d_1$ is the metric of $\bar{M}$ restricted to $U$, i.e.

$$d_1(z_1, z_2) = \sum_n 2^{-n} |E^{z_1} g_n - E^{z_2} g_n|,$$

with $0 \leq g_n \leq 1$ dense in $C(\bar{\Omega})$). Therefore, $R_\lambda^Z f$ is also uniformly continuous for $d_2$ on $U$, and hence it extends to a continuous function on $\bar{U}_\infty$. Thus we have shown that $R_\lambda^Z$ extends by continuity to map $C(\bar{U}_\infty)$ into itself.

# 76 | Foundations of the prediction process

If $f$ is only bounded and continuous on $U$ for the Ray topology, there exist $f_n \in C(\bar{U}_\infty)$, $f_n \uparrow f$ on $U$, and also $g_n \in C(\bar{U}_\infty)$, $g_n \downarrow f$ on $U$ (following Definition 2.5). Then by monotone convergence $R^Z_\lambda f_n \to R^Z_\lambda f$ and $R^Z_\lambda g_n \to R^Z_\lambda f$ pointwise on $U$. It follows that $R^Z_\lambda f$ is both lower and upper semicontinuous, hence it is continuous. This completes the proof of (2).

As for (3), by Lemma 2.21(2) and induction one has

$$P^z\{Z_t \text{ is r.c. with left limits } Z_{t-} \text{ in the } d_n\text{-topology}\} = 1$$

for each $n \geq 2$, implying the result.

Two easy consequences of Theorem 2.25 which were mentioned previously, but whose proofs depended on capacity theory, can now be given direct proofs as follows.

**Corollary 2.27** $P^z\{Z_t \in D \text{ for all } t \geq 0\} = 1$, $z \in M$ (i.e. Theorem 1.15).

**Proof** This is proved for Ray processes on p. 21 of Getoor (1975). The only point at which his proof relies on capacity is in using our Lemma 2.2 (extended to unbounded sequences, Exercise 2.1). By taking $U = M$ in Theorem 2.25 we see that $R^Z_\lambda$ becomes a Ray resolvent on $\bar{U}_\infty$ (it separates points because the metric is defined in terms of the range of $R^Z_\lambda$). Therefore we obtain a 'capacity-free' proof by following Getoor (1975). We omit further details.

**Corollary 2.28** *For any Borel demesne $U$, $U^*$ (of the Observation above) is a complete Borel demesne containing $U$.*

**Remark** We also obtain a direct proof of Lemma 2.11, without a section argument. Of course, using sections the result of Corollary 2.28 was already proved.

**Proof** We must show that, for $z \in U^*$,

$$P^z\{Z_t \in U^*, t \geq 0, \text{ and } Z_{t-} \in U^*, t > 0\} = 1,$$

where $Z_{t-}$ is the left limit in the prediction topology. Now, by definition of $U^*$, for $z \in U^*$,

$$1 = \lambda E^z \int_0^\infty e^{-\lambda t} I_U(Z_t)\, dt$$

$$= \lambda E^z \int_0^t e^{-\lambda s} I_U(Z_s)\, ds + \lambda E^z \left( E^z \int_t^\infty e^{-\lambda s} I_U(Z_s)\, ds \mid \mathscr{Z}_t \right)$$

$$= \lambda E^z \int_0^t e^{-\lambda s} I_U(Z_s)\, ds + \lambda E^z \left( e^{-\lambda t} \left( E^{Z_t} \int_0^\infty e^{-\lambda s} I_U(Z_s)\, ds \right) \right).$$

The first term on the right is at most $1 - e^{-\lambda t}$, and $\lambda E^{Z_t} \int_0^\infty e^{-\lambda s} I_U(Z_s)\,ds \leq 1$. It follows that, for each $t$, this last equals 1, $P^z$-a.s. But $\lambda R_\lambda^Z I_U(z)$ is $\lambda$-excessive, and, by Theorem 2.24 $Z_t$ is a right process on $D$. Therefore $\lambda E^{Z_t} \int_0^\infty e^{-\lambda s} I_U(Z_s)\,ds$ is right-continuous, $P^z$-a.s., and so equals 1. Since $P^z\{Z_t \in U\} = 1$ for a.e. $t$ by inversion of the transform, we have $P^z\{Z_t \in \bar{U}, \forall t \geq 0\} = 1$. Putting these together gives $P^z\{Z_t \in U^*, \forall t \geq 0\} = 1$, so $U^*$ is a Borel demesne.

To prove completeness, since the left-limit process of

$$\lambda e^{-\lambda t} E^{Z_t} \int_0^\infty e^{-\lambda s} I_U(Z_s)\,ds$$

is identically 1, $P^z$-a.s., it is enough to see that this process is given by substitution of $Z_{t-}$ for $Z_t$. Identifying $Z_t$ with its prediction process by Theorem 2.22, this becomes the left-limit process of $\lambda e^{-\lambda t} E^{\varphi(\tilde{Z}_t)} \int_0^\infty e^{-\lambda s} I_U(\tilde{Z}_0) \circ \Theta_s^Z\,ds$, and the interchangeability of left limits follows from the last paragraph of the proof of Theorem 2.4, completing the argument.

We turn now to another form of completeness of $U^*$: that of completeness under formation of probability entrance laws. In the special case $U = U^* = M$, this will answer in the affirmative Conjecture 2.10 of Knight (1981a, Essay I, p. 31). Let us first note explicitly that the class of processes definable by initial distributions on $U^*$ is the same as that definable by initial distributions on $U^* \cap D$. Indeed, for $\mu$ on $U^*$ we can define

$$v(dz) = P^\mu\{Z_0 \in dz\} = \int_{U^*} P^m\{Z_0 \in dz\}\mu(dm).$$

Since

$$1 = P^\mu\left\{\lambda \int_0^\infty e^{-\lambda t} I_U(Z_s)\,ds = 1\right\}$$

$$= E^\mu P^\mu\left\{\lambda \int_0^\infty e^{-\lambda t} I_U(Z_s)\,ds = 1 \mid \mathscr{Z}_0^\circ\right\}$$

$$= E^\mu\left(P^{Z_0}\left\{\lambda \int_0^\infty e^{-\lambda t} I_U(Z_s)\,ds = 1\right\}\right),$$

right-continuity of $Z_t$ at $t = 0$ gives $v(U^* \cap D) = 1$. Also since

$$P^\mu\{Z_0 \in A\} = P^v\{Z_0 \in A\}, \qquad A \in \mathscr{L}^\circ,$$

again by the Markov property we have $P^\mu = P^v$.

Another way of stating this observation is that any probability entrance law realizable by a distribution on $U^*$ is also realizable by a distribution on $U^* \cap D$, where *entrance law* is defined in the following.

## 78 | Foundations of the prediction process

**Definition 2.19** Let $\Omega_Z^+ = \{\text{r.c.l.l. paths } Z_t \in M \text{ defined only for } t > 0\}$ (where we do not assume existence of $Z_{0+}$). Let $\mathscr{L}_{(0,t)}^\circ = \sigma\{Z_s, 0 < s \leq t\}$ on $\Omega_Z^+$. A probability entrance law for $U$ is a probability $P$ on $\mathscr{L}_{(0,\infty)}^\circ$ such that

(1) $P\{Z_t \in U, \forall t > 0\} = 1$; and

(2) $P(Z_{t+s} \in A \mid \mathscr{L}_{(0,t]}^\circ) = q(s, Z_t, A); 0 < s, t; A \in \mathscr{M}$.

We note that, of course, (1) is equivalent to $P\{Z_{\varepsilon_n} \in U\} = 1$ for any sequence $\varepsilon_n \to 0+$.

The next theorem is the basic result on entrance laws.

**Theorem 2.29** *Every probability entrance law for a Borel demesne $U$ is realized by a unique probability $v$ on $U^* \cap D$.*

**Proof** Since $v$ is also the $P^v$-distribution of $Z_0$, uniqueness is clear from right-continuity of $Z_t$ at $t = 0$ for $P^v$. Turning to the existence, recall (by the above Observation) that $U^* = \{z: \lambda R_\lambda^Z I_U(z) = 1\} \cap \bar{U}$, so the existence of $v$ on $U^*$ has two parts. First, we obtain a $v$ on $\{z: \lambda R_\lambda^Z I_U(z) = 1\} \cap D$. This is a purely measure-theoretic condition, and can as well be treated by mapping $E$ back to $[0, 1]$ by a Borel isomorphism $\hat{\varphi}$. Then we will show, after mapping $v$ back to $M_E$, that it is carried on the closure $\bar{U}$ of $U$ in the prediction topology. (Actually, the same argument shows that it is on the closure of $U$ in the Ray topology of $M_E$, as well.)

Accordingly, we first replace the entrance law $P$, and demesne $U$, by their counterparts $P$ and $U$ in case $E = [0,1]$, obtained by a fixed but arbitrary isomorphism $\hat{\varphi}$. (We use the same notations: $(\Omega_Z, \mathscr{L}^\circ)$ of course depends on $E$, but $E$ is suppressed in the notation.) We next observe that for fixed $t > 0$, $Z_{t+s}, 0 \leq s$, is equivalent to the prediction process with initial distribution $P\{Z_t \in A\}$, and given $Z_t$ it is conditionally independent of $\mathscr{L}_{(0,t]}^\circ$. (In particular, $Z_t$ is concentrated on $U \cap D$ for $t > 0$.) This is by the usual Markov process reasoning, as in Lemma 2.8. Moreover, by Remark 3. of Theorem 2.14 it follows that $\Theta_t^{-1}\mathscr{L}^\circ$ and $\sigma\{\hat{\rho}(Z_s), s \geq t\}$ differ only by $P$-null sets for each $t$, and therefore $\mathscr{L}_{(0,\infty)}^\circ$ and $\sigma\{\hat{\rho}(Z_s), 0 < s\}$ also differ only by $P$-null sets.

Now we go back to the topology on $\Omega' (= \Omega'_{[0,1]})$ of Theorem 1.3, under which $(\Omega', \mathscr{F}')$ is a compact metric space with its Borel field, and $(M', \mathscr{M}')$ is likewise, with the topology of weak-* convergence of measures on $(\Omega', \mathscr{F}')$. According to Notation 1.4 and Lemma 1.7, $z_n \to z$ in $M'$ if and only if, for all $0 < \lambda \in \mathbb{Q}$ and $f_k$ in a uniformly dense set, $E^{z_n} \int_0^\infty e^{-\lambda s} f_k \circ \Theta_s \, ds \to E^z \int e^{-\lambda s} f_k \circ \Theta_s \, ds$. Now the $f_k$ are functions on $\Omega'$, but when we translate $\Omega'$ to the prediction space, we know that a probability $P^z(S')$ becomes $P^z\{\hat{\rho}(Z_{(\cdot)}) \in S'\}$. Therefore, we can obtain $P$-a.s. limits $Z_{0+}^* = \lim_{t \to 0+} Z_t$ on prediction space $\Omega_Z^+$ if and only if (writing $\theta$ for $\theta^Z$) we have

$$P\left\{\lim_{t\to 0+} E^{Z_t}\int_0^\infty e^{-\lambda s}f_k(\hat{\rho}(Z_{(\cdot)}))\circ\Theta_s\,ds\text{ exists}\right\} = 1.$$

On the other hand, this is the same as

$$P\left\{\lim_{t\to 0+} e^{-\lambda t}E\left(\int_0^\infty e^{-\lambda s}f_k(\hat{\rho}(Z_{(\cdot)}))\circ\Theta_s\,ds\circ\Theta_t \mid \mathscr{L}^\circ_{(0,t]}\right)\text{ exists}\right\},$$

and the same argument as for Lemma 1.6 shows that this is a supermartingale limit. So it follows that these limits exist, $P$-a.s., for each $k$ and $0 < \lambda \in \mathbb{Q}$. In other words, we can define $Z_0^*$ on $(\Omega_Z^+, \mathscr{L}^\circ_{(0,\infty)})$ in such a way that $P\{\lim_{t\to 0+} Z_t = Z_0^*\} = 1$, where the limit is in the compact space $M'$ of Definition 1.3. (Note, however, that $Z_0^* \in B$ is not excluded; $D$ is certainly not closed.) Moreover, Hunt's lemma immediately gives

$$E\left(\int_0^\infty \lambda e^{-\lambda s}f_k(\hat{\rho}(Z_{(\cdot)}))\circ\Theta_s\,ds \mid \mathscr{L}^\circ_{(0,0+)}\right) = E^{Z_0^*}\int_0^\infty \lambda e^{-\lambda s}f_k(\hat{\rho}(Z_{(\cdot)}))\circ\Theta_s\,ds,$$

and, letting $\lambda \to \infty$, recalling that $f_k \circ \Theta_s$ is uniformly continuous in $s$, uniformly on $\Omega'$, and the $f_k$ are dense in $C(\Omega')$, it follows that

$$P(\hat{\rho}(Z_{(\cdot)}) \in S' \mid \mathscr{L}^\circ_{(0,0+)}) = P^{Z_0^*}(S'), \qquad P\text{-a.s., } S' \in \mathscr{F}'.$$

Equivalently,

$$P(S^\circ \mid \mathscr{L}^\circ_{(0,0+)}) = P^{Z_0^*}(S^\circ), \qquad P\text{-a.s., } S^\circ \in \mathscr{L}^\circ_{(0,\infty)}.$$

Now let $\mu(dz)$ be the distribution of $Z_0^*$, and let $v(dz) = P^\mu\{Z_0 \in dz\}$, so that $v(dz)$ is indeed concentrated on $D$. Also, since clearly

$$E^{Z_0^*}\lambda\int_0^\infty e^{-\lambda t}I_U(Z_t)\,dt = 1$$

it follows that $v(dz)$ is concentrated on $\{\lambda R_\lambda^Z I_U(z) = 1\}$. Still retaining the case $E = [0,1]$, we now see that if we use $v(dz)$ as initial distribution for $Z_0$ on the subset of $\Omega_Z^+$ having right limits $Z_0$ at $t = 0$, we obtain a realization of the entrance law $P$ for the demesne $U$ on this space. We have then only to map $[0,1] \xrightarrow{\varphi^{-1}} E$, to obtain a realization of the original $P$ on a corresponding element of $\mathscr{L}^\circ_{(0,\infty)}$ for $E$, by an initial distribution $v$ concentrated on $\{\lambda R_\lambda^Z I_U(z) = 1\} \cap D$.

Now we turn to showing that $v$ is also concentrated on $\bar{U}$, the closure being in the prediction topology (or even the Ray topology of $M$). However, this is now very simple. We have already shown in the proof of Theorem 2.4 that right-continuity of $Z_t^{z\varphi^{-1}}$ in the compact topology of $M_{[0,1]}$ implies right-continuity of $Z_t^z$ in the prediction topology of $M$. Then in Theorem 2.25(3) we deduced from this that $Z_t$ is r.c.l.l. in the Ray topology of $M$. Here we need these proofs only at $t = 0$ to see that $P^z\{Z_0 \in \bar{U}\} = 1$, as required.

## 2.4 Compactifications and the Ray space

The space $U^*$ suffices for most purposes in studying the prediction process on $U$. However, a better space (and one which is closer to the literature of right processes) is the following.

**Definition 2.20** The Ray space $U_R$ of a Borel demesne is

$$U_R = \{z \in M : \lambda R_\lambda^z I_U(z) = 1\} \cap \bar{\bar{U}},$$

where $\bar{\bar{U}}$ is the closure of $U$ in the Ray topology of $M$.

Actually, we do not have an example in which $U_R \neq U^*$ as subsets of $M$, although the Ray and prediction topologies certainly may differ. Meanwhile, the next theorem is easy to prove.

**Theorem 2.30**

(1) $U_R$ is a complete demesne.

(2) $U_R \cap D = U^* \cap D$, i.e. $U_R$ and $U^*$ differ at most in their elements of $B$.

(3) Theorem 2.29 remains true with $U_R$ in place of $U^*$.

**Proof** Part (1) follows by Theorem 2.25(3) from Corollary 2.28. Both parts (2) and (3) follow from the fact, already observed in the proof of Theorem 2.29, that if $z \in U^* \cap D$ then $P^z\{\lim_{t \to 0+} Z_t = z$ in the Ray topology$\} = 1$. This implies $(U^* \cap D) \subset (U_R \cap D)$, and the converse is clear since the Ray topology is finer than the prediction topology (Definition 2.18).

**Remark** This gives also $U_R \subset U^*$.

A space analogous to $U_R$ appeared in Getoor (1975) in connection with the *Ray–Knight compactification* of a right process. There it is defined quite differently, and it is not even obvious that it is a subset of our $M$. Before we can establish the connection with the case of right processes, it is appropriate to introduce formally the compactification $\bar{U}_\infty$ (see the proof of Theorem 2.25).

In the proof of Theorem 2.25, we introduced a sequence of metrics $d_n(x_1, x_2)$, $2 \leq n$, on $U$ such that $d_{n+1}$ is stronger than $d_n$ for each $n$. Let us complete the list by including as $d_1(x, y)$ a metric of $U$ in the prediction topology, defined for the compactification $\bar{M}_E$ generated by compactifying $\Omega'_E$ to $\bar{\Omega}'_E$ as a subset of $\times_1^\infty \Omega'_{[0,1]}$, following Definition 2.5. Then (in view of Lemma 2.21 and its proof) we may assume that $d_1$ is weaker than $d_2$. We also introduced a compactification corresponding to $d_\infty = \sum_{n=2}^\infty 2^{-n} d_n$, in which we can just as well include a term for $n = 1$.

**Definition 2.21** The compactification of $U$ for the metric $d_\infty$ is denoted $\bar{U}_\infty$, and is called the Ray compactification of $U$.

A first observation is that $\bar{U}_\infty$ contains $U_R$, in the obvious sense that elements of $U_R$ are in the closure of $U$ in $M$ for the metric $d_\infty$. But since $d_\infty$ is bounded (by 1), $\bar{U}_\infty$ is a compact metric space of diameter less than or equal to 1. This shows that $\bar{U}_\infty$ is not in general contained in $M$ (in particular, if $U = M$ then $\bar{U}_\infty \neq M$ since $M$ is not compact in the Ray topology, as examples given below make clear).

Let us discuss in some detail the structure of $\bar{U}_\infty$. For each $n$, we may introduce (as in the proof of Theorem 2.25) a compactification $\bar{U}_n$ of $U$ in the metric $d_n$ (or equivalently in the metric $\sum_{k=1}^n 2^{-k} d_k$). To go from $\bar{U}_n$ to $\bar{U}_{n+1}$ we introduce a countable family $\{f_{k,n+1}\} \subset b^+(U)$ which can be extended to elements of $C(\bar{U}_{n+1})$ (but not necessarily of $C(\bar{U}_n)$). This family includes a set which is uniformly dense in $R_\lambda^Z C(\bar{U}_n)$ (or more precisely, the part of the range $\lambda R_\lambda^Z$, $0 < \lambda$, obtained from $f \in C(\bar{U}_n)$ with $0 \leq f \leq 1$). By Lemma 2.26 this dense subset of the range is obtained from a dense subset of $C(\bar{U}_n)$, where all functions can be considered as functions only on $U$. However, since $d_{n+1}$ is stronger than $d_n$, all $f \in C(\bar{U}_n)$ also extend into $C(\bar{U}_{n+1})$. Thus we can extend $R_\lambda^Z$ by continuity in such a way that it maps $C(\bar{U}_n) \to C(\bar{U}_{n+1})$, as functions on $\bar{U}_{n+1}$.

Conversely, to go from $\bar{U}_{n+1}$ back to $\bar{U}_n$, we have to form the equivalence classes for the equivalence relation $x_1 \equiv x_2$ iff $f_{k,j}(x_1) = f_{k,j}(x_2)$ for all $k$ and $j \leq n$. Since the $f_{k,j}$ are continuous on $\bar{U}_{n+1}$, it is easy to see directly that the quotient topology is compact. Now to describe the structure of $\bar{U}_\infty$, we can consider it as a projective limit of the spaces $\bar{U}_n$ under the projection mappings $\Theta_{n,n+1} : \bar{U}_{n+1} \to \bar{U}_n$ which map $u_{n+1} \in \bar{U}_{n+1}$ into the equivalence class $\{u : f_{k,j}(u) = f_{k,j}(u_{n+1}), \text{ all } k \text{ and } j \leq n\}$, considered as an element of $\bar{U}_n$. Then an element $u_\infty \in \bar{U}_\infty$ consists of a sequence $(u_1, u_2, \ldots)$, $u_n \in \bar{U}_n$, such that $\Theta_{n,n+1}(u_{n+1}) = u_n$ for all $n$. In other words, we can identify $\bar{U}_\infty$ as the set of all such sequences $u_\infty$, with the product topology. Indeed, the maps $\Theta_{n,n+1}$ are continuous, so that if a sequence $u_\infty^{(k)}$ converges in the product topology, as elements of $\times_1^\infty \bar{U}_n$, then the limit is again an element of $\bar{U}_\infty$. Besides, it is easy to see that $\bar{U}_\infty$ is compact, and that any $f \in C(\bar{U}_n)$ extends uniquely to an $f_\infty \in C(\bar{U}_\infty)$ by simply defining $f_\infty(u_1, u_2, \ldots) = f(u_n)$.

Going back to our introduction of Ray resolvents (Definition 2.17 and Equation (2.2) before Corollary 2.20) we have easily

**Theorem 2.31** *For $f \in C(\bar{U}_\infty)$, define $R_\lambda^\infty f$ on $\bar{U}_\infty$ as the extension by continuity of $R_\lambda^Z(f|U)$ on $U$, where $(f|U)$ denotes the restriction of $f$ to $U$. Then $(f|U) \in C_b(U)$, and $R_\lambda^\infty$ defines a Ray resolvent on $\bar{U}_\infty$. Moreover, $R_\lambda^\infty f(x) = R_\lambda^Z(f|U_R)(x)$ for $f \in b(\bar{\mathscr{U}}_\infty)$ and $x \in U_R$, when $R_\lambda^\infty$ is extended by the Riesz representation theorem to a kernel on the Borel sets $\bar{\mathscr{U}}_\infty$. Thus $R_\lambda^\infty$ and $R_\lambda^Z$ agree on $U_R$.*

82 | Foundations of the prediction process

**Proof** First we note that $U$ is a Borel subset of $\bar{U}_\infty$. Indeed, $f \in C(\bar{U}_\infty)$ restricts to a continuous function on $U_R$ of Definition 2.20 by definition of the Ray topology. Then it is clear that the injection map $U_R \to \bar{U}_\infty$ is one-to-one and Borel, so by Lusin's theorem $U_R \in \bar{\mathcal{U}}_\infty$, and $U \in \bar{\mathcal{U}}_\infty$ follows.

We noted above that $R_\lambda^Z$ maps $C(\bar{U}_n) \to C(\bar{U}_{n+1})$, in the sense of continuous extensions, for each $n$. Therefore, if $f \in C(\bar{U}_\infty)$ then $R_\lambda^Z(f|U)$ extends to an element of $C(\bar{U}_{n+1})$ for every $n$, i.e. to an element of $C(\bar{U}_\infty)$. Moreover, the metric $d_\infty$ is defined entirely from functions in the range of $R_\lambda^Z$, as is clear again by the Remark after the proof of Lemma 2.15, so $R_\lambda^\infty$ separates points in $\bar{U}_\infty$. Finally, the resolvent equation (2.2) cited above extends to $f \in C(\bar{U}_\infty)$ just as before, and since the positivity and boundedness are obvious, we have only to define kernels $R_\lambda^\infty(x, dy)$ by the Riesz representation theorem to obtain a Ray resolvent on $b(\bar{U}_\infty)$.

We will not give any proof of Ray's theorem asserting the existence of a unique semigroup $T_{t+}^\infty f(x)$, right-continuous in $t$ for $f \in C(\bar{U}_\infty)$ at each $x \in \bar{U}_\infty$, with $R_\lambda^\infty f(x) = \int_0^\infty e^{-\lambda t} T_{t+} f(x)\, dx$. Various analytical proofs are known (Ray 1959; Meyer 1966; Getoor 1975) and we have already used the result for Theorem 2.19(2). The purpose here is to make some comparison of $T_{t+}^\infty$ with the semigroup $T_{t+}$ of $Z_t$ on $U_R$, which of course needs no further existence proof since it is part of Theorem 1.14. Obviously we have $T_{t+}^\infty(f|U_R)(x) = T_{t+}(f|U_R)(x)$ for $f \in b(\bar{\mathcal{U}}^\infty)$, or for $f \in b(\mathcal{M})$, if $x \in U_R$, since both $T_{t+}^\infty$ and $T_{t+}$ have the same resolvent on $U_R$ and the same right-continuity for $(f|U_R)$, $f \in C(\bar{U}_\infty)$ (the latter because $Z_t$ is right-continuous in the Ray topology). Therefore we can state simply the following.

**Theorem 2.32** *For an initial distribution on $U_R$, the r.c.l.l. Ray process with semigroup $T_{t+}^\infty$ may be realized by $Z_t$ on $(\Omega_Z, \mathcal{L}_t^\circ)$ with the same initial distribution and transition semigroup $T_{t+}$.*

(Of course, we can equally use the canonical subspace $\Omega_{U_R}$, as remarked following Definition 2.14.)

From this we see that the Ray non-branch points of $U_R$ are the points $U_R \cap D$. But since every entrance law for $U$ (or, what is obviously the same, for $U_R$, since $U$ and $U_R$ clearly have the same entrance laws) is given by an initial distribution on $U_R$, it is immediately clear that there can be no Ray non-branch points which define entrance laws for $U$ other than those in $U_R \cap D$. Let us introduce the following definition.

**Definition 2.22** The entrance space of $U$ in $\bar{U}_\infty$ is

$$U_e = \{x \in \bar{U}_\infty : \lambda R_\lambda^\infty I_U(x) = 1, \lambda > 0\}.$$

This would be the formal analogue of the Ray space of Getoor (1975, Definition (15.5)) for the case of right processes. Of course, this $U_e$ contains $U_R$, and (as just pointed out) it has the same non-branch points as $U_R$. But we will see by examples that, in general, it contains branch points not in $U_R$. Let us consider the following examples in some detail, as they illustrate several different points.

**Example 1** Let $E = R$ and let $T_t$ be the semigroup of ordinary Brownian motion. Then $T_t = T_{t+}$, and the process identifies with $Z_t$. (It is a Borel right process.) If we extend $R$ to $\bar{R}_1$, the one-point compactification by $\infty$, we obtain a Ray resolvent by defining $T_t f(\infty) = \lim_{x \to \pm\infty} T_t f(x) = f(\infty)$, $f \in C(\bar{R}_1)$, so that $\infty$ becomes a *trap*. Then $\varphi(\bar{R}_1)$ is compact in the space of probability measures on $\Omega'_E$ with $E = \bar{R}_1$, and it is readily seen that $\varphi(\bar{R}_1) = \varphi(\bar{R}_1)_R = \varphi(\bar{R}_1)_\infty$, i.e. our extensions are equal for $T_t$ on $\bar{R}_1$. Then it is easy to see that on $R$ the prediction and Ray topologies both coincide with the original topology of $R$, and $\varphi(R) = \varphi(R)_R = \varphi(R)_e$. This last is equally true if we compactify $R$ by $\{\pm\infty\}$. (They do not depend on the choice of $\bar{E}$.) On the other hand, $\varphi(\bar{R})_\infty$ will correspond to $\bar{R}_2$, the two-point compactification, in this case, so the Ray compactification depends on the choice of $\bar{E}$. (In fact, even the compactification of $M$ for the prediction topology depends on $\bar{E}$.)

Similar considerations apply if, instead of only $\varphi(R)$, we include measures corresponding to initial distributions on $R$. Non-point initial distributions, of course, define branch points of $M$. If we include them all, then we obtain $U = U_R = U_e$ once more, because it is known that all entrance laws correspond to initial distributions on $R$ for Brownian motion. In between, the strong Feller property of $T_t$ shows that convergence in $M$, either in the prediction or the Ray topologies, is the same as weak convergence of initial distributions on $R$ (allowing only limits of mass 1) so $U_R = U_e \equiv$ {weak closure of the initial distributions}. We may leave the details of the verification of this to the interested reader.

**Example 2** Let $E$ be uncountable, and let $U = M_E$. We wish to show that $M$ is not compact in the Ray topology, hence $M \neq \bar{M}_\infty$ (we drop the subscript $E$). First, we will show that for any initial distribution $\mu$ on $D$, there is a point $z \in \bar{M}_\infty$ for which the Ray process with semigroup $T^\infty_{t+}$ starting at $z$ has distribution $\mu$ at $t = 0$. In other words,

$$R^\infty_\lambda f(z) = \int_D R^z_\lambda f(m) \, d\mu(m) \quad \text{for all } f \in b(\bar{M}_\infty).$$

To this effect, let $\hat{\varphi}$ be an isomorphism of $(D, \mathcal{M}|D)$ onto $(E, \mathcal{E})$, guaranteed by Theorem 1.1. Now we consider a sequence $z_n \in M$ defined as follows. The process whose probability is $z_n$ has initial distribution $\hat{\varphi}\mu$ on $E$, and waits at its starting point $e$ for $0 \le t < 1/n$. Then it evolves according to the probability $\hat{\phi}^{-1}(e)$, i.e. $z_n(\Theta_{1/n}^{-1}(S') \mid \mathscr{F}'_{1/n}) = \hat{\phi}^{-1}(e)(S')$, $S' \in \mathscr{F}'$. Then, as $n \to \infty$ the contribution to $R_\lambda^Z f(z_n)$ from $0 < t < 1/n$ tends to 0 for $f \in b(\bar{M}_\infty)$, hence $z_n \to z$ in the Ray topology where $z$ is the Ray branch point with distribution $\mu$ on $D$. Note that at $t = 1/n$ the prediction process $Z_{t-}^{z_n}$ has distribution $\mu$ on $D$, and $Z_t^{z_n} = Z_{t-}^{z_n}$ so there is no branch point. Now there exist measures $\mu$ such that $z \notin M$, i.e. no corresponding probability on $\mathscr{F}'$ exists. For example, suppose $\mu$ is concentrated on processes which all have the same value for $t \le 1$, and differ, say, on $\mathscr{F}'_{1+\varepsilon}$ for any $\varepsilon > 0$. Then a corresponding $z \in M$ would have to share this common value on $\mathscr{F}'_1$. This would imply $z \in D$, so $z$ is not even a branch point. (Necessary and sufficient conditions for there to exist a $z \in M$ with $P^\mu = P^z$ are given in Knight (1979, pp. 385–405, Theorem 1.2).)

**Example 3** Let $E = \{b_n, 0, \pm 1/n;\ 0 < n\}$ where $b_n$ and $\pm 1/n$ are discrete, $b_n \to 0$, $\pm 1/n \to 0$. We consider a Markov process defined from a fixed random variable $\hat{e}$: $P\{\hat{e} > t\} = e^{-t}$, by

$$P^{\pm 1/n}\left\{X_t = \pm\frac{1}{n},\ 0 \le t < \hat{e};\ X_t = \pm 1,\ \hat{e} \le t\right\} = 1,$$

the signs $+$ or $-$ being the same throughout; $\pm 1$ are traps. Also

$$P^{b_n}\left\{X_0 = +\frac{1}{n}\right\} = P^{b_n}\left\{X_0 = -\frac{1}{n}\right\} = \tfrac{1}{2},$$

so that $b_n$ are branch points. Finally, let

$$P^0\{X_t = 0,\ 0 \le t < \hat{e};\ X_t = 1,\ t \ge \hat{e}\} = P^0\{X_t = 0,\ 0 \le t < \hat{e};\ X_t = -1,\ t \ge \hat{e}\}$$
$$= \tfrac{1}{2}.$$

It is easy to see that the Criterion in Section 2.3 is satisfied, so $\tilde{X}_t$ provides a realization of the process and $T_t = T_{t+}$. Now for $f \in C(E)$, $f(\pm 1/n) \to f(0)$, and consequently

$$R_\lambda f(b_n) = \frac{1}{2}\left(R_\lambda f\left(\frac{1}{n}\right) + R_\lambda f\left(-\frac{1}{n}\right)\right) \to R_\lambda f(0),$$

and $b_n \to 0$ in the prediction topology. But

$$R_\lambda f\left(-\frac{1}{n}\right) = (\lambda + 1)^{-1}\left(f\left(-\frac{1}{n}\right) + \lambda^{-1} f(-1)\right),$$

so that, as $n \to \infty$, in the prediction topology, $P^{1/n} \to P^{0+}$ and $P^{-1/n} \to P^{0-}$ for two new points $0\pm$ such that $R_\lambda f(0\pm) = (\lambda + 1)^{-1}(f(0) + \lambda^{-1} f(\pm 1))$, respectively. Thus it can be seen that if $U$ is the demesne corresponding to

$\{P^x, x \in E\}$, then in the prediction topology

$$\bar{U} = \left\{ P^x, x \in \left\{ b_n, 0, \pm\frac{1}{n}, 0+, 0- \right\} \right\},$$

which is compact, while $U^* = U$ since $0+$ and $0-$ do not entrance into $U$. However, in the Ray topology we need to impose continuity of $R_\lambda f$ for $f \in C(\bar{U})$, and then $R_\lambda f(b_n) \to \frac{1}{2}(R_\lambda f(0+) + R_\lambda f(0-))$ which need not be equal to $R_\lambda f(0)$ for such $f$. Consequently, there is an additional Ray branch point $b_\infty$ with

$$P^{b_\infty}\{Z_0 = 0+\} = P^{b_\infty}\{Z_0 = 0-\} = \tfrac{1}{2},$$

and

$$\bar{U}_\infty = \left\{ b_n, b_\infty, 0, \frac{1}{n}, 0+, 0- \right\}$$

with $b_n \to b_\infty$, $1/n \to 0+$, $-1/n \to 0-$, so $\bar{U}_\infty$ is compact in the Ray topology (note that $b_\infty \notin M$). Since $b_\infty$ leads into $0\pm$, not into $E$, we still have the same $U_R = U$, but with a different topology from $U^*$.

**Example 4** Let us consider Example 3 a little differently, by beginning with $\bar{U}$ in the prediction topology. Now $\bar{U}$ is seen to be a complete demesne, so $\bar{U} = \bar{U}^*$. However, in the Ray topology we have $\bar{U}_e \neq \bar{U}^*$, although each point of $\bar{U}_e$ defines an entrance law into $\bar{U}$. This illustrates the need to intersect with $M$ in order to define the Ray space $\bar{U}_R = \bar{U}$.

**Example 5** Finally, let us give a Borel right process for which the topology generated by $R_\lambda f$, $f \in C(\bar{E})$ is not the prediction topology. Let

$$E = \left\{ 0, \pm 1, \pm\frac{1}{n}, \pm\left(1 + \frac{1}{n}\right); 1 < n \right\}$$

with the Euclidean topology, and let $P^x\{\hat{e}_i > t\} = e^{-t}$, $1 \leq i$, where $\hat{e}_i$ are independent exponential random variables. Let $P^0\{X_t = 0, t < \hat{e}_1, X_t = \pm 1, t \geq \hat{e}_1\} = \frac{1}{2}$, respectively. However, let us define

$$P^{1/n}\left\{ X_t = \frac{1}{n}, t < \hat{e}_1; X_t = \pm\left(1 + \frac{1}{n}\right), \hat{e}_1 \leq t < \hat{e}_1 + \hat{e}_2; \right.$$

$$\left. X_t = \mp\left(1 + \frac{1}{n}\right), \hat{e}_1 + \hat{e}_2 \leq t < \hat{e}_1 + \hat{e}_2 + \hat{e}_3, \text{etc.} \right\} = \tfrac{1}{2},$$

so that $X_t$ alternates between $\pm(1 + 1/n)$ after time $\hat{e}_1$. This also defines $P^{\pm(1+1/n)}$, and $P^{-1/n}$ is defined by $P^{-1/n}(S) = P^{1/n}(-S)$, while $\pm 1$ are traps.

Now $E$ is compact, and for $f \in C(E)$ we have

$$R_\lambda f\left(\frac{1}{n}\right) = \frac{1}{\lambda+1} f\left(\frac{1}{n}\right) + \frac{1}{2\lambda(\lambda+1)}\left(f\left(1+\frac{1}{n}\right) + f\left(-1-\frac{1}{n}\right)\right).$$

As $n \to \infty$ this approaches $R_\lambda f(0)$, with a similar behaviour for $R_\lambda f(-1/n)$. Thus the resolvent $R_\lambda$ defines the original topology of $E$. However, for the prediction topology we must have continuity of $E^x(\int_0^\infty e^{-\lambda t} I_{[1,\infty)}(X_t)\,dt)^2$. Now for $x=0$ this is $(2\lambda^2(2\lambda+1))^{-1}$, while for $x=1/n$ routine but tedious computations give the result

$$\tfrac{1}{2}[(2\lambda^2(\lambda+2))^{-1}(1+(1+2\lambda)^{-1})] = 2(\lambda+1)((\lambda+2)(2\lambda+1))^{-1}.$$

Thus $\lim_{n\to\infty} 1/n \neq 0$ in the prediction topology of $E$.

**Remark** It would be interesting to pursue this further, to obtain an example in which the prediction topology differs from the Ray topology of $E$ for a Borel right process.

These examples indicate a rather complicated situation relating the various topologies and compactifications. But a simplification occurs in the most important case, described by the next theorem.

**Theorem 2.33** *Suppose $p(t, x, B)$ satisfies the Hypothesis in Section 2.2, and in addition we have $P^{\varphi(x)}\{Z_t^{\varphi(x)} = \varphi(\tilde{X}_t), \forall t \geq 0\} = 1$, $x \in E$, as under the Criterion in Section 2.3. Then, for $U = \varphi(E)$, we have $U_R = U_e$, i.e. the Ray space equals the entrance space in $\bar{U}_\infty$.*

**Proof** It is obvious that $U_R \subset U_e$, with the same (Ray) topology, and the only elements of $U_e$ not in $U_R$ are also not in $M$. So it is enough to show that $U_e \subset M$. Now it is clear that, for $x \in U_e$, $P^x$ defines a probability entrance law of $Z_t$ on $U$, so by Theorem 2.29 $P^x$ may be realized (except perhaps at $t=0$) by a unique initial distribution $\nu$ for $Z_t$ on $U^* \cap D$ (where $P^x$ refers to the Ray process on $\bar{U}_\infty$). We will show that any such law $P^\nu$ for $Z_t$ may also be realized by a unique $P^z$, $z \in M$ (where $z \in B$ if and only if $\nu$ is not a point mass). We need the following result.

**Lemma 2.34** *For $\nu$ as above, the process $\varphi^{-1}(Z_t)$, $t > 0$, is a $P^\nu$-strong Markov process on $E$ relative to its coordinate $\sigma$-fields, and it is right-continuous in the topology of $E$ induced by $\varphi$ from $M$.*

**Proof** This is an immediate consequence of the Markov properties and right-continuity of $Z_t$ for $P^\nu$.

Accordingly, it suffices to set $z(S') = P^v(\varphi^{-1}(Z_{(\cdot)}) \in S')$, $S' \in \mathscr{F}'$, and to show that $Z_t^z$ is equivalent in law to $Z_t$ for $P^v$. This follows if we show that $Z_t$ defines the prediction process of $\varphi^{-1}(Z_t)$ on our prediction space, that is,

$$P^v\{\varphi^{-1}(Z_{(\cdot)}) \circ \theta_t^Z \in S' \mid \varphi^{-1}(Z_{(s)}), s \le t\} = Z_{t-}(S'),$$

$t > 0$, $S' \in \mathscr{F}'$ (whence the equivalence for $Z_t$ follows from right-continuity). Coming back to our hypothesis, this asserts that, for all $x \in E$, the process $\tilde{X}_t$ on $\Omega'$ is a $P^{\varphi(x)}$-strong Markov process relative to $\mathscr{F}'_{t+}$, whose prediction process is given by $\varphi(\tilde{X}_t)$. In the language of prediction space, this implies that, for any $z \in U$, $Z_t$ is the $P^z$-prediction process of the coordinate process $\varphi^{-1}(Z_t)$. Therefore, by the Markov property of $Z_t$ for $P^v$, for any $0 < \varepsilon < t$, and $S' \in \mathscr{F}'$, we have

$$P^v\{\varphi^{-1}(Z(\cdot)) \circ \theta_t^Z \in S' \mid \varphi^{-1}(Z(s)), 0 < s \le t\}$$
$$= P^v\{\phi^{-1}(Z(\cdot)) \circ \theta_t^Z \in S' \mid Z(\varepsilon), \varphi^{-1}(Z(s)), \varepsilon \le s \le t\}$$
$$= Z_{t-}(S'),$$

as was to be shown.

In the situation of Theorem 2.33, it is not necessary to introduce $U = \varphi(E)$ in order to define the entrance space. Instead, one can work directly on $E$.

**Definition 2.23** The Rayification of the topology of $E$ is the topology induced by $\varphi$ from the Ray topology of $\varphi(E)$. The Ray compactification of $E$, written $\bar{E}_\infty$, is the compactification of $E$ under the metric $d_\infty \circ \varphi$ (i.e. the $\varphi$-inverse of the Ray compactification $\bar{U}_\infty$ (Definition 2.21)). Finally, the entrance space $E_e$ of $E$ is the $\varphi$-inverse of $U_e$ (by Definition 2.22; note that $\varphi$ extends to a homeomorphism of $\bar{E}_\infty \to \bar{U}_\infty$).

In this terminology, Theorem 2.33 and its proof imply that any entrance law for a Borel right process on $E$ is realized for a unique element $\varphi_{(z)}^{-1} \in E_e$ by the process $\tilde{X}_t$ for the measure $P^z$. Indeed, by Theorem 2.17 the transition function of $\varphi^{-1}(Z_t)$ in this case is the given $p(t, x, B)$.

We have not defined the Ray space $R$ of $E$ because that has already been done for right processes in Getoor (1975, Definition (15.5)). Naturally one would hope that this Ray space $R$ is the same as $E_e \, (= \varphi^{-1}(U_R)$, by Theorem 2.33). Unfortunately a proof of this is lacking (contrary to Knight (1984) which has an error on p. 650). Nevertheless, the two are closely related, and in known applications they can be used interchangeably. For the sake of completeness we conclude this chapter by establishing the precise connection.

**Theorem 2.35** *In the case of Borel right processes, the Ray space $R$ of $E$ is the same as $E_e$ when $E$ is first given the topology (and metric) induced by $\{R_\lambda f, f \in C(\bar{E}), \lambda > 0\}$, (i.e. we use a countable dense subset, as in Definition 2.1).*

## 88 | Foundations of the prediction process

**Remark** It is shown in Getoor (1975, (15.17)) that $R$ does not depend on the choice of $\bar{E}$ (in a precise sense), hence the same is true of $E_e$ when $E$ is first re-topologized by $\{R_\lambda f, f \in C(\bar{E})\}$.

**Proof** We will show, more generally, that the Ray–Knight compactification of $E$, introduced in Walsh and Meyer (1971) is the same as $\bar{E}_\infty$ when $\bar{E}_\infty$ is defined from the new topology (and compactification) of $E$. By definition of $R$ this immediately implies the assertion, and also the fact that the Ray topology of Getoor (1975, p. 60) is the same as the Rayification of the topology on $E$ generated by $\{R_\lambda f, f \in C(\bar{E})\}$.

We must review how the Ray–Knight compactification is defined. One begins with the positive cone

$$\mathscr{C}_0 = \left\{ \sum_{i=1}^k R_{\lambda_i} f_i; f_i \in C^+(\bar{E}), 1 \leq k \right\}.$$

Letting $\Lambda$ denote the operation of closure under pairwise minima, one now defines inductively a sequence $\mathscr{C}_n$, $1 \geq n$, of positive convex cones, closed under minima, by setting

$$\mathscr{C}_{n+1} = \Lambda\left( \mathscr{C}_n + \left\{ \sum_{i=1}^k R_{\lambda_i} f_i; f_i \in \mathscr{C}_n, 1 \leq k \right\} \right).$$

Then the *Ray cone* is $\mathscr{C} = \bigcup_n \mathscr{C}_n$. It is shown that $\mathscr{C}$ contains a countable, uniformly dense subset, and the Ray–Knight compactification is the compact metric compactification defined from such a subset in the usual way (of Definition 2.1).

We can view the sequence $\mathscr{C}_n$, $1 \leq n$, in a more conventional manner by noting that the vector space $\mathscr{C}_n - \mathscr{C}_n$ is closed under both minima and maxima (since $f \vee g = (f + g) - (f \wedge g)$). Hence, by the lattice form of the Stone–Weierstrass theorem, if we compactify $E$ using a countable, dense subset of $\mathscr{C}_n - \mathscr{C}_n$, the subset remains dense in the continuous function on the compactification. Referring back to Definition 2.21 of $\bar{U}_\infty$, we used a sequence $\bar{U}_n$ of compactifications, such that $\bar{U}_1$ is a compactification for the prediction topology, and $\bar{U}_{n+1}$ is a compactification by a sequence containing both a uniformly dense subset of $C(\bar{U}_n)$ and of $R_\lambda^Z(C(\bar{U}_n))$. Translated back to $E$ by the map $\varphi$, the generators of $\bar{U}_1$ include a uniformly dense subset of $\mathscr{C}_0$, as shown at the start of the proof of Lemma 2.15, and since the topology generated by $\mathscr{C}_{n+1}$ is just that of $\mathscr{C}_n$ together with $\{R_\lambda f, f \in \mathscr{C}_n\}$ we see immediately by induction that the Ray compactification makes uniformly continuous the elements of $\mathscr{C}$. But the same remains true if we first re-topologize $E$ by $\{R_\lambda f, f \in C(\bar{E})\}$ (i.e. by the cone $\mathscr{C}_0$). To see this, we remark that by the resolvent equation, $R_\lambda f = \lim_{\mu \to \infty}(\mu - \lambda) R_\mu R_\lambda f$, so that elements of $\mathscr{C}_0$ remain uniformly continuous for the Rayification. The rest of the induction now applies as before. One notes, however, that the outcome is a weaker topology than in the previous case (at least formally, but we lack any proof of strict inequality, and the difference, if any, seems to be minor).

Turning to the converse, it emerges that the re-topologization of $E$ is needed for the proof. We first show that when $\bar{M}$ is defined from the re-topologized $E$, the metric on $E$ induced by $\varphi$ from that of $\bar{M}$ (following Definition 2.5) is weaker than that of the Ray–Knight compactification (generated by $\mathscr{C}$). The converse will then follow by iteration. Now according to the proof of Lemma 2.26, the metric on $E$ induced by $\varphi$ is generated by a countable family of the form $E^{\varphi(x)}T_m = R_\lambda(f_{n_1}g_{n_2})$ where $f_{n_1} \in C^+(\bar{E})$ and $g_{n_2}(x) = E^{\varphi(x)}T_{m-1}$, $1 \leq m$. Since $T_0 = 1$, we have $E^{\varphi(x)}T_1 = R_{\lambda_2}f_{n_1}(x)$, and then $E^{\varphi(x)}T_2 = R_{\lambda_3}(f_{n_1}(x)R_{\lambda_2}f_{n_2}(x))$, and so forth. But when we begin by re-topologizing $E$, we must use $f_{n_1}$ continuous on the compactification generated by $\{R_\lambda f, f \in C(\bar{E})\}$. It must be noted here that in defining the cone $\mathscr{C}$, one can as well begin with $\Lambda\mathscr{C}_0$ instead of $\mathscr{C}_0$, since clearly $\Lambda\mathscr{C}_0 \subset \mathscr{C}_1$. Then the vector space $\Lambda\mathscr{C}_0 - \Lambda\mathscr{C}_0$ is uniformly dense in the space $\{f_{n_1}\}$, and it follows that $\{R_\lambda f_{n_1}\}$ are in the uniform closure of $\mathscr{C}_2 - \mathscr{C}_2$. This last being closed under products, it follows that the terms $R_{\lambda_3}(f_{n_1}R_{\lambda_2}f_{n_2})$ are in the uniform closure of $\mathscr{C}_3 - \mathscr{C}_3$, and so on. By induction it follows that $E^{\varphi(x)}T_m$ is in the uniform closure of $\mathscr{C}_{m+1} - \mathscr{C}_{m+1}$ for every $m$. Hence the generators of the metric on $E$ induced by $\varphi$ are all in the uniform closure of $\mathscr{C} - \mathscr{C}$, which implies that the metric is weaker than that of the Ray–Knight compactification. But the uniform closure of $\mathscr{C} - \mathscr{C}$ is closed under $R_\lambda$ and under products. Therefore, we can repeat the argument starting with the $\varphi$-induced topology of $E$ (and the metric generators used above) and it follows that the generators of the compactification $\bar{U}_2$ are also in the uniform closure of $\mathscr{C} - \mathscr{C}$. Then by induction those of $\bar{U}_n$ are in the uniform closure of $\mathscr{C} - \mathscr{C}$ for every $n$, which implies that the metric of the Ray compactification, relative to the topology of $E$ generated by $\{R_\lambda f, f \in C^+(\bar{E})\}$, is weaker than, and hence equivalent to, that of the Ray–Knight compactification.

## Appendix

The following was used in the discussion of the Hypothesis in Section 2.2.

**Theorem 2.36** *Let $p(t, x, B)$ be a Markov transition function satisfying the Hypothesis in Section 2.2. Then the $\sigma$-fields on $E$ generated by $\{p(t, x, B), 0 < t, B \in \mathscr{E}\}$ and by $\{R_\lambda f(x), 0 < \lambda, f \in b(\mathscr{E})\}$ are the same.*

**Proof** For $\varepsilon > 0$ we have

$$R_\lambda f(x) = \int_0^\varepsilon e^{-\lambda s} T_s f(x) \, ds + e^{-\lambda \varepsilon} \int_E p(\varepsilon, x, dy) R_\lambda f(y) \, dy$$

where the second term on the right is clearly measurable over $\sigma\{p(\varepsilon, x, A), A \in \mathscr{E}\}$, and the first term on the right tends uniformly to 0 as $\varepsilon \to 0+$. This proves one inclusion. Conversely, a formula for inversion of the Laplace transform (Widder 1946, Chapter VII, Section 6) gives the inverse

$$\hat{p}(t, x, A) = \limsup_{k \to \infty} (-1)^k \frac{(kt^{-1})^{k+1}}{k!} \frac{d^k}{d\lambda^k} \int_0^\infty e^{-\lambda s} p(s, x, A) \, ds \bigg|_{\lambda = kt^{-1}}$$

for each $A \in \mathscr{E}$, in such a way that, by the uniqueness theorem for Laplace transforms, we have $\hat{p}(t, x, A) = p(t, x, A)$ for a.e. $t$. Also, for fixed $(t, A)$, $\hat{p}(t, x, A)$ is clearly measurable over $\sigma\{R_\lambda f\}$, and it is even $(\mathscr{B}^+ \times \sigma\{R_\lambda f\})$-measurable in $(t, x)$. Now for $B \in \mathscr{B}^+(0, t]$ and $C \in \mathscr{E}$ we have

$$\int_0^t \hat{p}(t - s, x, C) I_B(s) \, ds = \int_0^t p(t - s, x, C) I_B(s) \, ds \in \sigma\{R_\lambda f\}.$$

It follows that, for any simple function

$$g(s, y) = \sum_{k=1}^n a_k I_{B_k}(s) I_{C_k}(y),$$

we have

$$\int_0^t \int_E p(t - s, x, dy) g(s, y) \, ds \in \sigma\{R_\lambda f\}.$$

By the monotone class theorem, the same is true for bounded, $(\mathscr{B}^+ \times \mathscr{E})$-measurable $g$. Setting $g = p(s, y, A)$, the Chapman–Kolmogorov equation gives $tp(t, x, A) \in \sigma\{R_\lambda f\}$, as required.

# 3

# Gaussian processes and the prediction process in the wide sense

> It is clear then that chance is an incidental cause in the sphere of those actions for the sake of something which involve purpose. Intelligent reflection, then, and chance are in the same sphere, for purpose implies intelligent reflection.
>
> Aristotle, 4th Century B.C. Physics, Book II, 5.

## 3.1 The structure of a measurable Gaussian process

As we have already seen in the case of Markov processes (Theorem 2.13), if we make additional assumptions about a probability $z = P'$ on $(\Omega', \mathscr{F}')$ the corresponding $Z_t^z$ may give us special information about a measurable process whose law agrees with $P'$. In the present chapter, we will apply this method to real Gaussian processes. Essentially all of our results, however, are *linear* in the sense that the random variables involved are in the Hilbert subspace of $L^2(\Omega', \mathscr{F}', P')$ generated by the process itself, i.e. by $\{\int_0^t w(s)\,ds, t > 0, \text{ and } 1\}$. Moreover, they depend only on the mean and covariance structure of the process, which takes the following form.

**Definition 3.1** Provided that they exist, we set

(1) $\mu(\lambda) = E^z \int_0^\infty e^{-\lambda t} w(t)\,dt$, $\lambda > 0$; and

(2) $\sigma^2(\lambda_1, \lambda_2) = E^z[\{\int_0^\infty e^{-\lambda_1 s} w(s)\,ds - \mu(\lambda_1)\}\{\int_0^\infty e^{-\lambda_2 t} w(t)\,dt - \mu(\lambda_2)\}]$, $\lambda_1, \lambda_2 > 0$.

Therefore the results will have immediate generalizations to any $P'$ having the same mean and covariance structure.

As is well known (Doob 1953), if we begin with any real non-negative definite function $\Gamma(s, t)$ and an arbitrary real-valued $E(t)$, there exists a Gaussian process $X_t$ with $EX_t = E(t)$ and $E(X_s - E(s))(X_t - E(t)) = \Gamma(s, t)$, $0 \le s, t$, and this $X_t$ has a uniquely determined family of finite dimensional joint distributions. To apply the prediction process, however, we need to assume a bit more, as follows. Note that any boundedness condition as $t \to \infty$ is not restrictive for problems involving bounded $t < K$ since one can then replace the process $X_t$ by $X_K$ for all $t \ge K$, or by $e^{-\Lambda t} X_t$ for a fixed $\Lambda > 0$ when convenient.

## Assumption

(1) $E(t)$ is Borel measurable and $\int_0^\infty e^{-\lambda t} E^2(t)\, dt < \infty$ for $\lambda > 0$.

(2) $\Gamma(s, t)$ is the covariance of a measurable process (with mean 0) and $\int_0^\infty e^{-\lambda t} \Gamma(t, t)\, dt < \infty$, $\lambda > 0$.

For reasons of completeness concerning this assumption, let us prove its analytical counterpart.

## Theorem 3.1

(1) *A function $E(t)$ and a covariance $\Gamma(s, t)$ are the mean and covariance of some (Borel) measurable process if and only if $E(t)$ is Borel measurable, $\Gamma(s, t)$ is jointly Borel measurable in $(s, t)$, and $H$ is separable, where $H$ is the Hilbert space closure of any process $X_t$ having mean 0 and covariance $\Gamma$ (i.e. $\{\sum_{i=1}^n c_i X_{t_i}, t_1 < \cdots < t_n, n < \infty\}$ is dense in $H$ for the $L^2(\Omega, \mathscr{F}, P)$-norm).*

(2) *There exists a measureable process as in (1) if and only if there exists a measurable Gaussian process with the same mean and covariance.*

**Proof** If $X_t$ has finite mean $E(t)$, then $E(t)$ ($=\lim_{N\to\infty} E(-N \vee X_t \wedge N)$) is measurable if $X_t$ is so, by Fubini's theorem, and then $X(t) - E(t)$ is also measurable. Thus we can assume $E(t) = 0$ for the condition on $\Gamma(s, t)$. Then if $X_t$ is measurable, $(X_s, X_t)$ is $(\mathscr{B} \times \mathscr{B} \times \mathscr{F})$-measurable in $(s, t, w)$, and so $\Gamma(s, t)$ must be $(s, t)$-measurable by the same argument.

Now if $H$ is separable, since convergence in $H$ implies convergence in probability, $X_t$ takes its values in a separable part of $L^0$, and so, by a classical criterion (Dellacherie and Meyer 1975), $X_t$ has a measurable standard modification if and only if it is measurable as an $L^0$-valued-process. But the $L^2$-topology is finer, so $L^2$-measurability implies $L^0$-measurability, and $X_t$ is $L^2$-measurable when $\Gamma(s, t)$ is measurable. Thus the conditions of (1) are sufficient.

Conversely, if $X_t$ is measurable and $EX_t = 0$ then, for every $N$, $-N \vee X_t \wedge N$ is also measurable, and hence has its values in a separable part of $L^0$. But for bounded processes the $L^0$ and $L^2$-topologies coincide, and so $-N \vee X_t \wedge N$, for $N > 0$, has its values in a separable part $H_N$ of $L^2$. Then $\bigcup_N H_N$ has separable closure in $L^2$, and since $EX_t^2 = \Gamma(t, t)$ is finite, we have, by dominated convergence, $\lim_{N\to\infty} E\{(-N \vee X_t \wedge N) - X_t\}^2 = 0$. Thus $X_t$ is in the $L^2$ closure of $\bigcup_N H_N$, and the conditions of (1) are necessary.

Assertion (2) follows easily because the conditions of (1) depend only on $\Gamma(s, t)$.

For the present chapter, we denote $(\Omega'_E, \mathscr{F}'_E)$ of Definition 2.2 simply by $(\Omega', \mathscr{F}')$, with the understanding that $E = R$ throughout. In our approach, no use is made of spectral representations and there is no advantage in considering complex-valued processes. Analogously, we will write $(M, \mathcal{M})$ for $(M_R, \mathcal{M}_R)$, $\tilde{X}$ for $\tilde{X}_R$, $\tilde{\rho}$ for $\tilde{\rho}_R$, etc.

**Theorem 3.2** *Under the above Assumption there is a unique $P'$ ($=z$) on $(\Omega', \mathcal{F}')$ such that*

(1) $\mu(\lambda) = \int_0^\infty e^{-\lambda t} E(t) \, dt, \lambda > 0;$

(2) $\sigma^2(\lambda_1, \lambda_2) = \int_0^\infty \int_0^\infty e^{-(\lambda_1 s + \lambda_2 t)} \Gamma(s, t) \, ds \, dt; \lambda_i > 0;$

(3) *The family* $\{\int_0^\infty e^{-\lambda t} w(t) \, dt, \lambda > 0\}$ *is jointly Gaussian for $P'$.*

Two distinct pairs $E_i(t)$, $\Gamma_i(s, t)$ determine the same $P'$ if and only if $E_1(t) = E_2(t)$ for a.e. $t$ and $\Gamma_1(s, t) = \Gamma_2(s, t)$ for (Lebesgue-) a.e. $(s, t)$.

**Proof** If $X(t)$ is any measurable Gaussian process with mean $E(t)$ and covariance $\Gamma(s, t)$, we can define $P'(S') = P\{X(\cdot) \in S'\}, S' \in \mathcal{F}'$, and it follows easily from Fubini's theorem that (1) and (2) are satisfied. As to (3), since $X(t)$ is strongly measurable as an $L^2$-process it is a uniform limit of Gaussian simple functions, and since the $L^2$-integrals $\int_0^\infty e^{-\lambda t} X(t) \, dt$ exist, they are easily seen to form a Gaussian family. They also coincide, $P$-a.s., with the corresponding pathwise integrals (as one sees by computing the inner products with $\int_0^s X(u) \, du$, which by Fubini's theorem coincide in the two interpretations). The uniqueness assertions follow by writing

$$\int_0^\infty e^{-\lambda t} X(t) \, dt = \lambda^{-1} \int_0^\infty e^{-\lambda t} \left( \int_0^t X(u) \, du \right) dt,$$

so that, by inversion of the transform, the joint law of $\{\int_0^t X(u) \, du\}$ is determined by that of the transforms.

We can now introduce the basic objects with which we will be concerned in this chapter.

**Definition 3.2** A Gaussian process is a $P'$ on $(\Omega', \mathcal{F}')$, obtained as in Theorem 3.2 from some $(E(t), \Gamma(s, t))$ satisfying the above Assumption. The collection of such is denoted $M_G$.

We will eventually show that the class $M_G$ of Gaussian processes is a complete Borel demesne for the prediction process. This comes from the fact that their prediction processes have a particularly simple structure, as might be expected. Since $P'$ is really a probability for equivalence classes of $w(t)$ with $w_1 \equiv w_2$ when $w_1 = w_2$ a.e., it is not in keeping with our approach to assume $E(t)$ and $\Gamma(s, t)$ are completely known in advance. Rather, one should assume only an equivalence class, as in the second part of Theorem 3.2. The converse to Theorem 3.2 is as follows.

## 94 | Foundations of the prediction process

**Theorem 3.3** *If $z \ (=P')$ on $(\Omega', \mathscr{F}')$ is such that*

(1) $E' \int_0^\infty e^{-\lambda t} w^2(t) \, dt < \infty, \ \lambda > 0$; *and*

(2) *the family $\{\int_0^\infty e^{-\lambda t} w(t) \, dt, \ \lambda > 0\}$ is jointly Gaussian,*

*then there are $E_z, \Gamma_z$ satisfying the Assumption above, and defining $z$ as in Theorem 3.2. Thus $z \in M_G$.*

**Proof** Setting $E_z(t) = E'(\tilde{X}_t)$ when it exists, and $E_z(t) = 0$ elsewhere, we have

$$\int_0^\infty e^{-\lambda t} (E'|\tilde{X}_t|)^2 \, dt \leq E' \int_0^\infty e^{-\lambda t} \tilde{X}_t^2 \, dt < \infty,$$

because $w(t) = \tilde{X}_t$ for a.e. $t$ (and all $w \in \Omega'$) in view of Definition 2.6 and Lemma 2.6. Thus

$$E' \int_0^\infty e^{-\lambda t} (w(t) - E_z(t))^2 \, dt = E' \int_0^\infty e^{-\lambda t} w^2(t) \, dt - \int_0^\infty e^{-\lambda t} E_z^2(t) \, dt < \infty.$$

But, we have

$$E' \int_0^\infty e^{-\lambda t} (w(t) - E_z(t))^2 \, dt = E' \int_0^\infty e^{-\lambda t} (\tilde{X}_t - E_z(t))^2 \, dt$$

and $E_z(t) = E'\tilde{X}_t$ for a.e. $t$, so we can introduce (dropping the prime)

$$\Gamma_z(t, t) = \begin{cases} E(\tilde{X}_t - E\tilde{X}_t)^2 & \text{if finite} \\ 0 & \text{elsewhere,} \end{cases}$$

and clearly $\int_0^\infty e^{-\lambda t} \Gamma_z(t, t) \, dt < \infty, \ \lambda > 0$. Now let

$$\Gamma_z(s, t) = E(\tilde{X}_s - E\tilde{X}_s)(\tilde{X}_t - E\tilde{X}_t)$$

if both $s$ and $t$ belong to the first case, and set $\Gamma_z(s, t) = 0$ elsewhere. Then $\Gamma_z(s, t)$ is the covariance of the measurable process

$$X'_t = \begin{cases} \tilde{X}_t & \text{if } E(\tilde{X}_t - E\tilde{X}_t)^2 < \infty \\ 0 & \text{elsewhere,} \end{cases}$$

so it is clear that the Assumption is satisfied. Since $X'_t = \tilde{X}_t$, $P'$-a.s., except for a (Borel) null set of $t$, it is also clear that (1) and (2) of Theorem 3.2 are satisfied, completing the proof.

**Remark** Note that it is not asserted that either $\tilde{X}_t$ or $X'_t$ is a Gaussian process for $P'$ in the usual coordinate-wise sense. Only their integrals are Gaussian. On the other hand, by general theory there does exist a measurable Gaussian process with mean $E_z(t)$ and covariance $\Gamma_z(s, t)$ which induces the given $P'$ on $(\Omega', \mathscr{F}')$. Indeed by Theorem 3.1 any Gaussian process with $E_z(t)$

and $\Gamma_z(s, t)$ as mean and covariance will have a measurable standard modification.

It will sometimes by useful to know that there is a $\Gamma(s, t)$ corresponding to given $P' \in M_G$ which is continuous in $(s, t)$. Let us introduce here the following definition.

**Definition 3.3** Let $M_{G(c)} \subset M_G$ be the set of Gaussian processes with a continuous covariance.

Although $\Gamma(s, t)$ is not uniquely determined by $z \in M_G$, there can be at most one continuous $\Gamma$ for given $z$. Existence can be checked from a given $z$ as follows.

**Theorem 3.4** *We have $z \in M_{G(c)}$ if and only if $z \in M_G$ and*

$$\lim_{\Delta \to 0+} \Delta^{-2} \int_s^{s+\Delta} \int_t^{t+\Delta} \Gamma_z(u, v) \, du \, dv$$

*exists uniformly in bounded $(s, t)$ (for every $\Gamma_z$ corresponding to $z$).*

**Notation** The limit in this case will be denoted by $\Gamma_z^c(s, t)$. It is a continuous covariance, and it also generates $z$, using the same $E(t)$.

**Proof** Obviously the limit exists as asserted when $\Gamma_z$ is continuous, or when $\Gamma_z$ is equal a.e. to a continuous function. Conversely, since the double integral is continuous in $(s, t)$, the limit is likewise continuous if it exists uniformly. Moreover, the integral is the covariance of the process $\Delta^{-1} \int_t^{t+\Delta} \tilde{X}_u \, du$, which converges to $\tilde{X}_t$ in $L^2$ for a.e.t. If these covariances converge uniformly, then $\tilde{X}_t$ has a.e. covariance $\Gamma_z^c$ and hence if it is centred at the mean a.e., it extends by continuity in $L^2$ to a process whose covariance is $\Gamma_z^c$ as required.

Before commencing the study of $M_G$ and $M_{G(c)}$ as such, we investigate certain prediction demesnes determined by moment conditions. Besides the fact that some of these are needed for $M_G$, they are also of independent interest.

**Definition 3.4** For $r > 0$, let

$$M_r = \left\{ z \in M : E^z \int_0^\infty e^{-\lambda t} |w(t)|^r \, dt \quad \text{is finite for each } \lambda > 0 \right\}.$$

Subject to the following verification we call $M_r$ the *demesne of finite $r^{th}$ moments*.

**Theorem 3.5** *For each $r > 0$, $M_r$ is a complete Borel prediction demesne (Definition 2.14).*

## 96 | Foundations of the prediction process

**Proof** Since $\int_0^\infty e^{-\lambda t}(|w(t)|^r \wedge N)\,dt$, for each $N > 0$, is bounded and continuous on $\Omega'$, it follows by letting $N \to \infty$ that $E^z \int_0^\infty e^{-\lambda t}|w(t)|^r\,dt$ is $\mathcal{M}$-measurable. Hence $M_r \in \mathcal{M}$.

To continue, it is more in line with our earlier work to replace $w(t)$ by $\tilde{X}_t$ (Definition 2.6). Thus for $N < \infty$,

$$E^{Z_{\tilde{t}}^z} \int_0^\infty e^{-\lambda s}(|w_s|^r \wedge N)\,ds = E^{Z_{\tilde{t}}^z} \int_0^\infty e^{-\lambda s}(|\tilde{X}_s|^r \wedge N)\,ds$$

$$= E^z\left(\int_0^\infty e^{-\lambda s}(|\tilde{\rho}|^r \wedge N) \circ \theta_s\,ds \circ \theta_t | \mathscr{F}'_{t+}\right)$$

$$= e^{\lambda t} E^z\left(\int_0^\infty e^{-\lambda(s+t)}(|\tilde{\rho}|^r \wedge N)\theta_{s+t}\,ds | \mathscr{F}'_{t+}\right)$$

$$\leq e^{\lambda t} E^z\left(\int_0^\infty e^{-\lambda s}(|\tilde{\rho}|^r \circ \theta_s)\,ds | \mathscr{F}'_{t+}\right), \qquad P^z\text{-a.s.}$$

It follows exactly as in the proof of Theorem 2.4 that

$$e^{-\lambda t} E^{Z_{\tilde{t}}^z} \int_0^\infty e^{-\lambda s}(|\tilde{X}_s|^r \wedge N)\,ds$$

is a positive supermartingale for $(\mathscr{F}'_{t+}, P^z)$, and by Theorem 2.1 it is r.c.l.l. in $t$, $P^z$-a.s. for each $N$. By the above it is uniformly bounded by

$$E^z\left(\int_0^\infty e^{-\lambda s}(|\tilde{\rho}|^r \circ \theta_s)\,ds | \mathscr{F}'_{t+}\right), \qquad P\text{-a.s.,}$$

and letting $N \to \infty$ it follows by Exercise 2.1 that $E^{Z_{\tilde{t}}^z} \int_0^\infty e^{-\lambda s}|\tilde{X}_s|^r\,ds$ is likewise r.c.l.l., with left and right limits finite. This proves that $M_r$ is a Borel demesne. Now to finish the proof of completeness, observe that for finite $N$ the left limits $E^{Z_{\tilde{t}}^z} \int_0^\infty e^{-\lambda s}(|\tilde{X}_s|^r \wedge N)\,ds$ are given by replacing $Z_{\tilde{t}}^z$ by $Z_{\tilde{t}-}^z$ in view of Theorem 2.4. As $N \to \infty$, these tend to $E^{Z_{\tilde{t}-}^z} \int_0^\infty e^{-\lambda s}|\tilde{X}_s|^r\,ds$ by monotone convergence. On the other hand, this is also a positive $\lambda$-supermartingale, so a familiar maximal inequality (Dellacherie and Meyer 1980, VI, (1.2)) shows that the convergence is uniform in bounded $t$, $P^z$-a.s. Consequently, the above limits are also the left limits of $E^{Z_{\tilde{t}}^z} \int_0^\infty e^{-\lambda s}|\tilde{X}_s|^r\,ds$, are required.

In our study of Gaussian processes, we are mainly concerned with $M_2$.

**Corollary 3.6** *The collection of $P'$ satisfing (1) and (2) of Theorem 3.2 for some $E(t)$, $\Gamma(s, t)$ as in the above Assumption is precisely $M_2$. Hence it is a Borel prediction demesne.*

**Proof** For such $P'$, we have

$$E' \int_0^\infty e^{-\lambda t}(w^2(t) - (E(t))^2)\, dt = \int_0^\infty e^{-\lambda t}\Gamma(t,t)\, dt < \infty,$$

by the proof of Theorem 3.2 and (2) of the Assumption. By (1) of the same, this implies $P' \in M_2$. Conversely, for $P' \in M_2$ we have $\int_0^\infty e^{-\lambda t} E'(\tilde{X}_t)^2\, dt < \infty$, and hence, defining

$$\Gamma(s,t) = \begin{cases} \text{Cov}(\tilde{X}_s, \tilde{X}_t) & \text{on } \{(s,t): E'(\tilde{X}_s)^2 < \infty, E'(\tilde{X}_t)^2 < \infty\} \\ 0 & \text{elsewhere} \end{cases}$$

$$E(t) = \begin{cases} E'(\tilde{X}_t) & \text{if } E'|\tilde{X}_t| < \infty \\ 0 & \text{elsewhere} \end{cases}$$

we have the Assumption. Since the above exceptional set of $(s,t)$ is Lebesgue-null, and the expectation in defining $\sigma^2(\lambda_1, \lambda_2)$ is absolutely integrable, with the evident bound

$$\int_0^\infty \int_0^\infty e^{-(\lambda_1 s + \lambda_2 t)} \Gamma(s,t)\, ds dt \leq \int_0^\infty e^{-\lambda_1 s} \Gamma^{\frac{1}{2}}(s,s)\, ds \int_0^\infty e^{-\lambda_2 t} \Gamma^{\frac{1}{2}}(t,t)\, dt$$

$$\leq \frac{1}{\lambda_1 \lambda_2} \int_0^\infty e^{-\lambda_1 s} \Gamma(s,s)\, ds \int_0^\infty e^{-\lambda_2 t} \Gamma(t,t)\, dt$$

$$< \infty,$$

we can replace $w$ by $\tilde{X}$ in $\sigma^2(\lambda_1, \lambda_2)$ and apply Fubini's theorem to establish the identity (2) of Theorem 3.2.

Returning to the Gaussian case, let us first amplify its implications.

**Theorem 3.7** For $z \in M_G$, the set $\{\int_0^\infty f(t)w(t)\, dt;\ f \in L^2(e^{\lambda t}\, dt)$ for some $\lambda > 0\}$ is a jointly Gaussian family for $P^z$, with mean $\int_0^\infty f(t)E(t)\, dt$ and covariance $\int_0^\infty \int_0^\infty f_1(s)\Gamma(s,t)f_2(t)\, ds\, dt$.

**Proof** Since

$$E^z \int_0^\infty |f(t)w(t)|\, dt = E^z \int_0^\infty e^{\lambda t/2}|f(t)|\, e^{-\lambda t/2}(\tilde{X}(t))\, dt$$

$$\leq \left( \int_0^\infty e^{\lambda t} f^2(t)\, dt \int_0^\infty e^{-\lambda t} E'(\tilde{X}(t))^2\, dt \right)^{\frac{1}{2}}$$

$$< \infty,$$

Fubini's theorem yields the expectation $\int_0^\infty f(t)E(t)\,dt$ as asserted. Similarly, since

$$E^z \iint |f_1(s)w(s) - E(s))f_2(t)(w(t) - E(t))|\,ds\,dt$$

$$= \iint |f_1(s)f_2(t)|E^z|(\tilde{X}(s) - E(s))(\tilde{X}(t) - E(t))|\,ds\,dt$$

$$\leq \iint |f_1(s)f_2(t)|\Gamma^{\frac{1}{2}}(s,s)\Gamma^{\frac{1}{2}}(t,t)\,ds\,dt$$

$$\leq \left(\int e^{\lambda s}f_1^2(s)\,ds \int e^{-\lambda s}\Gamma(s,s)\,ds \int e^{\lambda t}f_2^2(t)\,dt \int e^{-\lambda t}\Gamma(t,t)\,dt\right)^{\frac{1}{2}}$$

$$< \infty,$$

the covariance assertion again follows. It remains to show that the family is Gaussian. By hypothesis, we know

$$\left\{\int_0^\infty e^{-\lambda t}w(t)\,dt, \lambda > 0\right\}$$

is Gaussian. If $\int_0^\infty e^{\lambda t}f^2(t)\,dt < \varepsilon$, then as above

$$E\left(\int_0^\infty f(t)w(t)\,dt\right)^2 \leq \left(\int e^{\lambda t}f^2(t)\,dt \int e^{-\lambda t}\Gamma(t,t)\,dt\right)$$

$$+ \left(\int e^{\lambda t}f^2(t)\,dt \int e^{-\lambda t}E^2(t)\,dt\right)$$

$$\leq \varepsilon\left(\int_0^\infty e^{-\lambda t}(\Gamma(t,t) + E^2(t))\,dt\right).$$

Therefore, the class of $f$ such that $\int_0^\infty f(t)w(t)\,dt$ is Gaussian contains $\{e^{-\lambda t}, \lambda > 0\}$, is linear, and is closed under passages to the limit in $L^2(e^{\lambda t}\,dt)$ for each $\lambda > 0$ (see below). By the Stone–Weierstrass theorem it contains functions of the form $f(t)e^{-\lambda t}$ for $f \in C_0(\mathbb{R}^+)$. Therefore, it contains all continuous $f$ with compact support. Their closure in $L^2(e^{\lambda t}\,dt)$ is $L^2(e^{\lambda t}\,dt)$, as required. Here we invoked the following well-known result.

**Lemma 3.8** *If a sequence* $X_n$ *of jointly Gaussian random vectors converges in probability then* $\lim_{n\to\infty} EX_n = \mu$ *and* $\lim_{n\to\infty} \text{Cov } X_n = (\sigma^2)$ *both exist. In this case the limit is Gaussian with mean* $\mu$ *and covariance* $(\sigma^2)$.

Since convergence in $L^2(\Omega', \mathscr{F}', P')$ implies convergence in probability, this completes the proof of Theorem 3.7.

To examine the prediction measure $P^z$ for $z \in M_G$, we require the following familiar lemma.

**Lemma 3.9** *Let $H_1 \subset H$ be Hilbert spaces of mean 0, jointly Gaussian random variables. Let $\mathscr{F}_1 \subset \mathscr{F}$ be the corresponding generated $\sigma$-fields, and let $\mathscr{E}(X|H_1)$ denote the Hilbert space projection of $X \in H$ onto $H_1$ (as in the Introduction). Then*

(1) $\mathscr{E}(X|H_1) = E(X|\mathscr{F}_1)$, *P-a.s.;*

(2) $X - \mathscr{E}(X|H_1)$ *is independent of $\mathscr{F}_1$.*

**Remark** In extending our work to the non-Gaussian case, it is important to recall that, without any Gaussian assumption (2) remains true if 'independent of $\mathscr{F}_1$' is replaced by 'orthogonal to $H_1$'.

The following lemma is an immediate consequence of this.

**Lemma 3.10** *For $z \in M_G$, and fixed $t$, $\{E^{Z^z(t)}(\int_0^\infty e^{-\lambda s} w(s)\,ds), \lambda > 0\}$ is jointly Gaussian, and $\{\int_0^\infty e^{-\lambda s} w(s+t)\,ds - E^{Z^z(t)} \int_0^\infty e^{-\lambda s} w(s)\,ds\}$ is independent of $\mathscr{F}'_{t+}$, and also Gaussian.*

**Proof** We may 'centre at the mean' by using here

$$X_t \stackrel{\text{def}}{=} \int_0^\infty e^{-\lambda s} w(s+t)\,ds - \int_0^\infty e^{-\lambda s} E_z(s+t)\,ds.$$

Then the conditions of Lemma 3.9 are satisfied with

$$H = L^2\left\{\int_0^\infty f(s)(w(s) - E_z(s))\,ds;\quad f \in L^2(e^{\lambda t}\,dt)\right\}$$

$$H_1 = \bigcap_{\varepsilon > 0} L^2\left\{\int_0^{t+\varepsilon} f(s)(w(s) - E_z(s))\,ds,\quad f \in L^2(e^{\lambda t}\,dt)\right\},$$

where $L^2\{\ \}$ denotes the linear closure of $\{\ \}$ in $L^2(\Omega', \mathscr{F}', P')$. It remains only to see that $\sigma(H_1)$ can be replaced by $\mathscr{F}'_{t+}$. First, we remark that

$$\mathscr{E}(X_t|H_1) = L^2 - \lim_{\varepsilon \to 0+} \mathscr{E}\left(X_t \Big| L^2\left\{\int_0^{t+\varepsilon} f(s)(w(s) - E(s))\,ds,\quad f \in L^2(e^{\lambda t}\,dt)\right\}\right),$$

and the collection of random variables on the right is Gaussian (Theorem 3.7) and generates $\mathscr{F}'_{t+\varepsilon}$ for each $\varepsilon > 0$, where $\bigcap_{\varepsilon > 0} \mathscr{F}'_{t+\varepsilon} = \mathscr{F}'_{t+}$. Thus, by martingale convergence,

$$\mathscr{E}(X_t|H_1) = L^2 - \lim_{\varepsilon \to 0+} E(X_t|\mathscr{F}'_{t+\varepsilon}) = E(X_t|\mathscr{F}'_{t+}),$$

and it follows that

$$E^{Z^z(t)} \int_0^\infty e^{-\lambda s}(w(s) - E_z(s+t))\,ds = L^2 \lim_{\varepsilon \to 0+} E(X_t | \mathscr{F}'_{t+\varepsilon}) = \mathscr{E}(X_t | H_1).$$

Hence

$$\left(X_t - E^{Z^z(t)} \int_0^\infty e^{-\lambda s}(w(s) - E_z(s+t))\,ds\right),$$

which is the $L^2$-limit of $(X_t - E(X_t | \mathscr{F}'_{t+\varepsilon}))$, each of which is independent of $\mathscr{F}'_{t+}$, is also independent of $\mathscr{F}'_{t+}$ as required.

By Lemma 3.10, for $z \in M_G$ we have $P^z\{Z(t) \in M_G\} = 1$ for each $t$. To show that $M_G$ is a complete Borel demesne, two problems remain: to show $M_G \in \mathscr{M}$, and to extend the probability 1 above to all $Z_{t-}$, $Z_t$ simultaneously. The first problem is straightforward. We have

$$M_G = M_2 \cap \left\{z: \int_0^\infty e^{-\lambda t} w(t)\,dt, \quad \lambda > 0 \text{ is jointly Gaussian}\right\},$$

where $M_2 \in \mathscr{M}$ is known (Theorem 3.5). Clearly $E^z \int_0^\infty e^{-\lambda t} w(t)\,dt \in \mathscr{M}$ (i.e. is $\mathscr{M}$-measurable) so we can centre at the mean, and show

$$M_G \cap \left\{z: E^z \int_0^\infty e^{-\lambda t} w(t)\,dt = 0, \quad \lambda > 0\right\} \in \mathscr{M}.$$

To this end, note that for any $0 < \lambda_1 < \cdots < \lambda_n$, the covariance

$$\mathrm{Cov}_z\left(\int_0^\infty e^{-\lambda_1 t} w(t)\,dt, \ldots, \int_0^\infty e^{-\lambda_n t} w(t)\,dt\right)$$

is an $\mathscr{M}$-measurable function of $z$ on $M_2$. Indeed, this is the limit as $N \to \infty$ of the matrix with components continuous in $z$:

$$\mathrm{Cov}_z\left(\int_0^\infty e^{-\lambda_k t}(-N \vee w(t) \wedge N)\,dt, \quad 1 \leq k \leq n\right).$$

Similarly, for fixed $\alpha_1, \ldots, \alpha_n$ the joint characteristic function

$$E^z \exp\left(i \sum_{k=1}^n \alpha_k \int_0^\infty e^{-\lambda_k t} w(t)\,dt\right)$$

is $\mathscr{M}$-measurable, and the Gaussian requirement may be stated by setting this equal to $\exp(-\frac{1}{2}\alpha\,\mathrm{Cov}_z\,\alpha')$ for all $\alpha$ with rational coordinates, and for all rational $0 < \lambda_1 < \cdots < \lambda_n$. Therefore, $M_G \in \mathscr{M}$.

The matter of extending $Z(t\pm) \in M_G$ to all $t$ simultaneously is a bit more subtle, and a preliminary insight into the problem may be given by considering an apparently similar problem in which, however, the inclusion does not extend; namely, let $M_2(c)$ denote the elements of $M_2$ with continuous covariance.

The criterion of Theorem 3.4 that

$$\lim_{\Delta \to 0+} \Delta^{-2} \int_s^{s+\Delta} \int_t^{t+\Delta} \Gamma_z(u, v) \, du \, dv$$

exists uniformly in bounded $(s, t)$, applies as before to show that $M_2(c) \in \mathcal{M}$. Also, after centring at the mean, the Remark before Lemma 3.10 shows that the variances of $\int_0^\infty e^{-\lambda s} w(s + t) \, ds - E^{Z^z(t)} \int_0^\infty e^{-\lambda s} w(s) \, ds$ are at most those of $\int_0^\infty e^{-\lambda s} w(s + t) \, ds$. Here we use the additional fact that, without any Gaussian assumption,

$$E^z\left(\int_0^\infty e^{-\lambda s} w(s+t)\, ds \Big| \mathcal{F}'_{t+}\right) = \mathcal{E}^z\left(\int e^{-\lambda s} w(s+t)\, ds \Big| H_z^*(t+)\right),$$

where we set $H_z^*(t+) = L^2(\mathcal{F}'_{t+}, P^z)$, so that the above difference is again a projection (onto $H_z^{*\perp}(t+)$). It follows by Jensen's inequality that

$$E^z \Gamma_{Z^z(t)}(s, s) \leq \Gamma_z(s + t, s + t).$$

But the most that can be shown by this reasoning is that, for each $t$, $P^{Z^z(t)}$ has a covariance whose $P^z$-expectation is continuous.

Indeed, it does not hold that $M_2(c)$ is a prediction demesne, as the following example shows.

**Example 1** Let $e$ be an exponential random variable, with $\lambda = 1$ (so that $P\{e > t\} = \exp(-t)$), and define a process $X_t = 0$ for $0 \leq t < e$; $X_t = \pm 1$ with probability $\frac{1}{2}$, independently of $e$, for $e \leq t < e + 1$; $X_t = X_e \pm 1$ with probability $\frac{1}{2}$, independently of the past, for $e + 1 \leq t$. Let $z$ be the prediction measure of $X_t$. Since $X$ has right-continuous paths, and as seen in Theorem 2.17, {r.c.} is universally measurable in $\Omega'$, it follows that $P^z\{r.c.\} = 1$, and so, for $P^z$, $\tilde{X}_t$ has the same law as $X_t$, and is also right-continuous, $P^z$-a.s. It is easy to see that $\tilde{X}_t$ has, not only for $z$, but also $P^{Z^z(t)}$ for each fixed $t$, a.s. continuous covariance. Indeed, this follows here because $\tilde{X}_t$ has no fixed times of discontinuity. However, let $T = e$ (a stopping time of $Z^z(t)$). Then, for $Z^z(T)$, $\tilde{X}_t$ has a fixed discontinuity at $t = 1$, at which its variance jumps from 0 to 1. Clearly $Z^z(T)$ has continuous covariance with probability 0. Hence $M_2(c)$ is not a demesne. However, it will be seen below that it becomes a demesne if we add the Gaussian requirement, i.e. $M_{G(c)}$ is a demesne, and so is $M_G$.

**Theorem 3.11** $M_G$ *and* $M_{G(c)}$ *are complete Borel prediction demesnes.*

**Proof** According to Lemma 3.10 we can construct the prediction probability $Z^z(t)$ for fixed $t$ as follows. First we construct for $P^z$-a.e. $Z^z(t)$ a measurable version of $\{E^{Z^z(t)} \tilde{X}(s), 0 < s\}$, which is possible since $\tilde{X}(s)$ is a measurable process with

$$E^{Z^z(t)} \int_0^\infty e^{-\lambda s}|\tilde{X}(s)|\, ds = E^{Z^z(t)} \int_0^\infty e^{-\lambda s}|w(s)|\, ds$$

$$\leq \frac{1}{\sqrt{\lambda}} E^z \left( \int_0^\infty e^{-\lambda s}(w(s+t))^2\, ds \Big| \mathscr{F}'_{t+} \right)$$

$$< \infty, \quad P^z\text{-a.s.}$$

This function of $s$ (which is of course a random function on $\Omega'$) is the expectation function $E(s)$ for $Z^z(t)$. To this we add a fixed Gaussian process, independent of $\mathscr{F}'_{t+}$, of mean 0 and variance less than that of $\{\int_0^\infty e^{-\lambda s} w(s+t)\, ds, \lambda > 0\}$ for $P^z$. We can thus construct $Z^z(t)$ for all rational $t > 0$, uniquely up to a $P^z$-null set. Our problem is to check that this prescription can be extended simultaneously to all $t$.

It is not hard to see how the expectation process extends. Indeed, letting

$$w^+(t) = \begin{cases} w(t) & \text{if } w(t) \geq 0 \\ 0 & \text{if } w(t) < 0 \end{cases}$$

and $w^-(t) = w^+(t) - w(t)$, the expressions $E^{Z^z(t)} \int_0^\infty e^{-\lambda s} w^+(s)\, ds$, $\lambda > 0$, are positive, $\lambda$-supermartingales (again as in Lemma 1.6), and as in the proof of Theorem 3.5 this is r.c.l.l. in $t$, with left limits given by substitution of $Z^z(t-)$. Subtracting the terms for $w^+$ and $w^-$, the same follows for $w$. Accordingly, to extend the Gaussian character to all $t$, it suffices to show that the covariance of $\{\int_0^\infty e^{-\lambda s} w(s)\, ds, \lambda > 0\}$ for $P^{Z^z(t)}$ is also r.c.l.l. in $t$, with left limits given by $Z^z(t-)$. Indeed, one then applies the following as a supplement to Lemma 3.8.

**Lemma 3.12** *A sequence* $(\mathbf{X}_n)$ *of jointly Gaussian random vectors converges in distribution if and only if* $\lim_{n\to\infty} E\mathbf{X}_n = \boldsymbol{\mu}$ *and* $\lim_{n\to\infty} \text{Cov } \mathbf{X}_n = (\boldsymbol{\sigma}^2)$ *exist. In this case the limit law is Gaussian* $(\boldsymbol{\mu}, (\boldsymbol{\sigma}^2))$.

**Proof** It is easy to see that Cov $\mathbf{X}_n$ must be bounded uniformly in $n$ to have convergence in distribution, since a component $\sigma^2_{n,k} \to \infty$ would cause probabilities of bounded sets to approach 0. Then one must also have $\boldsymbol{\mu}_n$ uniformly bounded, or the same contradiction ensues. Therefore, one can choose a subsequence $n_k \to \infty$ such that $\lim_{k\to\infty} \boldsymbol{\mu}_{n_k}$ and $\lim_{k\to\infty} \boldsymbol{\sigma}_{n_k}$ both exist. From the characteristic function of jointly Gaussian random variables it follows immediately that the limit law of $\mathbf{X}_{n_k}$ exists and is Gaussian with the corresponding limit mean and covariance. Then to have convergence in the distribution of $\mathbf{X}_n$, this limit must not depend on the subsequence $n_k$. It follows that $\lim_{n\to\infty} \boldsymbol{\mu}_n$ and $\lim_{n\to\infty} \boldsymbol{\sigma}_n^2$ exist, as asserted.

Now the covariance of $\{\int_0^\infty e^{-\lambda s} w(s)\, ds, \lambda > 0\}$ for $Z^z(t)$ is a.s. a constant for each $t$, because of the independence assertion in Lemma 3.10. Our problem thus reduces to showing that, for each $\lambda_1 < \lambda_2 < \cdots < \lambda_n$, these constant covariances are r.c.l.l. in $t$.

With a little thought one can convince oneself that this can be handled, much as in the case of the process of expectations, by writing $w = w^+ - w^-$ and using $\lambda$-supermartingales, but the details are tedious. A shorter, and at the same time more instructive, method is to appeal directly to the definition of the fact that $Z^z(t)$ is itself r.c.l.l. (Definitions 2.4 and 2.5). This means that for every bounded continuous function $f$ on $\Omega'$, $E^{Z^z(t)}f$ is r.c.l.l., where $\Omega'$ has the topology of (scaled) convergence in Lebesgue measure (Definition 2.2). Now fix $0 < \lambda_1 < \cdots < \lambda_n$, and let $f(x_1, \ldots, x_n)$ by any continuous function on $\mathbb{R}^n$ with compact support. Then the function on $\Omega'$

$$g_N(w) = f\left(\int_0^\infty e^{-\lambda_1 t} w_N(t)\, dt, \ldots, \int_0^\infty e^{-\lambda_n t} w_N(t)\, dt\right),$$

where $w_N(t) = -N \vee w(t) \wedge N$, is continuous on $\Omega'$ for each $0 < N < \infty$, and therefore $E^{Z^z(t)}g_N$ is r.c.l.l. in $t$. On the other hand, it follows as before that $E^{Z^z(t)} \int_0^\infty e^{-\lambda s}|w(s) - w^N(s)|\, ds$ is a $\lambda$-supermartingale, and is r.c.l.l. in $t$. (This reasoning is the same as for the expectation process.) For fixed $T$, we have

$$\lim_{N \to \infty} E^z E^{Z^z(T)} \int_0^\infty e^{-\lambda s}|w(s) - w_N(s)|\, ds$$

$$= \lim_{N \to \infty} E^z \int_0^\infty e^{-\lambda s}|w(s + T) - w_N(s + T)|\, ds$$

$$= 0 \quad \text{(since } z \in M_2 \subset M_1\text{)}.$$

and so a maximal inequality used before Theorem 3.5 implies that for $\varepsilon > 0$,

$$\lim_{N \to \infty} P^z\left\{\max_{t \leq T} E^{Z^z(t)} \int_0^\infty e^{-\lambda s}|w(s) - w_N(s)|\, ds > \varepsilon\right\} = 0.$$

Therefore, we have also (using once more the abbreviation $E^z f = zf$)

$$\lim_{N \to \infty} P^z\left\{\max_{t \leq T} |Z^z(t)(g_N) - Z^z(t)(g_\infty)| > \varepsilon\right\} = 0,$$

and it follows that $Z^z(t)(g_\infty)$ is also r.c.l.l., with left limits given by substitution of $Z^z(t)(g_\infty)$ is also r.c.l.l., with left limits given by substitution of $Z^z(t-)$. Finally, since the limits $Z^z(t)$ and $Z^z(t-)$ are already well-defined probabilities, convergence for $f(x_1, \ldots, x_n)$ with compact support implies convergence in distribution. According to Lemma 3.12, this implies (since $(\int_0^\infty e^{-\lambda_k s} w(s)\, ds$, $1 \leq k \leq n)$ is Gaussian for $Z^z(t)$ if $t$ is rational) that the limits $Z^z(t\pm)$ are Gaussian for all $t$, $P^z$-a.s., as was to be shown for $z \in M_G$.

Before specializing to $z \in M_{G(c)}$, we pause to mention another Gaussian process which has entered the picture, although so far only in a minor role. This is the *process of expectations* $(E^{Z^z(t)}(\int_0^\infty e^{-\lambda s} w(s)\, ds), \lambda > 0)$ where $t$ is fixed. Of course, its mean is $\int_0^\infty e^{-\lambda s} E_z(s+t)\, ds$ where $E_z(s+t)$ is the mean function of $z \in M_G$. The variance is left as an exercise.

**Exercise 3.1** Show that, for each $t$ and $z \in M_G$, the above process (namely, if $E_z \equiv 0$, the projection of the shifted Gaussian process $\int_0^\infty e^{-\lambda s} w(s+t)\, ds$, $\lambda > 0$, onto $H_z(t+)$ of Notation 3.5) is again represented by an element of $M_G$, and give an upper bound for its variance satisfying Assumption (2).

Another remark about the foregoing proof is that it does not use the fact that the covariance $\Gamma(u, v)$ of $Z^z(t)$ actually does not depend on $Z^z(t)$ (but may depend on $t$). The proof that $M_{G(c)}$ is a demesne, however, must depend on this fact, because this is the essential difference between $M_{G(c)}$ and $M_2(c)$ which, as we have seen, is not a demesne. The above proof actually shows that, for any $z \in M_2$, the joint laws $(\int_0^\infty e^{-\lambda_k s} w(s)\, ds, \lambda_1 < \cdots < \lambda_n)$ are r.c.l.l. in $t$ for $P^{Z^z(t)}$, for all choices of $\lambda_k$ (which may be assumed rational), but without the Gaussian condition convergence in law does not imply that of the covariance.

Returning to the proof that $M_{G(c)}$ is also a demesne, (Theorem 3.11), let $z \in M_{G(c)}$. We pointed out before Example 1 that for each $t$ the expectation $E^z \Gamma_{Z(t)}(s_1, s_2)$ is continuous in $(s_1, s_2)$ (writing $Z$ for $Z^z$). In the present case, since $\Gamma_{Z(t)}$ is $P^z$-a.s. constant, this constant covariance is continuous for each $t$. Let us denote it, for short, by $\Gamma_t^c(s_1, s_2)$, so that $\Gamma_t^c(s_1, s_2)$ is continuous in $(s_1, s_2)$. We know from the foregoing proof that the covariance of $Z^z(t)$ is r.c.l.l. in $t$, at least in the sense that each expression

$$\int_0^\infty \int_0^\infty e^{-(\lambda_1 s_2 + \lambda_s s_2)} \Gamma_{Z_z(t)}(s_1, s_2)\, ds_1\, ds_2$$

is r.c.l.l. in $t$, $P^z$-a.s. For all rational $r > 0$, these are $P^z$-a.s. the same as

$$\int_0^\infty \int_0^\infty e^{-(\lambda_1 s_2 + \lambda_2 s_2)} \Gamma_r^c(s_1, s_2)\, ds_1\, ds_2,$$

so it follows that, for each $t \geq 0$, $\int_0^\infty \int_0^\infty e^{-(\lambda_1 s_2 + \lambda_2 s_2)} \Gamma_t^c(s_1, s_2)\, ds_1\, ds_2$ is the right limit of its values for rational $r > t$. But this also means that the same expression is r.c.l.l. in $t$ (for all $\lambda_i > 0$). From this it follows that

$$P^z \left\{ \int_0^\infty \int_0^\infty e^{-(\lambda_1 s_2 + \lambda_2 s_2)} \Gamma_{Z_z(t)}(s_1, s_2)\, ds_1\, ds_2 \right.$$

$$= \int_0^\infty \int_0^\infty e^{-(\lambda_1 s_2 + \lambda_2 s_2)} \Gamma_t^c(s_1, s_2)\, ds_1\, ds_2$$

$$\left. \text{for all } t > 0, \text{ and all } \lambda_i > 0 \right\} = 1,$$

and in the same way the left limits $\Gamma^c_{t-}(s_1, s_2)$ ($=\Gamma_{Z^c_{t-}}(s_1, s_2)$, $P^z$-a.s.) are also continuous in $(s_1, s_2)$ and equal to $\Gamma_{Z^z(t-)}$ (up to equivalence in $(s_1, s_2)$) simultaneously in $t$, by the same reasoning. This completes the proof of Theorem 3.11.

The last part of the proof for $M_{G(c)}$ also applies to give an extra dividend about $M_G$. Indeed, continuity in $(s_1, s_2)$ was not used.

**Corollary 3.13** *For $z \in M_G$, let $\Gamma_t(s_1, s_2)$ (resp. $\Gamma_{t-}(s_1, s_2)$) denote the constant covariance which equals that of $Z^z(t)$ (resp. of $Z^z(t-)$), $P^z$-a.s. (we identify equivalence classes in $(s_1, s_2)$, equivalent up to $(ds_1 \times ds_2)$-nullsets). Then*

$$P^z\{\Gamma_{Z(t)} = \Gamma_t \text{ for all } t \geq 0, \quad \text{and} \quad \Gamma_{Z(t-)} = \Gamma_{t-} \text{ for all } t > 0\} = 1.$$

*Moreover $\Gamma_t$ is right-continuous in $t$, with left limits $\Gamma_{t-}$, in the sense that $\int_0^\infty \int_0^\infty e^{-(\lambda_1 s_2 + \lambda_2 s_2)} \Gamma_t(s_1, s_2) \, ds_1 \, ds_2$ is r.c.l.l. for all $\lambda_i > 0$, with analogous left limits using $\Gamma_{t-}$.*

Our next aim is to look more closely at the paths of $Z^z(t)$ for $z \in M_G$. It is clear from Lemma 3.12 that the times of discontinuity of $Z^z(t)$ are the combined discontinuities

$$\left\{t: \Gamma_{t-} \neq \Gamma_t \text{ or } E^{Z(t-)} \int_0^\infty e^{-\lambda s} w(s) \, ds \neq E^{Z(t)} \int_0^\infty e^{-\lambda s} w(s) \, ds \right.$$
$$\left. \text{for some } \lambda > 0 \right\}, \quad P^z\text{-a.s.}$$

However, to proceed further we need to show that the presence of $\{t: \Gamma_{t-} \neq \Gamma_t\}$ is superfluous, this set being $P^z$-a.s. contained in the other. At this juncture, it is time to 'shift' and transfer our field of opration to the demesne $M_G$ (or $M_{G(c)}$, when indicated). Thus instead of $(\Omega', \mathscr{F}')$ we substitute the canonical space $(\Omega_{M_G}, \mathscr{L}^0_t, \Theta^Z_t, P^z)$ consisting of all r.c.l.l. paths with values in $M_G$ (in the prediction topology of $M$) and the usual coordinate filtrations and translation operators. As we have seen (Theorem 2.14), the coordinate process $Z(t)$ is an r.c.l.l., strong Markov process, with moderately Markov left-limit process $Z(t-)$, both with the same Borel transition function $q(t, z, A)$ on $\mathbb{R}^+ \times M_G \times \mathscr{M}_G$. If we restrict further to $M_G \cap D$, we obtain a Borel right process (Corollary 2.27), but we may lose the left limits. The a.e. coordinate process $\tilde{X}_t$ is now replaced by $\hat{\rho}(Z_t)$, from Definition 2.7. Like $\tilde{X}_t$, $\hat{\rho}(Z_t)$ is not a Gaussian process in general, even for $P^z$ with $z \in M_G$, but it is Gaussian for a.e. $t$. It should also be remarked that even $Z_t$ itself is not always Gaussian in the sense of Definition 3.2 when the initial distribution $\mu$ is not a single point mass in $M_G$.

106 | Foundations of the prediction process

**Lemma 3.14** For $z \in M_G$,

$$P^z\{\{t: Z_{t-} \neq Z_t\} = \{t: R_\lambda^z \hat{\rho}(Z_{t-}) \neq R_\lambda^z \hat{\rho}(Z_t) \text{ for some } \lambda\}\} = 1.$$

This is just a restatement on $\Omega_{M_G}$ of the assertion described above. Indeed, from (2.1),

$$R_\lambda^z \hat{\rho}(z) = E^z \int_0^\infty e^{-\lambda t} \hat{\rho}(Z_t) \, dt = E^z \int_0^\infty e^{-\lambda t} w(t) \, dt, \qquad z \in M_G,$$

and our assertion is that the times of discontinuity of $Z(t)$ are $P^z$-a.s. the same as those of the Laplace transform of the expectation function $E(\cdot)$ for $P^{Z(t)}$. This of course will follow equally for any initial distribution $\mu$ on $M_G$, once it is shown for $z \in M_G$. The following may help to introduce the proof.

**Exercise 3.14** Show that, as a function of $t$,

$$\int_0^\infty e^{-\lambda s} (\hat{\rho}(Z(t+s)) - E_z(t+s)) \, ds, \qquad z \in M_2,$$

is $P^z$-continuous in the quadratic mean.

**Solution** For $t_1 < t_2$ the mean square difference of these expressions (writing $\Gamma$ for $\Gamma_z$) is

$$\int_0^\infty \int_0^\infty e^{-\lambda(s_1+s_2)} (\Gamma(t_1+s_1, t_1+s_2) + \Gamma(t_2+s_1, t_2+s_2)) \, ds_1 \, ds_2$$

$$- 2 \int_0^\infty \int_0^\infty e^{-\lambda(s_1+s_2)} \Gamma(t_1+s_1, t_2+s_2) \, ds_1 \, ds_2.$$

But, for example, we have

$$\int_0^\infty e^{-\lambda s_1} \int_0^\infty e^{-\lambda s_2} \Gamma(t_1+s_1, t_1+s_2) - \Gamma(t_1+s_1, t_2+s_2) \, ds_2 \, ds_1$$

$$= \int_0^\infty e^{-\lambda s_1} \int_{t_2-t_1}^\infty e^{-\lambda s_2} \Gamma(t_1+s_1, t_1+s_2) \, ds_2 \, ds_1 (1 - e^{+\lambda(t_2-t_1)})$$

$$+ \int_0^\infty e^{-\lambda s_1} \int_0^{t_2-t_1} e^{-\lambda s_2} \Gamma(t_1+s_1, t_1+s_2) \, ds_2 \, ds_1,$$

and, as $t_2 \to t_{1+}$, an easy application of $\Gamma^2(s,t) \leq \Gamma(s,s)\Gamma(t,t)$ (as in Corollary 3.6) shows that this has limit 0, as required.

**Exercise 3.15** Show that, for $z \in M_{G(c)}$, $\lim_{s \to t} \Gamma_s^c(s_1, s_2) = \Gamma_t^c(s_1, s_2)$, where the convergence is uniform in compact sets of $(s_1, s_2)$ for each $t \geq 0$. For

$z \in M_G$, show that

$$\lim_{s \to t+} \int_0^{s_1} \int_0^{s_2} \Gamma_s(u_1, u_2) \, du_1 \, du_2 = \int_0^{s_1} \int_0^{s_2} \Gamma_t(u_1, u_2) \, du_1 \, du_2, \quad \text{for } 0 \le s_1, s_2.$$

Hint: use Dini's theorem.

**Proof of Lemma 3.14** By Theorem 3.7, we have

$$\int_0^\infty f(s)(\hat{\rho}(Z_s) - E_z(s)) \, ds \in L^2(\Omega_{M_G}, \mathscr{L}_\infty^\circ, P^z)$$

for $f \in L^2(e^{\lambda t} \, dt)$, $\lambda > 0$, $z \in M_G$;

indeed, the analogue holds on $M_2$. It is consistent with our previous notation to introduce the Hilbert spaces $H_z(t)$ and $H_z(t+)$ generated by these integrals. In the present Gaussian case these are spaces of jointly Gaussian, mean 0, elements of $L^2(\Omega_{M_G}, \mathscr{L}_\infty^\circ, P^z)$ generating $\mathscr{L}_{t-}^\circ$ and $\mathscr{L}_t^\circ$, respectively (up to $P^z$-null sets).

**Notation 3.5** For any $z \in M_2$ and $0 < t \le \infty$, let

$$H_z(t) = L^2 \left\{ \int_0^\infty f(s)(\hat{\rho}(Z_s) - E_z(s)) \, ds; \ f \in L^2(e^{\lambda s} \, ds) \right.$$

$$\left. \text{for some } \lambda > 0 \text{ and } f(s) = 0, s > t \right\},$$

and let $H_z(t+) = \bigcap_{\varepsilon > 0} H_z(t + \varepsilon)$.

In the Gaussian case, it follows by Lemma 3.9 that for $X \in H_z(\infty)$ we have

$$\mathscr{E}^z(X | H_z(t)) = E^z(X | \mathscr{L}_{t-}^\circ) \quad \text{and} \quad \mathscr{E}^z(X | H_z(t+)) = E^z(X | \mathscr{L}_t^\circ), \quad P^z\text{-a.s.},$$

where $\mathscr{E}^z$ denotes projection in $H_z(\infty)$. We next assert that (for fixed $z$) $\Gamma_t^- \ne \Gamma_t$ holds if and only if $H_z(t) \ne H_z(t+)$. Indeed, from the above,

(1) $\mathscr{E}^z \left( \int_0^\infty e^{-\lambda s}(\hat{\rho}(Z(t+s)) - E_z(t+s)) \, ds \Big| H_z(t+) \right)$

$$= E^{Z(t)} \int_0^\infty e^{-\lambda s}(\hat{\rho}(Z(s)) - E_z(t+s)) \, ds;$$

and

(2) $\mathscr{E}^z \left( \int_0^\infty e^{-\lambda s}(\hat{\rho}(Z(t+s)) - E_z(t+s)) \, ds \Big| H_z(t) \right)$

$$= E^{Z(t-)} \int_0^\infty e^{-\lambda s}(\hat{\rho}(Z(s)) - E_z(t+s)) \, ds, \quad P^z\text{-a.s.}$$

Thus if $H_z(t+) = H_z(t)$, these expressions (1) and (2) are equal, and so the variances of their differences from $\int_0^\infty e^{-\lambda s}\hat\rho(Z(t+s))\,ds$, projected onto $H_z(t+)$ or $H_z(t)$, are also equal. In other words, $\Gamma_{t-} = \Gamma_t$. Conversely, suppose that $H_z(t+) \neq H_z(t)$. Then we claim that, for some $\lambda > 0$, $P\{(1) \neq (2)\} > 0$. Otherwise, by repeating the proof of Theorem 3.7 it would follow that

$$\mathscr{E}^z\left(\int_0^\infty f(s)(\hat\rho(Z(t+s)) - E_z(t+s))\,ds\,|\,H_z(t+)\right)$$
$$= \mathscr{E}^z\left(\int_0^\infty f(s)(\hat\rho(Z(t+s)) - E_z(t+s))\,ds\,|\,H_z(t)\right)$$

for all $f \in L^2(e^{\lambda t}\,dt)$, $\lambda > 0$. Since projections are linear operators, this implies of course the same assertion with $s$ in place of $t+s$, i.e. $H_z(t+) = H_z(t)$ in contradiction to our assumption. Therefore, since (1) and (2) are Gaussian, for some $\lambda$ they are unequal with probability 1, and therefore $P^z\{Z(t-) \neq Z(t)\} = 1$. Moreover, for this same $\lambda > 0$, we have

$$E^z((1) - (2))^2 = \int_0^\infty \int_0^\infty e^{-\lambda(s_1+s_2)}(\Gamma_{t-}(s_1, s_2) - \Gamma_t(s_1, s_2))\,ds_1\,ds_2,$$

because the term $\int_0^\infty e^{-\lambda s}(\hat\rho(Z(t+s)) - E_z(t+s))\,ds$ can be subtracted from both (1) and (2), and then the cross terms reduce to a square, because $H_z^\perp(t+) \subset H_z^\perp(t)$. Thus we must have $\Gamma_{t-} \neq \Gamma_t$, as asserted. We can summarize this discussion as follows. Letting $\{t_j\} = \{t: H_z(t) \neq H_z(t+)\}$ (which is countable because $H_z(\infty)$ is separable) we have $\Gamma_{t-} \neq \Gamma_t$ if and only if $t \in \{t_j\}$, and in this case $P^z\{Z(t-) \neq Z(t)\} = 1$, and for some $\lambda > 0$, $P^z\{R_\lambda^z\hat\rho(Z_{t-}) \neq R_\lambda^z\hat\rho(Z_t)\} = 1$. Conversely, if for some $t$ and $\lambda > 0$, $P^z\{R_\lambda^z(\hat\rho(Z_{t-})) \neq R_\lambda^z(\hat\rho(Z_t))\} > 0$, then this probability is 1 and $t \in \{t_j\}$. Thus, we have a complete characterization of the fixed discontinuities of $Z(t)$ for $z \in M_G$. On the other hand, for $t \notin \{t_j\}$ we have $Z(t-) = Z(t)$ if and only if

$$R_\lambda^z(\hat\rho(Z_{t-})) = R_\lambda^z(\hat\rho(Z_t)) \quad \text{for all } \lambda > 0,$$

this holding simultaneously in $t$ because $Z_t \in M_G$ for all $t$, $P^z$-a.s., and the elements of $M_G$ are determined uniquely by $(E(t), \Gamma(s, t))$ as in Theorem 3.2. In particular, this completes the proof of Lemma 3.14.

To describe the discontinuities of $Z(t)$ for $z \in M_G$, it remains only to consider the possibility of a stopping time $T$ with $P\{T \in \{t_j\}\} = 0$ and $P\{Z(T-) \neq Z(T)\} > 0$. To treat this possibility we introduce a family of martingales which is of fundamental importance for our whole approach to Gaussian processes, and for its sequel as well.

**Definition 3.6** For $z \in M_2$ and $\lambda > 0$, let

$$M_\lambda(t) = R_\lambda^z \hat{\rho}(Z(t)) - R_\lambda^z \hat{\rho}(Z(0)) + \int_0^t (\hat{\rho}(Z(s)) - \lambda R_\lambda^z \hat{\rho}(Z(s))) \, \mathrm{d}s.$$

Since $R_\lambda^z \hat{\rho}(Z(t))$ is r.c.l.l. $P^z$-a.s. (as in the proof of Theorem 3.11, writing $\hat{\rho} = \hat{\rho}^+ - \hat{\rho}^-$) the integral is well defined (on the exceptional null set we just set $M_\lambda(t) \equiv 0$). Then $M_\lambda(t_1 + t_2) = M_\lambda(t_1) + M_\lambda(t_2) \circ \Theta_{t_1}^Z$, $P^z$-a.s., and furthermore $M_\lambda(t)$ is r.c.l.l. with left limits given by substituting $Z(t-)$ for $Z(t)$, as shown at the start of Theorem 3.11. Thus $M_\lambda(t)$ is an *additive functional* of the prediction process on $M_G$. Actually, it is easy to check that these facts also hold on $M_2$, or even on $M_1$ if we omit the following square-integrability.

**Theorem 3.15**

(1) *For $z \in M_2$, $M_\lambda(t)$ is a square-integrable martingale additive functional of $Z(t)$ on $M_2$.*

(2) *For $z \in M_G$ it is also a process with independent Gaussian increments, and it is independent of $\mathscr{L}_0^\circ$ (recall that $\mathscr{L}_0^\circ \equiv \mathscr{L}_{0+}^\circ$).*

**Proof** We observe that, for $z \in M_G$, since the terms in $M_\lambda(t)$ will be in $H_z(\infty)$ under (1) (Notation 3.5), $M_\lambda(t)$ becomes Gaussian. Thus, to see that (2) follows from (1), it is only necessary to recall that a square-integrable martingale has orthogonal increments, and that for a mean 0 Gaussian process orthogonality implies independence. In particular, $M_\lambda(s)$ is independent of $\{R_\lambda^z \hat{\rho}(Z(0))\}$ which generates $\mathscr{L}_0^\circ$, so we need only to prove (1). Clearly we have $M_\lambda(t) \in \mathscr{L}_t$, where $\mathscr{L}_t$ denotes the usual intersection over $z \in M_2$ of the $P^z$-augmentations of $\mathscr{L}_t^\circ$, and then

$$R_\lambda^z \hat{\rho}(Z(t)) = E^z \left( \int_0^\infty e^{-\lambda s} \hat{\rho}(Z(s+t)) \, \mathrm{d}s \,\Big|\, \mathscr{L}_t \right).$$

For the martingale property, it suffices that

$$E^z(M_\lambda(t_2) - M_\lambda(t_1) | \mathscr{L}_{t_1}) = 0, \quad t_1 < t_2,$$

which, for the additive functional, becomes $E^{Z(t_1)} M_\lambda(t_2 - t_1) = 0$, so it is enough to check

$$E^z \lambda \int_0^t R_\lambda^Z \hat{\rho}(Z(s)) \, ds$$

$$= E^z \lambda \int_0^t \int_0^\infty e^{-\lambda u} \hat{\rho}(Z(s+u)) \, du \, ds$$

$$= E^z \left[ \lambda \int_0^\infty e^{-\lambda u} \left( \int_u^{u+t} \hat{\rho}(Z(v)) \, dv \right) du \right]$$

$$= E^z \int_0^t \hat{\rho}(Z(u)) \, du + E^z \int_0^\infty e^{-\lambda u} (\hat{\rho}(Z(u+t)) - \hat{\rho}(Z(u))) \, du,$$

where the interchange of integrals is justified because, for $z \in M_2 \subset M_1$, the integrand is absolutely (double) integrable, $P^z$-a.s., and the integration by parts follows because $\lim_{u \to \infty} E^z e^{-\lambda u} \int_u^{u+t} |\hat{\rho}(Z(v))| \, dv = 0$, by finiteness of $E^z \int_0^\infty e^{-\lambda u} |\hat{\rho}(Z(u))| \, du$. Substitution of the above into the definition of $M_\lambda(t)$ yields $E^z M_\lambda(t) = 0$, as asserted. It remains to check that $E^z M_\lambda^2(t) < \infty$. By Jensen's inequality

$$E^z (R_\lambda^Z \hat{\rho}(Z(t)))^2 \leq E^z \left( \int_0^\infty e^{-\lambda s} \hat{\rho}(Z(s+t)) \, ds \right)^2$$

$$\leq \frac{1}{\lambda} E^z \int_0^\infty e^{-\lambda s} (\hat{\rho}(Z(s+t)))^2 \, ds$$

$$\leq \frac{e^{\lambda t}}{\lambda} E^z \int_0^\infty e^{-\lambda s} (\hat{\rho}(Z(s)))^2 \, ds.$$

The last integral is finite for $z \in M_2$, and now it is clear that $M_\lambda(t)$ is a sum of four terms, each of which has finite mean square. This completes the proof of (1). As a final remark on (2), we note that, since $E^z M_\lambda(t) = 0$, we can as well centre at expectations termwise in the definition of $M_\lambda(t)$, to see that $M_\lambda(t) \in H_z(t+)$ (see also Theorem 3.7).

**Remark** For $z \in M_G$, if $E_z(\cdot) = 0$ we can write

$$R_\lambda^Z \hat{\rho}(Z(t)) = \mathscr{E}^z \left( \int_0^\infty e^{-\lambda s} \hat{\rho}(Z(s+t)) \, ds \, | \, H_z(t+) \right) \quad \text{for each } t, \quad P^z\text{-a.s.}$$

The right side is only unique up to a $P^z$-null set, however, so we cannot assert that it is r.c.l.l. except by choosing $R_\lambda^Z \hat{\rho}(Z(t))$ by its more detailed definition. Nevertheless, it shows how to define the *wide-sense* analogue of Theorem 3.15 on $M_2$. Namely, if we make this replacement through Definition 3.6, and then interpret the term

$$\int_0^t \hat{\rho}(Z(s)) - \lambda \mathscr{E}^z \left( \int_0^\infty e^{-\lambda u} \hat{\rho}(Z(u+s)) \, du \, | \, H_z(s+) \right) ds$$

as an integral in quadratic mean, then by general principles we can assert that it becomes a wide-sense square-integrable martingale, alias, a process with orthogonal increments. Indeed, this orthogonality property, which holds in the Gaussian case, depends only on the covariance structure $(E(t) = 0, \Gamma(s, t))$, and hence it also holds for $z \in M_2$. This wide-sense extension of our prediction theory, in which $Z(t)$ would be replaced by the family of projections of the linear future $\int_0^\infty e^{-\lambda u} \hat{\rho}(Z(u+t))\, du$ onto $H_z(t+)$, seems to be of lesser interest than the strict sense form, but we will comment on it occasionally.

Using our martingales $M_\lambda(t)$, we can now show

**Theorem 3.16** *For $z \in M_G$, the set of times of discontinuity of $Z(t)$ is, with probability 1, precisely the countable set*

$$\{t_j\} = \{t > 0 : H_z(t) \neq H_z(t+)\} = \{t > 0 : \Gamma_{t-} \neq \Gamma_t\}.$$

**Proof** We have seen in Lemma 3.14 that the discontinuity times include the above, and equal the set on which $R_\lambda^Z \hat{\rho}(Z(t-)) \neq R_\lambda^Z \hat{\rho}(Z(t))$ for some $\lambda > 0$, where, as usual, we can assume $\lambda$ is rational. (Note that since $R_\lambda^Z \hat{\rho}(z)$ is a Laplace transform for $z \in M_2$, if $R_\lambda^Z \hat{\rho}(Z(t-)) = R_\lambda^Z \hat{\rho}(Z(t))$ for all rational $\lambda > 0$, then they are equal for all $\lambda > 0$.) Now the discontinuity times of $R_\lambda^Z \hat{\rho}(Z(t))$ are the same as those of the Gaussian martingale $M_\lambda(t)$, so it is enough to review certain well-known facts about r.c.l.l. Gaussian processes with independent increments. Namely, the assertion of Theorem 3.16 is an immediate consequence of the following Lemma.

**Lemma 3.17** *Let $G(t)$ be an r.c.l.l. Gaussian process with mean 0 and independent increments, $G(0) = 0$. Let $\sigma^2(t) = EG^2(t)$ denote the variance, so that $\sigma^2(t)$ is r.c.l.l. and non-decreasing, with a countable set $\{t_j\}$ of discontinuities. The process $G_d(t) = \sum_{t_j \leq t} (G(t_j) - G(t_j-))$ is well defined up to a P-null set, and the series, suitably reordered, converges uniformly in bounded t. Then $G_d(t)$ is r.c.l.l. and Gaussian with independent increments and $\{t_j\}$ is its total set of discontinuities, P-a.s., while $G_c(t) = G(t) - G_d(t)$ is independent of $G_d(t)$ and has continuous paths and variance $\sigma_c^2(t) = \sigma^2(t) - \sigma_d^2(t)$, where $\sigma_d^2(t) = \sum_{t_j \leq t} (\sigma^2(t_j) - \sigma^2(t_j-))$. If we set $\tau_c(t) = \inf\{s : \sigma_c^2(s) > t\}$, the process $W(t) = G_c(\tau_c(t))$ is a standard Brownian motion stopped at time $t = \sigma_c^2(\infty)$.*

**Proof** With a view to more general situations which are to occur in subsequent chapters, it is best to prove this by martingale methods. Of course, $G(t)$ is an r.c.l.l. martingale relative to its generated filtration, and so, if $I(s)$ denotes the indicator function of the set $\{t_j\}$, the $L^2$-Lebesgue–Stieltjes integral $\int_0^t I(s)\, dG(s)$ is well defined for each $t$, and is Gaussian with independent increments in $t$, so it is a martingale in $t$. Its variance is

$\int_0^t I(s) \, d\sigma^2(s) = \sigma_d^2(t)$. Now, for $\varepsilon > 0$, let $I_\varepsilon(s)$ be the indicator function of $\{t_j(\varepsilon)\} = \{t_j: \sigma^2(t_j) - \sigma^2(t_j-) > \varepsilon\}$. It is clear that

$$\int_0^t I_\varepsilon(s) \, dG(s) = \sum_{t_j(\varepsilon) \leq t} G(t_j(\varepsilon)) - G(t_j(\varepsilon) -),$$

and by a martingale maximal inequality it follows that there are $\varepsilon_n \to 0+$ with

$$\lim_{n \to \infty} \int_0^t I_{\varepsilon_n}(s) \, dG(s) = \int_0^t I(s) \, dG(s), \quad P\text{-a.s.}$$

uniformly in bounded $t$. We take this as a definition of the series appearing in Lemma 3.17. Then, as the uniform limit of r.c.l.l. processes, $G_d(t)$ is r.c.l.l., and the Gaussian property and independent increments also transfer to the limit. It is similarly clear that the discontinuity times of $G_d(t)$ are precisely $\{t_j\}$, since each $t_j$ is a discontinuity of $\int_0^t I_\varepsilon(s) \, dG(s)$ for all small $\varepsilon$, and the discontinuity size does not depend on $\varepsilon$. Also, this integral being independent of $\int_0^t (1 - I_\varepsilon(s)) \, dG(s)$ for each $\varepsilon$, it follows that $G_d(t)$ and $G_c(t)$ are mutually independent, and $G_c(t)$ has continuous variance $\sigma_c^2(t) = \sigma^2(t) - \sigma_d^2(t)$. Again, since $G_c(t)$ is r.c.l.l., and $\tau_c(t)$ is right-continuous, we see that $W(t)$ is also r.c.l.l., and the variance of $W(t)$ is

$$\sigma_c^2(\tau(t)) = \begin{cases} t & \text{for } t < \sigma_c^2(\infty) \\ \sigma_c^2(\infty) & \text{for } t > \sigma_c^2(\infty) \end{cases}$$

But then it is a well-known fact that the r.c.l.l. Gaussian process $W(t)$ with independent increments, as right limit of its values for rational $t$, must be a stopped Brownian motion (and thus continuous). So $G_c(t) = W(\sigma_c^2(t))$ also has continuous paths, completing the proof.

Now to finish the proof of Theorem 3.16, we know that the discontinuity times are the same as those of $\{M_\lambda(t), 0 < \lambda \text{ rational}\}$. Each $M_\lambda(t)$ satisfies the hypotheses of Lemma 3.17, so its times of discontinuity are all fixed, and contained in the combined set $\{t_j\}$ of the assertion, as we remarked at the end of the proof of Lemma 3.14. Again, these are all probability 1 discontinuity times by the Gaussian hypothesis so the proof is complete.

The family $\{M_\lambda(t)\}$ has an important role in applying $Z(t)$ to Gaussian processes (other than as an aid to proving Theorem 3.16). The following result follows from a more general theorem, due originally to Knight (1983a) and Meyer (1983) which we relegate to the appendix at the end of the chapter, for later use.

**Theorem 3.18** *For $z \in M_G$ with $H_z(0+) = 0$, (see the Remark following Corollary 3.19). the Hilbert space closure of $\{M_n(s), s \leq t, n \geq N\}$, for each $N \geq 1$, is precisely $H_z(t+)$ (Notation 3.5).*

**Proof** According to the appendix, for $z \in M_2$ we have

$$\bar{\sigma}\{R_n^Z\hat{\rho}(Z(0)) + M_n(s), s < t, n \geq N\} = \bar{\sigma}\left\{\int_0^s \hat{\rho}(Z(u))\,du, s < t\right\}$$

where $\bar{\sigma}\{\cdot\}$ denotes augmentation of $\sigma\{\cdot\}$ by all $P$-null sets. By Theorem 2.7, the right side is $P^z$-equivalent to $\mathscr{L}_{t-}^\circ$. Now if $H_z(0+) = \mathbf{0}$, $R_n^Z\hat{\rho}(Z(0))$ is a.s. constant, and thus, even without the Gaussian assumption we have $\bar{\sigma}\{M_n(s), s < t, n \geq N\} = \bar{\sigma}(\mathscr{L}_{t-}^\circ)$. Clearly

$$\bar{\sigma}\{M_n(s), s \leq t, n \geq N\} = \bar{\sigma}\{M_n(s), s < t, R_n^Z\hat{\rho}(Z(t)), n \geq N\}.$$

It follows by the Stone–Weierstrass theorem, as in the appendix, that the right side contains $\sigma\{R_\lambda^Z\hat{\rho}(Z(t)), \lambda > 0\}$. But, for $z \in M_G$, we know already from the proof of Theorem 3.11 that $\bar{\sigma}\{R_\lambda^Z\hat{\rho}(Z(t)), \lambda > 0\} = \bar{\sigma}(Z(t))$, since this determines the mean $E_{Z(t)}(s)$. Thus it follows that

$$\bar{\sigma}\{M_n(s), s \leq t, n \geq N\} = \mathscr{L}_t^z$$

with the superscript $z$ denoting the usual $P^z$-augmentation of $\mathscr{L}_t^\circ$. But since $\mathscr{L}_t^z$ is right-continuous in $t$, this is the same as $\bar{\sigma}(H_z(t+))$. Thus the Hilbert space closure of $\{M_n(s), s \leq t, n \geq N\}$, which is contained in $H_z(t+)$, generates the same $\sigma$-field as $H_z(t+)$. Unless it equals $H_z(t+)$, there is an element of $H_z(t+)$ (other than $\mathbf{0}$) orthogonal to the closure. Since these are mean 0, Gaussian spaces, it follows from Lemma 3.9(2) that this element is independent of the closure. But then it generates a $\sigma$-field independent of $\bar{\sigma}(H_z(t+))$, which is a contradiction. This proves Theorem 3.18.

As a corollary, we can state the wide-sense analogue (see the Remark following Theorem 3.15).

**Corollary 3.19** *For $z \in M_2$, with $E_z(\cdot) \equiv 0$, let $\hat{\mathscr{M}}_\lambda(t)$ be defined like $M_\lambda(t)$ but replacing $R_\lambda^Z\hat{\rho}(Z(s))$, $0 \leq s \leq t$, by $\mathscr{E}^z(\int_0^\infty e^{-\lambda u}\hat{\rho}(Z(u+s))\,du \mid H_z(s+))$. Then if $H_z(0+) = \mathbf{0}$, the Hilbert space closure of the set $\{\hat{\mathscr{M}}_m(s), s \leq t, m \geq N\}$ is again $H_z(t+)$.*

**Proof** One only has to realize: (a) that for $z \in M_G$ this reduces to Theorem 3.18, and (b) the identity of the two Hilbert spaces depends only on the covariance structure $\Gamma_z$: it asserts the existence of a sequence in the linear span of $\{\hat{\mathscr{M}}_n(s), s \leq t, n \geq N\}$ which converges to a prescribed element of $H_z(t+)$ in the sense of the norm defined by $\Gamma_z$. Clearly this holds for the $z \in M_2$ if and only if it holds for the $z \in M_G$ with the same $\Gamma_z$ (and we take $E_z(t) \equiv 0$).

**Remark** The assumption $H_z(0+) = \mathbf{0}$ can, of course, always be arranged by replacing the given (centred) process by its projection onto $H_z^\perp(0+)$. In the Gaussian case, this means using the process with mean $E(t) = 0$ and the fixed covariance of $Z(0)$ (which we denoted by $\Gamma_0(s, t)$ above). In the general case, it holds (a.s.) for the process $Z^z(0)$ because $Z^z(0) \in D$ (by Definition 1.7 and Theorem 1.15). In the Gaussian case, where $\mathscr{X}_0^0$ is $P^z$-independent of $H_z^\perp(0+)$, $z \in D$ is both necessary and sufficient for $H_z(0+) = 0$. Note that for $z \in M_G$, the Hilbert space generated by $\{M_n(t), 0 < t, n \geq N\}$ is always $H_z^\perp(0+)$.

We next use $\{M_n(t), n \geq N\}$ to obtain a canonical representation of the process, of the type due to P. Lévy, H. Cramer, and T. Hida in varying degrees of generality. At least formally our approach is little more than an orthogonalization of the sequence $M_n(t)$. In the Gaussian case, of course, this implies independence, and hence a representation in terms of independent processes of the type discussed in Lemma 3.17. This evidently is as simple a basis for representing the process as could be hoped for. Its practical usefulness, however, depends on the degree to which it can be explicitly carried out in terms of the observable data. Our method relies on knowing $\{R_n^z \hat{\rho}(Z(t))\}$ as a function of $t$. This is tantamount to assuming a known solution of the prediction problem of projecting the linear future at time $t$ onto the past $H_z(t+)$. This problem has been the object of a vast literature in applied mathematics, and we do not attempt to portray it in any detail. In principle, it is just the problem of obtaining a projection onto a Hilbert subspace. It has been found (and we give examples below) that in many cases the family $M_n(t)$ provides a particularly convenient way of utilizing these projections, once they are known, to obtain a canonical representation of the process. This is especially true in the stationary case (for reasons which will appear) and of course the stationary case is by all odds the prevalent case for applied work. In the general case the canonical form, when defined abstractly, is by no means unique, and there is no way to prove rigorously that one method of obtaining such is 'better' than another. Even in our rather particularized approach, the outcome will depend on the value of $N$ from Theorem 3.18. As we do not have any significant example to indicate a preference of one value of $N$ over another (and all of this will become rather trivial in the stationary case) we will just assume the choice $N = 1$ to lighten the notation. The other possibilities are completely analogous.

The orthogonalization procedure is carried out in terms of the inner products $E^z(M_m(t)M_n(t))$, but our primary assumed data is rather the quantities $R_n^z \hat{\rho}(Z(t))$. It is thus of interest to express the former in terms of the inner products of the latter. We have the following result, but the proof is tedious, and we do not recommend it.

**Theorem 3.20** *Let $(m, n)_z^t$ denote $E^z(R_m^z \hat{\rho}(Z(t)) R_n^z \hat{\rho} Z(t)))$. Then for $z \in M_G$, assuming that $E_z(\cdot) = 0$,*

$$E^z(M_m(t)M_n(t)) = (m, n)_z^t - (m, n)_z^0 - (n + m)\int_0^t (m, n)_z^u \, du$$

$$+ \int_0^t \left(\int_0^\infty (e^{-ns} + e^{-ms})\Gamma_z(u, u + s)\, ds\right) du.$$

**Proof (sketch)** For simplicity of notation we confine ourselves to the case $n = m$. By orthogonality

$$E^z M_n^2(t) = \sum_{k=1}^K E^z(\Delta_k M_n)^2, \tag{3.1}$$

where $\Delta_k M_n(\cdot) = M_n(a_k) - M_n(a_{k-1})$ with $a_k = kK^{-1}t$ $(= a_k(K, t))$. We now let $K \to \infty$ and do a calculation in quadratic mean, based on the following lemma.

**Lemma 3.21** $R_n^Z \hat{\rho}(Z(t))$ *is right-continuous and bounded in* $t \leq K < \infty$ *in the $L^2$-norm, for each $z \in M_G$.*

**Proof** By Exercise 3.14, $\int_0^\infty e^{-\lambda s}\hat{\rho}(Z(t + s))\, ds$ is $P^z$-continuous in quadratic mean, and routine observation, as in Theorem 3.5, gives the bound $e^{\lambda t}E^z(\int_0^\infty e^{-\lambda s}|\hat{\rho}(Z(s))|\, ds)^2$, which is finite, as in Corollary 3.6. Now recalling that $R_n^Z(Z(t)) = \mathscr{E}^z(\int_0^\infty e^{-\lambda s}\hat{\rho}(Z(t+s))\, ds | H_z(t+))$, we have

$$(E^z\{R_n^Z\hat{\rho}(Z(t + \Delta)) - R_n^Z\hat{\rho}(Z(t))\}^2)^{\frac{1}{2}}$$

$$= \left(E^z\left(\mathscr{E}^z\left\{\int_0^\infty e^{-\lambda s}\{\hat{\rho}(Z(t + \Delta + s)) - \hat{\rho}(Z(t + s))\}\, ds | H_z(t+)\right\}\right)^2\right)^{\frac{1}{2}}$$

$$+ \left(E\left(\mathscr{E}^z\left\{\int_0^\infty e^{-\lambda s}\hat{\rho}(Z(t + \Delta + s))\, ds | H_z(t + \Delta +) \cap H_z^\perp(t+)\right\}\right)^2\right)^{\frac{1}{2}}.$$

As $\Delta \to 0+$, the first term on the right tends to 0 because it is bounded by the $L^2$-norm with the projection $\mathscr{E}$ omitted, while in the second term, we can replace $Z(t + \Delta + s)$ by $Z(t + s)$ as in (3.3) below, and then use the fact that $H_z(t + \Delta) \cap H_z^\perp(t+)$ decreases to $\mathbf{0}$ as $\Delta \to 0+$. Thus the lemma is proved.

Coming back to the sum (3.1), if we square out the increments $\Delta_k M_n$ and let $K \to \infty$, it is easy to see that the sums involving the continuous compensators $\int_{a_{k-1}}^{a_k} (\hat{\rho}(Z(s)) - nR_n^Z\hat{\rho}(Z(s)))\, ds$ tend to 0 in $L^2$ because of the bound $E^z(\int_a^b \hat{\rho}(Z(s))\, ds)^2 \leq (b - a)E^z\int_a^b (\hat{\rho}(Z(s)))^2\, ds$ applied to each increment, and Cauchy's inequality applied to the cross-product terms. So we are left with

$$\lim_{K \to \infty} \sum_{k=1}^K E^z(\Delta_k R_n^Z \hat{\rho}(Z(\cdot)))^2 = E^z M_n^2(t), \tag{3.2}$$

in place of (3.1). Now we will require

**116** | Foundations of the prediction process

**Lemma 3.22**

$$L^2\text{-}\lim_{K \to \infty} K \cdot \left[ \mathscr{E}^z(\Delta_k R_n^Z \hat{\rho}(Z(\cdot))|H_z(a_{k-1}+)) \right.$$
$$\left. - \left(\frac{nt}{K} R_n^Z \hat{\rho}(Z(a_{k-1})) - \mathscr{E}^z \int_0^{tK^{-1}} e^{-\lambda s}\hat{\rho}(Z(a_{k-1}+s))\,ds\right)|H_z(a_{k-1}+)\right] = 0,$$

uniformly in $k \leq K$ (for each $t$).

**Proof** We have for $\varepsilon > 0$, in the sense of $L^2$,

$$\int_0^\infty e^{-\lambda s}\hat{\rho}(Z(t+\varepsilon+s))\,ds - \int_0^\infty e^{-\lambda s}\hat{\rho}(Z(t+s))\,ds$$

$$= (e^{\lambda\varepsilon} - 1)\int_\varepsilon^\infty e^{-\lambda s}\hat{\rho}(Z(t+s))\,ds - \int_0^\varepsilon e^{-\lambda s}\hat{\rho}(Z(t+s))\,ds \quad (3.3)$$

$$= \varepsilon\lambda \int_0^\infty e^{-\lambda s}\hat{\rho}(Z(t+s))\,ds - \int_0^\varepsilon e^{-\lambda s}\hat{\rho}(Z(t+s))\,ds + o(\varepsilon).$$

Setting $\lambda = n$, $\varepsilon = tK^{-1}$, $t = a_{k-1}$, and projecting onto $H_z(a_{k-1}+)$, the norm of the $o(\varepsilon)$ is only decreased, and it is easy to see that it is uniform in $k \leq K$ as required.

This implies first that in calculating (3.2) we can replace $\Delta_k R_n^Z \hat{\rho}(Z(\cdot))$ by $R_n^Z \hat{\rho}(Z(a_k)) - \mathscr{E}^z(\int_0^\infty e^{-ns}\hat{\rho}(Z(a_k+s))\,ds|H_z(a_{k-1}+)$ (that is, we can subtract $\mathscr{E}^z(\Delta_k R_n^Z \hat{\rho}(Z(\cdot))|H_z(a_{k-1}+))$) with an error whose norm square is of order $o(K^{-1})$ and hence is negligible for the summation. Next, by iteration of conditional expectation, and then reordering, (3.2) becomes

$$\lim_{K \to \infty} \sum_{k=1}^K E^z(R_n^Z\hat{\rho}(Z((a_k)))^2 - E^z\left(\mathscr{E}^z\int_0^\infty e^{-ns}\hat{\rho}(Z(a_k+s))\,ds|H_z(a_{k-1}+)\right)^2$$

$$= \lim_{K \to \infty} -\sum_{k=1}^{K-1}\left[E^z\left(\mathscr{E}^z\left(\int_0^\infty e^{-ns}\hat{\rho}(Z(a_{k+1}+s))\,ds|H_z(a_k+)\right)\right)^2\right.$$
$$\left. - E^z(R_n^Z\hat{\rho}(Z(a_k))^2)\right] + ((n,n)_n^t - (n,n)_0^t).$$

Here a term in the summation equals

$$E^z\left[\mathscr{E}^z\left(\int_0^\infty e^{-ns}(\hat{\rho}(Z(a_{k+1}+s)) - \hat{\rho}(Z(a_k+s)))|H_z(a_k+)\right)\right.$$
$$\left. \times \mathscr{E}^z\left(\int_0^\infty e^{-ns}(\hat{\rho}(Z(a_{k+1}+s)) + \hat{\rho}(Z(a_k+s)))|H_z(a_k+)\right)\right]$$

and, by Lemma 3.22 again, the second factor may be replaced by $2R_n^Z\hat{\rho}(Z(a_k))$ with sufficiently small error. Therefore, in the sense of convergence in $L^1$ (via Cauchy's and Schwartz's inequalities) the limit becomes

$$\lim_{K\to\infty} -\sum_{k=1}^{K-1} \left[\frac{2t}{K}(n(R_n^Z\hat{\rho}(Z(a_k)))^2) - 2R_n^Z\hat{\rho}(Z(a_k))\right.$$
$$\left. \times \mathscr{E}^z \int_0^{tK^{-1}} e^{-ns}\hat{\rho}(Z(a_k+s))\,ds\,|\,H_z(a_k+)\right].$$

Now by Lemma 3.21, we can write

$$R_n^Z\hat{\rho}(Z(a_k))\mathscr{E}^z \int_0^{tK^{-1}} e^{-ns}\hat{\rho}(Z(a_k+s))\,ds\,|\,H_z(a_k+)$$
$$= \int_{a_k}^{a_{k+1}} R_n^Z\hat{\rho}(Z(s))\hat{\rho}(Z(s))\,ds$$

plus another term whose sum converges to 0 in $L^1$ by the dominated convergence theorem, and therefore the limit equals

$$-2\int_0^t n(R_n^Z\hat{\rho}(Z(s)))^2 - R_n^Z\hat{\rho}(Z(s))\hat{\rho}(Z(s)))\,ds.$$

Since the convergence is in $L^1$ we can take expectation with respect to $P^z$, and putting back the term $(n, n)_z^t - (n, n)_0^t$ yields the assertion of Theorem 3.20 for $n = m$. The situation for $n \neq m$ is analogous.

**Remark** The analogous formulas for the *quadratic variations* of the martingales $M_n(t)$ in the non-Gaussian case require the Ito stochastic integral. Above, the computation is basically just a calculation in quadratic mean (although it is tedious). Of course, if we replace $M_n(t)$ by $\hat{\mathscr{M}}_n(t)$ of Corollary 3.19 in the non-Gaussian case, the present result and proof remain valid for $z \in M_2$ with mean 0 if we also replace $(m, n)_z^t$ by the corresponding

$$E^z\left(\mathscr{E}^z\left(\int_0^\infty e^{-ms}\hat{\rho}(Z(t+s))\,ds\,|\,H_z(t+)\right)\right.$$
$$\left.\times \mathscr{E}^z\left(\int_0^\infty e^{-ns}\hat{\rho}(Z(t+s))\,ds\,|\,H_z(t+)\right)\right).$$

On the other hand, the present result goes through with no change for any $z \in M_2$ with mean 0. Indeed, the only use of the Gaussian assumption was to replace, when convenient, a conditional expectation by an $L^2$-projection. But for $z \in M_2$, as we noted in the discussion preceding Example 1 as well as in the Introduction, we still have $R_\lambda^Z\hat{\rho}(Z(t)) = \mathscr{E}^z(\int_0^\infty e^{-\lambda s}w(s+t)\,ds\,|\,H_z^*(t+))$ where $H_z^*(t+)$ denotes the larger Hilbert space $L^2(\mathscr{L}_t^\circ, P^z)$ in our prediction space set-up. This projection serves to carry over the proof for the non-Gaussian case with no further change.

**Exercise 3.20** Using this result, show that for $t > 0$ and $z \in M_2$ with mean 0,

$$L^2\text{-}\lim_{n \to \infty} \int_0^t (\hat{\rho}(Z(s)) - nR_n^z\hat{\rho}(Z(s)))\,ds = 0.$$

Conclude that, for $z \in M_2$,

$$\bar{\sigma}\{R_\lambda^z \hat{\rho}(Z(s)), \lambda > 0, s < t\} = \mathscr{L}_{t-}^z.$$

Hint: show that $\lim_{n \to \infty}(n, n)_z^t = 0$, and that $\lim_{n \to \infty} E^z M_n^2(t) = 0$. Thus the conclusion follows by definition of $M_n(t)$.

We come now to the Cramér–Hida representation.

**Theorem 3.23** For $z \in M_G$, if $H_z(0+) = 0$ (i.e. $z \in D$) there is a sequence $\{Y_n, n < N + 1\}$, for some $N \leq \infty$, of independent, mean 0, r.c.l.l. Gaussian processes with independent increments, $Y_n(0) = 0$, such that

(1) $L_z^2\{Y_n(s); n < N + 1, s \leq t\} = H_z(t+)$ for all $t$, where the left side denotes the linear closure in $L^2(P^z)$; and

(2) $d\sigma_1^2 \gg d\sigma_2^2 \gg \cdots \gg d\sigma_n^2 \gg \cdots$, in the sense of absolute continuity of measures, where $\sigma_n^2(t) = E(Y_n^2(t))$.

For any two such sequences, the absolute continuity classes of the $d\sigma_n^2$ are the same for each $n$. The index of multiplicity $\Phi_z(t) = \inf\{n: \sigma_{n+1}^2(t) = 0\}$ is thus uniquely defined.

**Discussion**

1. It is shown further by Cramér (1964) that for every sequence of measures $d\sigma_n^2$ satisfying (2) there is a Gaussian process $X(t)$ with continuous covariance (i.e. a $z \in M_{G(c)}$) whose spectral sequence (2) has the same absolute continuity class as $d\sigma_n^2(t)$ for every $n$. Since we can replace $Y_n(t)$ by $\int_0^t f_n(s)\,dY_n(s)$, for any $f_n$ for which $d(\int_0^t f_n^2(s)\,d\sigma_n^2(s))$ is mutually absolutely continuous with respect to $d\sigma_n^2$, without negating (1) or (2), we can assert that any such measures $d\sigma_n^2$, finite on each $(0, t]$, may always be fitted by some $z \in M_{G(c)}$, subject only to conditions (2). On the other hand, the greatest novelty in our proof below is that it provides a particular way of constructing a sequence $Y_n$, which has practical advantages in certain cases. Thus it does not lead to an arbitrary sequence $d\sigma_n^2$ for given absolute continuity classes, but to a particular one. Moreover, in most cases of any practical importance one has $\Phi_z(t) = 1$ for all $t$, and hence there is only a single process $Y_1(t)$.

2. From a statistical point of view, it is quite important to distinguish

which features of our development depend on knowing the law $P^z$, and which do not. The construction of our basic data $R_n^Z \hat{\rho}(Z^z(t))$ depends, of course, on $P^z$. On the other hand, if $E_z \equiv 0$ it follows from the proof of the appendix that the determination of the elements $\int_0^s \hat{\rho}(Z(u)) \, du$, $s < t$, of $H_z(t+)$ from $\{M_n(s), s < t\}$ does not depend on $z$. There is a fixed functional dependence between them, which of course carries over to their linear span in $\bigcap_{z \in M_2} H_z(t+)$. This is equally true for Theorem 3.18 (the Gaussian case) and for Corollary 3.19 (the $M_2$ case). Thus, the 'process' $\int_0^s \hat{\rho}(Z(u)) \, du$ is determined by $\{M_n(s), s < t\}$ in a manner free of the law $P^z$. In the Gaussian case with continuous paths the same is true of the inner products $E^z(M_m(t)M_n(t))$ of Theorem 3.20. One need only show that the limit (3.2) can be defined free of $z$. We will see later (in Chapter 5, Theorem 5.5) that, in a more general (non-linear) context, the corresponding martingales have quadratic variations $\langle M_n, M_n \rangle_t$ which, in the absence of accessible discontinuities, can be calculated from the sample paths. In the Gaussian case all discontinuities are accessible, so we must assume $Z(t)$ is continuous. Then $M_n(t)$ is a process with independent increments whose quadratic variation is simply the variance $E^z M_n^2(t)$, which is what is given in Theorem 3.20. So we see that $\langle M_n, M_n \rangle_t$ is non-random in the continuous Gaussian case, which means that $E^z M_n^2(t)$ (and analogously $E^z(M_m(t)M_n(t))$) can be obtained by observing the quadratic variations, which are $P^z$-a.s. constant, and do not depend explicitly on $z$ once the $R_n^Z \hat{\rho}(Z(t))$ are known. It follows that the processes $Y_1, Y_2, \ldots$ of Theorem 3.23 can also be determined, under the same proviso, without knowing $z$. Moreover, an important use of these processes is to express in terms of $(Y_n)$ more immediately useful quantities such as $R_n^Z \hat{\rho}(Z(t))$, and here the coefficients of the inversion for $(M_n)$ also depend only on $\langle M_n, M_m \rangle$, and hence are $P^z$-a.s. determinable by observation of $R_n^Z \hat{\rho}(Z_t)$. On the other hand, these constant coefficients are not determined by observations of $Y_1, Y_2, \ldots$, so one does not have a canonical representation by processes with independent Gaussian increments which can be implemented without knowing $z$. In particular, knowledge of $d\sigma_1^2, d\sigma_2^2, \ldots$, even for our particular choice thereof, does not determine $z$.

**Proof of Theorem 3.23** The proof is quite easy. We begin by orthogonalizing the sequence $M_n$. Let $N_1 = M_1$, and inductively for $1 \leq n$ let

$$N_{n+1}(t) = M_{n+1}(t) - \sum_{k=1}^{n} \int_0^t \frac{dE^z(M_{n+1}(s)N_k(s))}{dE^z N_k^2(s)} \, dN_k(s). \quad (3.4)$$

To see that the integrand is well defined, one need only use the fact that the projection of $M_{n+1}(t)$ onto the $L_z^2$-subspace generated by $(N_k(s), s \leq t)$ has the form $\int_0^t F(t, s) \, dN_k(s)$ for some fixed $F(t, s)$ with $\int_0^t F^2(t, s) \, dE^z N_k^2(s) < \infty$. Indeed, all elements of the subspace have this form, since it is closed in $L_z^2$. Then, from

$$0 = E^z((M_{n+1}(t) - \int_0^t F(t,s)\,dN_k(s))N_k(t_1))$$

$$= E^z(M_{n+1}(t_1)N_k(t_1)) - \int_0^{t_1} F(t,s)\,dE^z N_k^2(s), \quad t_1 \leq t,$$

we obtain the integrand equal to $F$ by differentiation with respect to $t_1$, in the Radon–Nikodyn sense. It is clear that the sequence $(N_n)$ are independent Gaussian processes with independent increments (where we interpret $0/0 = 0$), and they generate the same spaces $H_z(t+)$ as the $(M_n)$. Now we introduce measures $d\mu_n(t) = dE^z N_n^2(t)$, and for each $n$ we introduce a Lebesgue decomposition

$$d\mu_1 = dv_1,$$

$$d\mu_{n+1}(t) = = f_{n+1}(t)\left(\sum_{k=1}^n d\mu_k(t)\right) + dv_{n+1}(t),$$

where $dv_{n+1}$ is singular with respect to $\sum_{k=1}^n d\mu_k$, and $f_{n+1}(t)$ is a Radon–Nikodym derivative of the part of $d\mu_{n+1}$ absolutely continuous with respect to $\sum_{k=1}^n d\mu_k$. Let $S_{n+1}$ be a Borel set with $dv_{n+1}(R^+) = dv_{n+1}(S_{n+1})$ and $\sum_{k=1}^n d\mu_k(S_{n+1}) = 0$, $1 \leq n$. The existence of such $S_{n+1}$ is an immediate consequence of the definition of singularity of measures. Since $d\mu_k(S_{n+1}) = 0$ for $k < n+1$, the conditions of $S_{n+1}$ remain valid if we replace $S_{n+1}$ by $\bigcup_{n+1}^\infty S_m$ for $1 \leq n$, thus we can assume $S_2 \supset S_3 \supset \cdots$. Next, since $v_n \leq \mu_n$, and $\mu_n(S_{n+1}) = 0$, we have $v_n(S_{n+1}) = 0$, and so we can replace $S_n$ by $S_n \cap S_{n+1}^c$, $2 \leq n$, to obtain a *disjoint* sequence also satisfying the conditions. Denoting this sequence again by $S_2, S_3, \ldots$, and setting $S_1 = (\bigcup_2^\infty S_n)^c$, we define our first process $Y_1$ on the interval $0 \leq t \leq 1$ as follows (in which terms with $\mu_n(1) = 0$ are interpreted at 0):

$$Y_1(t) = \int_0^t I_{S_1}(s)\,dN_1(s) + \sum_{n=2}^\infty 2^n \sqrt{\mu_n^{-1}(1)} \int_0^t I_{S_n}(s)\,dN_n(s). \quad (3.5)$$

Writing $E$ for $E^z$, we have

$$EY_1^2(t) = EN_1^2(t) + \sum_{n=2}^\infty 2^{-2n}\mu_n^{-1}(1)\int_0^t I_{S_n}(s)\,dEN_n^2(s)$$

$$< EN_1^2(t) + \tfrac{1}{3} < \infty \quad \text{for } 0 \leq t \leq 1,$$

and setting $\sigma_1^2(t) = EY_1^2(t)$ we have

$$d\sigma_1^2(t) = d\mu_1(t) + \sum_{n=2}^{\infty} 2^{-2n}\mu_n^{-1}(1) I_{S_n}(t) \, d\mu_n(t)$$

$$= d\mu_1(t) + \sum_{n=2}^{\infty} 2^{-2n}\mu_n^{-1}(1) I_{S_n}(t) \left\{ dv_n(t) + f_n(t) \sum_{k=1}^{n-1} d\mu_k(t) \right\}$$

$$= d\mu_1(t) + \sum_{n=2}^{\infty} 2^{-2n}\mu_n^{-1}(1) \, dv_n(t).$$

Thus $dv_n(t) \ll d\sigma_1^2(t)$ for every $n$, and it follows by induction that also $d\mu_n(t) \ll d\sigma_1^2(t)$ for every $n$. Thus we have a process $Y_1(t)$, $0 \le t \le 1$, which has independent Gaussian increments, is in $H^z(t+)$ for each $t \le 1$, and whose variance measure has the minimal absolute continuity class dominating that of each $d\mu_n(t)$. Hence it also dominates that of each $dM_n(t)$, since, by (3.4), $M_n(t)$ has the form $N_n(t) + \sum_{k=1}^{n-1} \int_0^1 c_k(s) \, dN_k(s)$. Conversely, it is clear that for some sufficiently small positive constants $l_n$ we have $\sum_{n=1}^{\infty} l_n M_n^2(1) < \infty$, and then we have $d\sigma_1^2(t) \equiv dE^z(\sum_{n=1}^{\infty} l_n M_n^2(t))$ in the sense of mutual absolute continuity. Indeed, this defines $d\sigma_1^2(t)$ uniquely up to mutual absolute continuity.

We will indicate the construction of $Y_2(t)$, $0 \le t \le 1$, before the induction step. In short, we simply use the new sequence

$N_{k,2}(t) = (N_k$ minus its projection onto the Hilbert space generated by $Y_1$ in the same time interval), $k \ge 2$,

and repeat the construction of $Y_1$ with the new sequence. Then, since the sets $S_k$ are disjoint and the $N_k$ are orthogonal, the projection of $N_k$ onto the space of $Y_1$ is the process

$$\int_0^t \frac{dE(N_k(s) Y_1(s))}{dE Y_1^2(s)} \, dY_1(s) = \int_0^t I_{S_k}(s) \, dN_k(s),$$

hence the new sequence is simply

$$\int_0^t I_{S_k^c}(s) \, dN_k(s) = \int_0^t I_{\bigcup_1^{k-1} S_j}(s) \, dN_k(s), \qquad k \ge 2,$$

and each of the variance measures is $d\sigma_1^2$-absolutely continuous. Then

$Y_2(t) = \int_0^t I_{S_1}(s) \, dN_2(s) + $ (terms in $dN_k(s)$, $k > 2$, which are supported on an $I_{S_1}(s) \, d\mu_2(s)$-null subset of $\bigcup_{j=1}^{k-1} S_j$).

It follows quite easily that $N_2(t)$ is in the space generated by $Y_1(s)$ and $Y_2(s)$, $s \le t$. Indeed $N_1(t)$ is in the space of $Y_1(s)$, $s \le t$, and since $S_2$ is $d\mu_1$-singular, $\int_0^t I_{S_2}(s) \, dN_2(s)$ is also in the space of $\{Y_1(s), s \le t\}$. Since $\int_0^t I_{S_1}(s) \, dN_2(s)$ $(= \int_0^t I_{S_2^c}(s) \, dN_2(s))$ is in the space of $\{Y_2(s), s \le t\}$, we have $N_2(t)$ in the combined space. Also, of course, $d\sigma_2^2 \ll d\sigma_1^2$.

The induction step is analogous, and need not be detailed. For $N < t \leq N+1$, one has merely to repeat the argument using $(M_n(t) - M_n(N))$ to construct $(Y_n(t) - Y_n(N))$, so the proof of Theorem 3.23 is complete.

**Corollary 3.24** *For $z \in M_2$, the statement of Theorem 3.23 remains true if, instead of $z \in D$, we only require $H_z(0+) = 0$, but, instead of $\{Y_n, n < t+1\}$, we obtain an orthogonal sequence $\{\hat{\mathcal{Y}}_n, n < N+1\}$ of processes with orthogonal increments. The definition of multiplicity $\Phi_z(t)$ remains unchanged.*

**Proof** We have only to replace $M_n(t)$ by their wide-sense analogues $\hat{\mathcal{M}}_n(t)$ from Corollary 3.19, after centring at the mean.

**Remark** A process $z$ which, at $t = 0$, chooses with $p = \tfrac{1}{2}$ either a standard Brownian motion $B(t)$ or a standard compensated Poisson process $P(t) - t$, provides an example where $H_z(0+) = \mathbf{0}$ but $z \notin D$.

**Corollary 3.25** *For $z \in M_2$ with mean 0 and $H_z(0+) = \mathbf{0}$, there are real-valued, Borel measurable $K_{n,k}(s)$, $1 \leq k \leq n$, such that*

$$\hat{\mathcal{M}}_n(t) = \sum_{k=1}^{n} \int_0^t K_{n,k}(s)\, d\hat{\mathcal{Y}}_k(s), \qquad 1 \leq n, \qquad P^z\text{-a.s.}$$

*for each $t$, where the integral is defined in the space $L^2$ in the sense of Stieltjes–Lebesgue.*

**Discussion** The existence of such $K_{n,k}$ follows immediately from the previous theorem and corollary if we permit $k < \infty$ and a $t$-dependence $K_{n,k}(s,t)$, because every element of $H_z(t+)$ has such a representation. To see this, note that the family of such representations, with $\sum_k \int_0^t K_{n,k}^2(s,t)\, d\sigma_k^2(s) < \infty$, forms a Hilbert space containing $\{\hat{\mathcal{Y}}_n(s), s \leq t, 1 \leq n\}$. On the other hand, since the definition of $\{\hat{\mathcal{Y}}_n(t) - \hat{\mathcal{Y}}_n(t_0), t_0 < t\}$ only requires $(\hat{\mathcal{M}}_k(t) - \hat{\mathcal{M}}_k(t_0)$, $k \leq n$, $t_0 < t)$, one sees that the inversion for $\hat{\mathcal{M}}_n$ is free of $t$ and $\hat{\mathcal{Y}}_k$, $k > n$, as asserted. An important point here is that the $K_{n,k}$ may be actually calculated without excessive difficulty. Since $\hat{\mathcal{M}}_1 = \hat{\mathcal{N}}_1$ (where we use $\hat{\mathcal{N}}_k$ for wide-sense $N_k$) (3.5) implies immediately $\hat{\mathcal{M}}_1(t) = \int_0^t I_{S_1}(s)\, d\hat{\mathcal{Y}}_1(s)$. If $\hat{\mathcal{Y}}_k = 0$, $k \geq 2$, (as is usually the case!) one has similarly

$$\hat{\mathcal{N}}_n(t) = 2^n \mu_n^{1/2}(1) \int_0^t I_{S_n}(s)\, d\hat{\mathcal{Y}}_1(s), \qquad t \leq 1,$$

and to express $\hat{\mathcal{M}}_n$ one uses (3.4). If $\hat{\mathcal{Y}}_2 \neq 0$, then to find $\hat{\mathcal{N}}_2(t)$ one must add $\int_0^t I_{S_1}(s)\, d\hat{\mathcal{Y}}_2(s)$ to the above case $n = 2$, and so forth. It is important to stress this inversion to some extent, because it will be seen that in the case $z \in M_{G(c)}$, where there is a well-defined coordinate process for each $t$, representation of $(\hat{\mathcal{M}}_n)$ in terms of $(\hat{\mathcal{Y}}_n)$ leads to the proper canonical representation of Lévy, Cramér, and Hida for the coordinate process.

## 3.2 Specialization to continuous covariances

In the preceding part of this chapter, we have developed an abstract theory of Gaussian and $L^2$ processes in connection with the prediction process. In the remainder of the chapter we indicate how the theory applies in more specialized situations, and ultimately in those of the applications. Our sequence of specializations goes as follows: first, we specialize to $\{z \in M_{G(c)}\}$; next, to $\{z \in M_{G(c)}$ with multiplicity $\Phi_z(t) \leq 1\}$; and finally to $\{z \in M_{G(c)}$ with stationary covariance $\Gamma = \Gamma(s - t)\}$. In the interests of brevity, we do not elaborate on many details or go into much depth. Some additional steps in this direction (by no means exhaustive) are in the author's papers (Knight 1983a,b).

**Definition 3.7** For $z \in M_{G(c)}$, let

$$X_z^c(t) = L^2\text{-}\lim_{\varepsilon \to 0} \varepsilon^{-1} \int_t^{t+\varepsilon} (\hat{\rho}(Z(s)) - E_z(s))\, ds, \qquad 0 \leq t.$$

The existence of this limit follows easily from continuity of $\Gamma_z^c$ (Theorem 3.4), and it is clear that $\Gamma_z^c$ is the covariance of $X_z^c(t)$, which also has mean 0. Since $X_z^c(t)$ is $L^2$-continuous, it is $L^2$-Riemannian integrable, and thus clearly the Hilbert space closure of $\{X_z^c(s), s \leq t\}$ is $H_z(t)$, of Notation 3.5.

Caution: it must not be supposed that $X_z^c$ has a standard modification with continuous paths. For the dichotomy which occurs here, see for example Ito and Nisio (1968). Of course, the existence of the limit $X_z^c$ itself does not even require that $z$ be Gaussian.

As in Corollary 3.25, we have immediately

**Theorem 3.26*** *For $z \in D \cap M_{G(c)}$ there are jointly measurable $F_n(t, s), 0 < s < t$, $1 \leq n$, such that, in the $L^2$ sense,*

$$X_z^c(t) = \sum_{n=1}^{\Phi_z(t)} \int_0^t F_n(t, s)\, dY_n(s), \qquad 0 < t. \tag{3.6}$$

**Remarks** Since $H_z(t) = \bigvee_{\varepsilon > 0} H_z(t - \varepsilon)$, it is easy to see that one can replace $Y_n(s)$ by $Y_n(t-)$ for $s = t$ in the integrals (i.e. integrate only up to t−). If $H_z(t) \neq H_z(t+)$, then $X_z^c(t)$ does not record the discontinuities

$$\Delta Y_n(t) = Y_n(t) - Y_n(t-).$$

We call (3.6) a *proper canonical representation* of $X_z^c(t)$ in terms of $(Y_n)$, following Lévy (1956, 1957) and Hida (1960).

---

* In Knight (1983a) there is a mistake in the proof of right-continuity in $t$ ($\geq s$) of $F_n(t, s)$, and this (minor) point remains open.

A problem which naturally arises is how to obtain the $F_n(s, t)$ from the $K_{n,k}(s)$ of Corollary 3.25. The key to this is in the following.

**Theorem 3.27** *For $z \in D \cap M_{G(c)}$, with K from Corollary 3.25 and F from Theorem 3.26,*

$$K_{n,k}(s) = \int_0^\infty e^{-nu} F_k(u + s, s) \, du, \qquad k \leq n, \tag{3.7}$$

*for $d\sigma_k^2$-a.e. s.*

**Discussion** More generally, as a definition of $K_{\lambda,k}$, for each $t$ and $\lambda > 0$, the following two equations are equivalent:

$$M_\lambda(t) = \sum_{k=1}^\infty \int_0^t K_{\lambda,k}(s) \, dY_k(s), \tag{3.8}$$

$$K_{\lambda,k}(s) = \int_0^\infty e^{-\lambda u} F_k(u + s, s) \, du \qquad \text{for } d\sigma_k^2\text{-a.e. } s \leq t.$$

In fact, as shown later in Lemma 5.9(1), if $Y(t) \in H_z(t+)$ is any process with orthogonal increments, the same identity connects the coefficients $K_\lambda(s)$ and $F(t, s)$ of the projections of $M_\lambda(t)$ and $X_z^c(t)$ onto the space of $\{Y(s), s \leq t\}$. One should note, moreover, that $K_\lambda$ and $F$ are only unique up to $d\sigma^2$-null sets of $s$, where $\sigma^2(t) = E^z Y^2(t)$. One may check, however, that they may be chosen $(\mathcal{B}^+ \times \mathcal{B}^+)$-measurable. Then, by the inversion formula of the Laplace transform, $K_{\lambda,k}(s)$ ($0 \leq s, \lambda > 0$) determines $F_k(u + s, s)$ for $(du \times d\sigma_k^2)$-a.e. $(u, s)$. On the other hand, setting $F_k(t, s) = 0$ for $s \geq t$, as functions of $s$ the $F_k$ are continuous in $t$ in the sense of $L^2(d\sigma_k^2)$. Indeed, because $X_c^z$ is $L^2$-continuous we have

$$0 = \lim_{\varepsilon \to 0} \int_0^\infty (F_k(t + \varepsilon, s) - F_k(t, s))^2 \, d\sigma_k^2(s).$$

By Fubini's theorem, for Lebesgue-a.e. $t$, $K_\lambda$ determines $F_k(t, s)$ for $d\sigma_k^2$-a.e. $s < t$. Therefore $\int_{t_1}^{t_2} (\int_{s_1}^{s_2} F_k(t, s) \, d\sigma_k^2(s)) \, dt$, $s_2 < t_1$, are determined. But the $dt$ integrand is continuous in $t$, since the $L^2(d\sigma_k^2)$-continuity implies $L^1(d\sigma_k^2)$-continuity. So we conclude that $\int_{s_1}^{s_2} F_k(t, s) \, d\sigma_k^2(s)$ is determined for all $s_1 < s_2 < t$. This is just the statement that $F_k(t, s)$ is uniquely determined within its equivalence class of $s < t$, for each $t$, by $K_{\lambda,k}$, $\lambda > 0$, $s \leq t$. The same is true of $K_{n,k}$, $n \geq 1$, by a Stone–Weierstrass argument, but (3.8) is easier to invert. Thus, in practice, we use (3.8) as a means of obtaining the prediction coefficients $F_k$ of $X_z^c$.

**Proof of Equation (3.8)** We first establish the existence of

$$\int_0^t \left( \int_0^\infty e^{-\lambda u} F_k(t + u, s) \, du \right) dY_k(s),$$

for which it suffices that the transform be well defined for $d\sigma_k^2$-a.e. $u$, and that we have $\int_0^t (\int_0^\infty e^{-\lambda u} F_k(t+u,s)\,du)^2\,d\sigma_k^2(s) < \infty$. Now by hypothesis for $M_2$,

$$\infty > E^z \int_0^\infty e^{-\lambda u} (X_z^c(u))^2\,du$$

$$\geq \int_0^\infty e^{-\lambda u} E^z \left( \int_0^t F_k(u,s)\,dY_k(s) \right)^2 du$$

$$= \int_0^\infty e^{-\lambda u} \int_0^t F_k^2(u,s)\,d\sigma_k^2(s)\,du$$

$$= \int_0^t \left( \int_0^\infty e^{-\lambda u} F_k^2(u,s)\,du \right) d\sigma_k^2(s)$$

$$\geq \lambda \int_0^t \left( \int_0^\infty e^{-\lambda u} |F_k(u,s)|\,du \right)^2 d\sigma_k^2(s)$$

$$\geq e^{2\lambda t} \lambda \int_0^t \left( \int_0^\infty e^{-\lambda u} |F_k(t+u,s)|\,du \right)^2 d\sigma_k^2(s).$$

Next, we observe the important fact that since $E^z M_\lambda(t) = 0$, and $M_\lambda$ is linear in $\hat\rho(Z(s))$, it will not be changed if we replace $\hat\rho(Z(s))$ by $\hat\rho(Z(s)) - E_z(s)$ throughout, which is equivalent to using, in place of $\hat\rho(Z(s))$, a measurable version of the process $X_z^c(s)$. Therefore we have the $L^2$ identity

$$R_\lambda^z \hat\rho(Z(t)) - \int_0^\infty e^{-\lambda s} E_z(s+t)\,ds = \int_0^\infty e^{-\lambda u} \mathscr{E}^z(X_z^c(t+u)|H_z(t+))\,du$$

$$= \int_0^\infty e^{-\lambda u} \left( \sum_{k=1}^\infty \int_0^t F_k(t+u,s)\,dY_k(s) \right) du$$

$$= \sum_{k=1}^\infty \int_0^t \left( \int_0^\infty e^{-\lambda u} F_k(t+u,s)\,du \right) dY_k(s),$$

where the interchange of summation is justified by the independence of the $Y_k$ processes, and the interchange of integration by the foregoing estimate. Note that it suffices to justify

$$E^z \left( Y_k(\rho) \left( \int_0^\infty e^{-\lambda u} \left( \int_0^t F_k(t+u,s)\,dY_k(s) \right) du \right) \right)$$

$$= E^z \left( Y_k(\rho) \int_0^t \left( \int_0^\infty e^{-\lambda u} F_k(t+u,s)\,du \right) dY_k(s) \right)$$

for $0 < \rho \leq t$, which is straightforward since $\int_0^t F_k(t+u,s)\,dY_k(s)$ is $L^2$-continuous in $u$.

It is worthwhile stating this separately.

**Lemma 3.28** *For $z \in D \cap M_{G(c)}$ (or only for $z \in M_2$ with $H_z(0+) = 0$ and continuous covariance, if we replace $Y_k$ by $\mathcal{Y}_k$ from Corollary 3.24), if we centre at the mean (i.e. take $E_z(t) = 0$),*

$$R_\lambda^z \hat{\rho}(Z(t)) = \sum_{k=1}^{\infty} \int_0^t h_{k,\lambda}(t,s) \, dY_k(s),$$

where

$$h_{k,\lambda}(t,s) = \int_0^\infty e^{-\lambda u} F_k(t+u, s) \, du.$$

*In other words, the coefficients of the prediction $R_\lambda^z \hat{\rho}(Z(t))$ (if $E_z(t) = 0$) are expressed from those of the process $X_z^c$, which are determined by those of $M_n(t)$, as in the Theorem.*

Now returning to $M_\lambda$ itself, we can as well calculate it assuming $E_z(t) = 0$. Letting $M_{k,\lambda}(t)$ denote its projection onto the space of $Y_k(s)$, $s \leq t$, we have (using Definition 3.6)

$$M_{k,\lambda}(t) = \int_0^t h_{k,\lambda}(t,s) \, dY_k(s) + \int_0^t \left( \int_0^v (F_k(v,s) - \lambda h_{k,\lambda}(v,s)) \, dY_k(s) \right) dv$$

$$= \int_0^t \left( \int_0^\infty e^{-\lambda u} F_k(t+u, s) \, du \right) dY_k(s)$$

$$+ \lambda \int_0^t \left[ \int_0^v \int_0^\infty e^{-\lambda u} \{ F_k(v,s) - F_k(v+u, s) \} \, du \, dY_k(s) \right] dv.$$

Collecting the coefficient of $dY_k(s)$, we get

$$\int_0^\infty e^{-\lambda u} F_k(t+u, s) \, du + \lambda \int_s^t \left( \int_0^\infty e^{-\lambda u} \{ F_k(v,s) - F_k(v+u, s) \} \, du \right) dv.$$

Now integration by parts gives

$$\int_0^\infty e^{-\lambda u} \left( \int_0^t (F_k(v,s) - F_k(v+u, s)) \, dv \right) du$$

$$= -\lambda^{-1} \int_0^\infty e^{-\lambda u} \frac{d}{du} \int_s^t F_k(v+u, s) \, dv \, du$$

$$= -\lambda^{-1} \int_0^\infty e^{-\lambda u} \frac{d}{du} \int_{s+u}^{t+u} F_k(w, s) \, dw \, du$$

$$= -\lambda^{-1} \int_0^\infty e^{-\lambda u} (F_k(t+u, s) - F_k(s+u, s)) \, du,$$

## Gaussian processes and the prediction process | 127

and substitution into $M_{\lambda,k}(t)$ gives $\int_0^t (\int_0^\infty e^{-\lambda u} F_k(s+u, s)\, du)\, dY_k(s)$, as required. We leave it to the reader to check that all the integrands used above are absolutely integrable, so that the interchange is justified. The stochastic reordering of $s$ and $v$ is also justified by routine analysis based on the estimate

$$\int_0^t \int_0^v \left( \int_0^\infty e^{-\lambda u} \{F_k(v,s) - F_k(v+u, s)\}\, du \right)^2 d\sigma_k^2(s)\, dv < \infty.$$

We also note the following wide-sense analogue.

**Corollary 3.29** *The results of Theorems 3.26 and 3.27 hold for $z \in M_2$ with $H_z(0+) = 0$ and continuous covariance, if we replace $Y_k$ by $\tilde{\mathcal{Y}}_k$ from Corollary 3.24, after centring at the mean.*

We next take up the case $\Phi_z(t) \leq 1$, where the multiplicity $\Phi_z(t)$ was introduced in Theorem 3.23 and, for the wide-sense case, in Corollary 3.24. It should be noted carefully that $\Phi_z(t)$ has not been defined unless $H_z(0+) = \mathbf{0}$. In short, we consider the family $H_z(t+)$ on an open interval $(0, \infty)$, without permitting a jump at $t = 0$. This is necessary for $\{M_n(s), s \leq t\}$ to generate $H_z(t+)$, since $M_n(0) = 0$. In the Gaussian case, even if $z \notin D$ (and hence $H_z(0+) \neq \mathbf{0}$), this causes little problem because $H_z(t+)$ is generated by $\{M_n(s), s \leq t\}$ together with $H_z(0+)$, and for $z \in M_G$ these last are independent of $\{M_n(s)\}$. Thus Theorem 3.23 and Corollary 3.24 provide, without change, a basis $(Y_n)$ for $H_z^\perp(0+)$. However, the restriction to $z \in D$ does imply that the requirement $\{\Phi_z(t) \leq 1, t > 0\}$ cannot define a *complete* demesne unless we require also $Z(t-) \in D$ for all $t$, which would be the same as requiring $Z(t)$ to be continuous, by Theorem 3.16. We will see later, in Chapter 4, that this continuity restriction does give a complete demesne. Meanwhile, we have the following result.

**Theorem 3.30** $M_G \cap \{z \in D: \Phi_z(\cdot) \leq 1\}$ *is a Borel demesne.*

**Proof** We will show more generally that

$$P^z\{\Phi_{Z(t)}(s-t) \leq \Phi_z(s) \quad \text{for all } 0 < t \leq s\} = 1.$$

Since evidently $\{z \in M_G \cap D: \Phi_z(s) < k\} = \{\sigma_k^2(s) = 0\} \in \mathcal{M}$, this suffices for the proof. For fixed $z$, we see by the Markov property of $Z(t)$ that, for each $t$, the process $M_n(t+s) - M_n(t) = M_n(s) \circ \Theta_t^z$ has conditional law given $\mathcal{L}_t^\circ$ equal $P$-a.s. to the law of $M_n(s)$ for $Z(t)$. But, for $z \in M_G$, $(M_n(t+s) - M_n(t))$ is independent of $\mathcal{L}_t^\circ$, and hence we can assert that the multiplicity $\Phi_{Z(t)}(s)$ is $P^z$-a.e. for all $s$ that of $(M_n(t+s) - M_n(t))$ for $P^z$, which is certainly not greater than $\Phi_z(s+t)$. This establishes only that

$$1 = P^z\{\Phi_{Z(t)}(s-t) \leq \Phi_z(s), \quad s > 0 \text{ and rational } t < s\}. \quad (3.9)$$

128 | Foundations of the prediction process

But $\int_0^\infty e^{-\lambda s} M_n(s)\,ds$ is a Gaussian family of random variables for $z \in M_G$, so it follows from the fact that $M_G$ is a demesne (Theorem 3.11), together with the convergence Lemma 3.8, that the mixed moments

$$E^{Z(t)}\left(\int_0^\infty e^{-\lambda s} M_m(s)\,ds \int_0^\infty e^{-\lambda u} M_n(u)\,du\right)$$

are right-continuous in $t$, $P^z$-a.s. This expression reduces by change of variable to $2\lambda^{-1} E^{Z(t)} \int_0^\infty e^{-2\lambda v}(M_m(v)M_n(v))\,dv$. For each $t$, by mutual independence of increments of $(M_n(t+v))$ and $\mathscr{F}_t^\circ$, this is $P^z$-a.e. equal to

$$2\lambda^{-1} E^z \int_0^\infty e^{-2\lambda v}(M_m(v+t)M_n(v+t) - M_m(t)M_n(t))\,dv,$$

so by right-continuity this equality holds for all $t$. Then the $P^z$ joint law of $(M_n(t+s) - M_n(t), 1 \leq n, s > 0)$ suffices for all $t$ simultaneously to determine the multiplicities $\Phi_{Z(t)}(s)$, and this extends (3.9) to all $t$, as required.

The reason for singling out $\Phi_z \leq 1$ is not only for simplicity, but mainly because it applies to all cases of any real interest. Of course $\Phi_z(t) = 0$ means that the process reduces to its mean $E(s)$, $0 < s \leq t$ (at least for a.e. $s$), so the actual case of interest is $\Phi_z(t) = 1$ for all $t$. This case was the main concern of P. Lévy in his path-breaking studies of canonical representations of Gaussian processes. When $\Phi_z(t) \leq 1$ is known or can be assumed (and frequently this can be tested by the methods of Lévy (1957) or Hida (1960, Theorem 1.7)) it is not necessary to go through the construction (3.5) of $Y_1(t)$. All that is required is to find some $\lambda_1 > 0$ (or a finite set $0 < \lambda_1 < \cdots < \lambda_n$) such that the measure $d\sigma_1^2(t)$ is equivalent to $dE^z M_{\lambda_1}^2(t)$ (or only to $dE^z(\sum_{k=1}^n c_k M_{\lambda_k}(t))^2$ for suitable $c_k$). Then $M_{\lambda_1}$ (or $\sum_{k=1}^n c_k M_{\lambda_k}(t)$) can replace $Y_1$ for the representation theory. This is not difficult, in view of the next theorem.

**Theorem 3.31** *If $z \in M_g \cap D$ and $\Phi_z(\cdot) \leq 1$, then $dE^z M_\lambda^2(t)$ is equivalent (mutually absolutely continuous) with $d\sigma_1^2(t)$, except for at most a closed countable set of $\lambda$. To choose such a $\lambda$ ($=\lambda_1$) it suffices to make $dE^z M_{\lambda_1}^2(t)$ equivalent to $d(\int_a^b E^z M_\lambda^2(t)\,d\lambda)$, for any $0 \leq a < b$.*

**Proof** By Theorem 3.27 and Equation (3.8), since $Y_2 \equiv 0$, we have

$$M_\lambda(t) = \int_0^t \left(\int_0^\infty e^{-\lambda u} F_1(u+s, s)\,du\right) dY_1(s),$$

hence the equality of measures $dE^z M_\lambda^2(t) = (\int_0^\infty e^{-\lambda u} F_1(u+t, t)\,du)^2\,d\sigma_1^2(t)$. It follows that the Laplace transform on the right exists for $d\sigma_1^2$-a.e. $t$, and the measure is equivalent to $d\sigma_1^2$ if and only if

$$d\sigma_1^2\left\{t: \int_0^\infty e^{-\lambda u} F_1(u+t, t)\,du = 0\right\} = 0.$$

Now for each $t$ such that the transform exists, either

$$\left\{\lambda: \int_0^\infty e^{-\lambda u} F_1(u+t, t)\, du = 0\right\}$$

is countable, or else $F_1(u+t, t) = 0$ for Lebesgue a.e. $u > 0$, because the transform is analytic. Let $A = \{t: F_1(u+t, t) = 0 \text{ for a.e. } u > 0\}$, and write

$$X_z^c(s) = \int_0^s I_{A^c}(u) F_1(s, u)\, dY_1(u) + \int_0^s I_A(u) F_1(s, u)\, dY_1(u). \quad (3.10)$$

Since

$$0 = \int_0^\infty \int_u^\infty I_A(u) F_1^2(s, u)\, ds\, d\sigma_1^2(u) = \int_0^\infty \left(\int_0^s I_A(u) F_1^2(s, u)\, d\sigma_1^2(u)\right) ds,$$

it follows by Fubini's theorem that the second term on the right of (3.10) is $0$ ($P^z$-a.s.) for Lebesgue-a.e. $s$. But since $X_z^c(s)$ is $L^2$-continuous so is this term, and hence it vanishes identically in $s$. This implies that $d\sigma_1^2(A) = 0$, because we know that $\int_s^t I_A(u)\, dY_1(u) \in H_z(t+)$ for all $t$, and $H_z(t)$ is generated by $\{X_z^c(s), s \leq t\}$. Indeed by (3.10)

$$X_z^c(s) = \int_0^s I_{A^c}(u) F_1(s, u)\, dY_1(u),$$

which is independent of $\int_0^t I_A(u)\, dY_1(u)$ for all $t$, hence we see that the latter is independent of $H_z(\infty)$, and hence is $\mathbf{0}$. Thus $d\sigma_1^2(A) = 0$.

Next, set $\hat{F}(t, s) = F_1(t, s)$ if $\int_0^\infty e^{-\lambda u} F_1(u+t, t)\, du$ exists for all $\lambda > 0$ and $t \notin A$, and set $\hat{F} = 0$ elsewhere. Then, except for a $d\sigma_1^2$-null set of $t$, the closed set $B(t) = \{\lambda: \int_0^\infty e^{-\lambda u} \hat{F}(u+t, t)\, du = 0\}$ is at most countable. It follows that, for every continuous (i.e. non-atomic) Borel measure $v(d\lambda)$,

$$0 = \int_0^\infty \left(\int_0^\infty I_{B(t)}(\lambda)\, d\sigma_1^2(t)\right) v(d\lambda).$$

This implies that

$$d\sigma_1^2\{t: \lambda \in B(t)\} = \int_0^\infty I_{B(t)}(\lambda)\, d\sigma_1^2(t) = 0$$

except for at most a countable set of $\lambda$. Indeed, the function $\int_0^\infty I_{B(t)}(\lambda)\, d\sigma_1^2(t)$ is upper-semicontinuous in $\lambda$ (since the $B(t)$ are closed), hence the set where it is $\geq \varepsilon > 0$ is closed. If the set were uncountable, being closed it would support a monotone continuous mapping of the usual Cantor set, and hence a non-trivial continuous measure $v(d\lambda)$. This completes the proof.

**Corollary 3.32** With $\lambda_1$ as in Theorem 3.31 and $M_{\lambda_1}$ in place of $Y_1$, the representation of $M_\lambda$ in (3.8) becomes

$$M_\lambda(t) = \int_0^t K_\lambda(s) \, dM_{\lambda_1}(s),$$

where

$$K_\lambda(s) = \frac{dE^z(M_\lambda(s) \, dM_{\lambda_1}(s))}{dE^z(M_{\lambda_1}(s) M_{\lambda_1}(s))}$$

*from Theorem 3.20.*

**Proof** Since such a representation is known to exist, we need only write the projection of $M_\lambda$ onto the space of $M_{\lambda_1}$, as we did in orthogonalizing $M_n$ in (3.4).

Before turning to the stationary case, let us discuss both an example (treated from another point of view by Lévy (1956)) and an exercise in the non-stationary case. The former will illustrate the discussion following Theorem 3.27, to the effect that (3.8) applies (and so does its proof) to any choice of $Y_1$, and in particular to $Y_1 = \sum_{k=1}^n c_k M_{\lambda_k}$ or $Y_1 = M_{\lambda_1}$ from Theorem 3.31.

**Example 2** Let $B(t)$ be a given Brownian motion ($B(0) = 0$), and define a process

$$X(t) = \int_0^t (-3t + 4u) \, dB(u) \tag{3.11}$$

(as in Lévy (1956, (3.6.7))). Of course $X(t)$ is measurable, mean 0, Gaussian, and we can identify it in distribution with $X_z^c(t)$ on our prediction space. It may then appear that the proper canonical representation of $X(t)$ should have the same form as its definition, but this is not so. The trouble lies in the fact that this $B(t)$ is not in the Hilbert space generated by $\{X(s), s \leq t\}$. Proceeding to our method, the covariance of $X$ is easily calculated: $\Gamma(s, t) = 3s^2 t - \frac{2}{3}s^3$, $0 \leq s \leq t$, and by the 0-1 Law for $B(t)$ it follows that $H_z(0+) = \mathbf{0}$, so all assumptions of Theorem 3.27 are met. The initial hurdle in applying our method is to find the prediction $R_\lambda^z(\hat\rho(Z_t)) = E^z(\int_0^\infty e^{-\lambda s} X_z^c(t+s) \, ds | X_z^c(u), u \leq t)$, where $X_z^c(t)$ has the law of $X(t)$ on prediction space. It turns out that

$$E^z(X(t+s) | X(u), u \leq t) = (1 + 2st^{-1}) X_t - 2st^{-2} \int_0^t X_u \, du,$$

a fact which is easy to check because it suffices that

$$0 = E^z(X(v)(X(t+s) - E^z(X(t+s) | X(u), u \leq t))) \qquad \text{for } 0 < v \leq t,$$

but since this is our real starting-point we omit the derivation. From this it follows that

$$R^z_\lambda \rho(Z(t)) = \lambda^{-1}(1 + 2(\lambda t)^{-1})X^c_z(t) + 2(\lambda t)^{-2}\int_0^t X^c_z(s)\,ds.$$

(We note that $X^c_z(t)$ can be chosen continuous here—see Definitions 2.6 and 2.7.) Now an integration by parts (justified by appeal to $\Gamma$) yields easily

$$M_\lambda(t) = \lambda^{-1}(1 + 2(\lambda t)^{-1})X^c_z(t) - 2((\lambda t)^{-2} + (\lambda t)^{-1})\int_0^t X^c_z(s)\,ds.$$

From here, calculation using $\Gamma$ gives us

$$E^z(M_\lambda(t)M_\mu(t)) = (3\lambda\mu)^{-1}t^3 + (\lambda\mu)^{-1}(\lambda^{-1} + \mu^{-1})t^2 + 4(\lambda\mu)^{-2}t,$$

and using this, we can write the projection of $M_\mu(t)$ onto the space of $\{M_\lambda(s), s \le t\}$, and check that it has the same $L^2$-norm as $M_\mu(t)$. Therefore, we must have $\Phi_z(t) = 1$, and we can use any $M_\lambda$ in place of $Y_1$ for a canonical representation, since all $d\sigma^2_\lambda$ are equivalent.

Now fixing $\lambda_1$, we have for (3.8) since there is only a single $k$

$$K_\lambda(s) = \frac{dE^z(M_\lambda M_{\lambda_1}(s))}{dE^z M^2_{\lambda_1}(s)},$$

and this transform in $\lambda$ is easy to invert giving*

$$F(u + t, t) = \frac{(t^2\lambda_1^{-1} + 2t\lambda_1^{-2}) + (2t\lambda_1^{-1} + 4\lambda_1^{-2})u}{t^2\lambda_1^{-2} + 4t\lambda_1^{-3} + 4\lambda_1^{-4}}.$$

Therefore, by change of variables,

$$X^c_z(t) = \int_0^t F(t, s)\,dM_{\lambda_1}(s)$$

$$= \int_0^t \frac{(2\lambda_1^{-2} + s\lambda_1^{-1})(2t - s)}{(s^2\lambda_1^{-2} + 4s\lambda_1^{-3} + 4\lambda_1^{-4})}\,dM_{\lambda_1}(s)$$

$$= \int_0^t (2t - s)(2\lambda_1^{-2} + s\lambda_1^{-1})^{-1}\,dM_{\lambda_1}(s).$$

But $(2\lambda_1^{-2} + s\lambda_1^{-1})^{-1}\,dM_{\lambda_1}(s)$ is just another Brownian motion $B^*(s)$, so our proper canonical representation is

$$X^c_z(t) = \int_0^t (2t - s)\,dB^*(s), \tag{3.12}$$

---

* An even easier way is to project $X^c_z(t)$ directly onto $dM_{\lambda_1}(s)$, but the transform approach has other consequences for the sequel (Chapter 5).

132 | Foundations of the prediction process

in contrast to the original (3.11). It should be observed as obvious that this $B^*(t)$ is in the space of $X_z^c(s)$, $s \leq t$ (along with $M_\lambda(t)$). As soon as this is known, both $B^*$ and the coefficient $2t - s$ (for a.e. s) are uniquely determined, the latter by projection of $X_z^c(t)$ onto the space of $(B^*(s), s \leq t)$. The same remarks apply, of course, to $X(t)$ and $(B(s), s \leq t)$ when $B$ is given in advance to satisfy the canonical representation.

**Remark** By Hida (1960, Theorem 1.7) a representation of the form $X(t) = \int_0^t F(t, u) \, dB(u)$ is proper canonical if and only if $\int_0^t F(t, u) f(u) \, du = 0$ and $\int_0^t |f(s)| \, ds < \infty$ for all $t$ imply $f = 0$ a.e. Thus it is easy to check that (3.12) is, and (3.11) is not, proper canonical. The point of the example, however, is to find (3.12) when it is not known in advance.

A mean 0 Gaussian process $z \in M_{G(c)} \cap D$ is 'Markovian' if and only if $E^z(X_z^c(t + s) | \mathscr{X}_t^0) = \varphi(s + t, t) X_z^c(t)$, $P^z$-a.s. for some fixed $\varphi$ and $0 < s$. That this implies the usual conditional independence of $X_z^c(t + s)$ and $\mathscr{X}_t^0$ given $X_z^c(t)$ follows by Lemma 3.9. It may also be written $\mathscr{E}^z(X_z^c(t + s) | H_z(t+)) = \varphi(s + t, t) X_z^c(t)$, and in this form it defines a 'wide-sense Markov process' when only $z \in M_2 \cap D$ with continuous covariance is assumed.

**Exercise 3.31** For a Gaussian Markov process $z \in M_{G(c)} \cap D$, assuming $\Gamma_z(t, t) \neq 0$ and $E_z(\cdot) = 0$, show that if we set $c(\lambda, t) = \int_0^\infty e^{-\lambda u} \varphi(t + u, u) \, du$, then $c(\lambda, t)$ is continuous in $t$, and

$$M_\lambda(t) = c(\lambda, t) X_z^c(t) + \int_0^t X_z^c(s)(1 - \lambda c(\lambda, s)) \, ds, \qquad P^z\text{-a.s.} \quad (3.13)$$

Hence we can choose $X_z^c(t)$ to be r.c.l.l. in such a way that (3.13) becomes true for all $t$. Next, assume $c(\lambda, s) > \delta > 0$ for all $s \leq t$, and that $dc(\lambda, s)/ds$ is continuous. Show that

$$dM_\lambda(t) = \left\{ \frac{d}{dt} c(\lambda, t) + 1 - \lambda c(\lambda, t) \right\} X_z^c(t) + c(\lambda, t) \, dX_z^c(t),$$

in a sense of $L^2$, and solve for $X_z^c(t)$ to get the proper canonical representation

$$X_z^c(t) = \frac{\int_0^t \exp(-\lambda u + \int_0^u (c(\lambda, v))^{-1} \, dv) \, dM_\lambda(u)}{c(\lambda, t) \cdot \exp(-\lambda t + \int_0^t (c(\lambda, v))^{-1} \, dv)}. \quad (3.14)$$

Conversely, without assuming the existence of $dc(\lambda, t)/dt$ but only $c(\lambda, s) > \delta > 0$, $s \leq t$, show that if we define a process $Y(t)$ by the right side of (3.14), then (3.13) follows with $Y(t)$ in place of $X_z^c(t)$. But the equation

$$0 = c(\lambda, t) X(t) + \int_0^t X(s)(1 - \lambda c(\lambda, s)) \, ds, \qquad X(0) = 0,$$

implies $X(t) = 0$ in $L^2$, hence $Y(t) = X_z^c(t)$ and (3.14) still provides the proper canonical representation.

Consider finally the special case that $\varphi(s + t, t)$ is a function of $s$ alone, say $\varphi(s)$. Then $c(\lambda, t) = c(\lambda)$, and assuming $c(\lambda) \neq 0$ show that (3.14) becomes

$$X_z^c(t) = \left( \int_0^t \exp((\lambda - c^{-1}(\lambda))(t - u)) \, dM_\lambda(u) \right) c^{-1}(\lambda). \quad (3.15)$$

Show, besides, that $c(\lambda) \neq 0$ follows from (3.13) and the fact that $M_\lambda$ has independent increments, unless $X_z^c \equiv 0$. From (3.15) deduce that $X_z^c(t) \exp(c^{-1}(\lambda) - \lambda)t$ does not depend on $\lambda$, since it has independent increments, with mean 0, and therefore $c^{-1}(\lambda) - \lambda = \beta$ does not depend on $\lambda$. It follows that (dropping the ornaments)

$$dX(t) = -\beta X(t) + (\lambda + \beta) \, dM_\lambda(t), \quad (3.16)$$

where $(\lambda + \beta) \, dM_\lambda$ does not depend on $\lambda$.

Show, conversely, that for any mean 0, $L^2$-continuous Gaussian process $y$ with independent increments in place of $(\lambda + \beta)M_\lambda(t)$, (3.16) defines a unique, mean 0, Gaussian Markov process $z$ starting at 0, with $z \in M_G \cap D$ and $\varphi = e^{-\beta s}$. If, in addition, $y$ is a Brownian motion, then $X(t)$ is an *Ornstein–Uhlenbeck velocity process with parameter $\beta$, starting at 0*.

Returning to (3.14), note that for suitable $F_\lambda(t)$ and $G_\lambda(u)$ we have $F_\lambda(t)X_z^c(t) = \int_0^t G_\lambda(u) \, dM_\lambda(u)$, where the right side has independent increments; in the wide-sense analogue, its increments are orthogonal.

It can be shown (Hida (1960), for example) that such an equation is necessary and sufficient for Gaussian $X_z^c$ to be Markovian, under negligible assumptions. For more information on Gauss–Markov processes, see for instance Doob (1949) and Hida (1960).

As a final topic for the present chapter, we examine how our theory applies to stationary Gaussian processes. An apology is made in advance to those readers who insist on spectral representations. Our theory is entirely in the time (not spectral) domain. At least this has a compensating advantage of not 'revealing' any periodicities which actually may have no physical meaning.

A covariance $\Gamma$ is called stationary if it is a function of $|t - s|$; we will write $\Gamma(t - s)$ in this case. To any stationary covariance there is associated a mean 0, stationary Gaussian process on $0 \leq t < \infty$, and also one on $-\infty < t < \infty$. To best apply the prediction process, it is necessary to permit $-\infty < t < \infty$. Otherwise, the prediction process will not be stationary, and the stationarity assumption loses force. As a preliminary it is therefore necessary to discuss the prediction theory in case $-\infty < t < \infty$. One may assert that everything goes through just as before except for one difference. The states of the prediction process are, by definition, measures on $\{w'(t), 0 \leq t\}$. This does not change when we consider a given $P'$ on $\{w'(t), -\infty < t < \infty\}$, thus $P'$ itself is not a permissible state for $Z^{P'}(t)$. The process $Z^{P'}(t)$ is now defined, $-\infty < t < \infty$, by $Z^P_t(S) = P'(\Theta_t^{-1}S|\mathscr{F}'_{t+})$, together with right-continuity in an appropriate topology as before, but now $\Omega'$ becomes $\Omega'_{-\infty,\infty} = \{w'(t), -\infty < t < \infty\}$ and $\Theta_s w'(s) = w'(t+s)$ is defined for $-\infty < t < \infty$, so that $\Theta_t^{-1} w'(s) = w'(s-t)$ is always unique. Here, of course, we use $\mathscr{F}'_t = \sigma\{\int_s^t f(w(u))\,du, -\infty < s < t, f \in b(\mathscr{E})\}$, and the atoms of $\mathscr{F}'$ are equivalence classes, just as in Definition 1.1. But in defining $Z^{P'}_t(S)$ we only allow $S \in \mathscr{F}'_{[0,\infty)} = \sigma\{\int_0^s f(w(u))\,du, 0 < s, f \in b(\mathscr{E})\}$. Therefore, the state space of $Z^{P'}_t$ is the same as before. The same holds true when we transfer to prediction space, which now consists of all r.c.l.l. paths $w_Z(t)$, $-\infty < t < \infty$.

Except for these obvious modifications, the rest of our theory goes through without change. In particular, $Z_t$ is a strong Markov process, $-\infty < t < \infty$, with the same transition function $q(t, z, A)$, $t \geq 0$, as before. To see this most readily, one can observe that for $t > t_0$ this can be reduced to the former case by a change of time parameter such as

$$\tau(t) = \begin{cases} \dfrac{\pi}{2} + \arctan(t - t_0) & t < t_0 \\[1em] t - t_0 + \dfrac{\pi}{2} & t > t_0 \end{cases}$$

mapping $(-\infty, \infty) \to (0, \infty)$, with $\tau(t_0) = \pi/2$. Under the mapping $w'(t) \leftrightarrow w'(\tau(t))$, $-\infty < t < \infty$, the given $\sigma$-field $\mathscr{F}'_{t+}$ is mapped onto the given $\mathscr{F}'_{\tau(t)+}$ for the positive parameter, while for $t > t_0$ the map $\tau(t)$ is linear in such a way that, clearly, $Z^{P'}_t(w')$ is mapped $Z^{\tau(P')}_{\tau(t)}(w'(\tau))$, where $\tau(P')(S) = P'(\tau^{-1}(S))$; $\tau^{-1}(S) = \{w' : w'(\tau) \in S\}$. Again, the linearity of $\tau$ shows that not only the Markov properties, but also the precise transition function $q$, are valid for $t > t_0$. Since $t_0$ is arbitrary, this extends to $-\infty < t < \infty$.

Returning now to $\Gamma(t-s)$, in order to define a stationary Gaussian $z$ on $\Omega'_{-\infty,\infty}$ we require measurability of the process, as before. But in the stationary case, we may state the following well-known theorem.

**Theorem 3.33** $\Gamma(t-s)$ *is covariance of a measurable (mean 0) process if and only if it is continuous.*

**Proof** This reduces to a form (due to von Neumann) of Stone's theorem on the spectral representation of semigroups (here, the maps $T_t: X_s \to X_{s+t}$) of unitary operators on a separable Hilbert space. We sketch the argument. By Theorem 22.4.3 of Hille and Phillips (1937), such a semigroup is continuous in the strong operator topology if it is strongly measurable. Of course strong continuity implies weak continuity which implies continuity of $\Gamma(t - s)$. On the other side, by Hille and Phillips (1937, Theorem 3.5.3), strong measurability is equivalent to weak measurability and almost separably valuedness. Weak measurability is easily equivalent to measurability of $\Gamma$, and the almost separability is implied by the separability condition of Theorem 3.1(1) above, and hence by the existence of a corresponding measurable process.

**Remark** This result is included only for completeness. One can avoid it by taking continuity of $\Gamma$ as a hypothesis in the stationary case. Actually, our basic theory is concerned with equivalence classes of $\Gamma$ up to $(ds \times dt)$-null sets, and so the hypothesis could be modified accordingly. But it follows from Theorem 3.33 that the equivalence classes contain at most a single element of the form $\Gamma(s - t)$.

It should be emphasized at the outset that, although $P'$ corresponding to $E(\cdot) = 0$ and $\Gamma(s - t)$ defines a stationary Gaussian process $X^c(t)$, its prediction process $Z(t)$ has its values in the space of non-stationary probabilities. Indeed, $Z(t) \in D$, together with stationarity (defined, to be sure, only for the positive time) would require that $Z(t)$ is $P'$-a.s. a random 'point mass', since $H_{Z(t)}(0+) = \mathbf{0}$ and stationary would imply $\Gamma_{Z(t)}(0) = 0 = \Gamma_{Z(t)}(s, s)$ for all $s > 0$. This is in contrast to the behaviour of $Z(t)$ in $t$. The next theorem is obvious.

**Theorem 3.34** *The prediction process of a stationary process is stationary.*

**Remark** With the usual definition of (strict) stationarity, namely that $P'(\Theta_t S) = P'(S)$ for all $t$ and $S$, this result holds without any Gaussian hypothesis. Indeed, the stationarity of $Z^{P'}(t)$ in distribution follows from the invariance of the defining conditional probabilities under $\Theta_t$, which is immediate from the definition. The unrestricted stationarity of a joint distribution then follows from the Markov property in the obvious way, but the question arises of characterizing the subspace of $Z(t)$ for stationary Gaussian $P'$, and of determining whether it is a demesne. This is an open problem.

The development of $M_\lambda(t)$ and $H_z(t+)$, going back to Notation 3.5, requires only slight modification when $-\infty < t < \infty$ is permitted. For $H_{P'}(t)$ we take as generators $\{\int_s^t f(u)(\hat{\rho}Z(u)) - E_{P'}(u)) \, du, -\infty < s < t, f \in L^2(du)\}$ and $H_{P'}(t+) = \bigcap_{\varepsilon > 0} H_{P'}(t + \varepsilon)$ as before. (Since we form the $L^2$-linear closure, the precise set of generators is quite arbitrary.) We do not define $M_\lambda(t)$, but rather the increments $M_\lambda(t) - M_\lambda(s)$, $-\infty < s < t < \infty$, for which Definition 3.6 obviously suffices: $M_\lambda(t) = M_\lambda(t) - M(0)$. Another minor adjustment is needed in stating Theorem 3.18, representing $H_z(t+)$ in terms of $\{M_n(t) - M_n(s), s \leq t, n \geq N\}$ (where we already note a minor change). Namely, the requirement $H_z(0+) = \mathbf{0}$ must be replaced by $H_{P'}(-\infty) = \mathbf{0}$, where $H_{P'}(-\infty)$ $(= \bigcap_t H_{P'}(t))$ is the *deterministic component*. To see this, we note that the proof of Theorem 3.18 shows that for each $s < t$, $H_{P'}^{\perp}(s+) \cap H_{P'}(t+)$ is generated by $\{M_n(u) - M_n(s), s < u \le t, n > N\}$. Therefore, letting $s \to -\infty$ it follows readily that $H_{P'}^{\perp}(-\infty) \cap H_{P'}(t+)$ is generated by $\{M_n(t) - M_n(s), s < t, n \ge N\}$. To represent $H_{P'}(t+)$ in terms of this last, it is necessary and sufficient that $H_{P'}(-\infty) = 0$.

**Theorem 3.35** *For stationary Gaussian $P'$ with $H_{P'}(-\infty) = 0$, we have the following.*

(1) $H_{P'}(t+) = $ [Hilbert space closure of $\{M_n(t) - M_n(s), s < t, n \ge 1\}$].

(2) *Theorem 3.20 extends to define* $E^{P'}(M_m(t) - M_m(s))(M_n(t) - M_n(s))$, $-\infty < s < t < \infty$.

(3) *Theorem 3.23 extends to define a sequence* $\{Y_n(t) - Y_n(s), -\infty < s < t < \infty, n < N+1\}$ *(for some $N \le \infty$) of independent Gaussian processes with independent increments and* $d\sigma_1^2 \gg d\sigma_2^2 \gg \cdots$, *having the analogous uniqueness and representation properties. The multiplicity*

$$\Phi_{P'} = \inf\{n : \sigma_{n+1}^2(t) - \sigma_{n+1}^2(s) = 0, -\infty < s < t\}$$

*is constant in t. The wide-sense form (Corollaries 3.24 and 3.25) likewise extends.*

(4) *The definition of $X_{P'}^c(t)$ extends from Definition 3.7 (note that $E_{P'}(t)$ is now a constant, which may be set equal to 0), and the existence of a proper canonical representation (Theorem 3.26) becomes*

$$X_{P'}^c = \sum_{n=1}^{\Phi_{P'}} \int_{-\infty}^{t} F_n(t, s) \, dY_n(s), \qquad (3.17)$$

Before extending our modifications to Theorem 3.27, we will show that in the present stationary case much more than (4) is true.

**Theorem 3.36** *For stationary Gaussian $P'$ with $H_{P'}(-\infty) = 0$, either $\Phi_{P'} = 0$ and the process is a constant $E_{P'}$ for all t, or $\Phi_{P'} = 1$. Moreover, $F(t, s)$ ($= F(t - s)$) depends only on $t - s$, and $Y(t_0 + s) - Y(t_0)$ is a Brownian motion $B(s)$ (we can absorb the variance constant into F). Thus Equation (3.17) can be written*

$$X_{P'}^c(t) = \int_{-\infty}^{t} F(t - s) \, dB(s), \qquad B(t) - B(s) \in H_{P'}(t). \qquad (3.18)$$

*The analogous fact for $P'$ stationary in the wide sense ($\Gamma = \Gamma(t - s)$, $E_{P'}(t)$ constant) simply replaces $B(t) - B(s)$ by a process with orthogonal increments and $d\sigma^2(t) = dt$ (more explicitly, by $\sigma_1^{-1}(Y_1(t) - Y_1(s))$, when Theorem 3.23 is used).*

**Proof** We first observe that, for every $\lambda$, either $M_\lambda(t)$ ($= M_\lambda(t) - M_\lambda(0)$) is identically 0 or else $\sigma^{-1}M_\lambda(t)$ is a two-sided Brownian motion ($\sigma^{-1}M_\lambda(t)$ and $\sigma^{-1}M_\lambda(-t)$ are independent Brownian motions) where $\sigma^2 = EM_\lambda^2(1) \neq 0$. This is a consequence of stationarity, which implies that $d\sigma^2$ is a $\sigma$-finite measure which is translation invariant, hence it is a constant $\sigma^2$ times Lebesgue measure. The same argument applied to the inner products $E^{P'}(M_{\lambda_1} M_{\lambda_2}(t))$ shows that they have the form $c(\lambda_1, \lambda_2)|t|$, and this in turn implies that $(c_n Y_n, n < N+1)$ are independent Brownian motions, or identically 0, where $c_n = E^{-\frac{1}{2}} Y_n^2(1)$ with $0/0 = 0$. Since the construction of $(Y_n(t+s) - Y_n(t))$ from $(M_n(t+s) - M_n(t))$, $s > 0$, is the same for all $t$, we have

$$Y_n(t+s) - Y_n(0) = (Y_n(t) - Y_n(0)) + (Y_n(s) - Y(0)) \circ \Theta_t^Z,$$

$P'$-a.s. for each $t$ and $s > 0$, which extends to all $s > 0$ by right-continuity. From this, we see that in the representation (3.26) we can define $F_n(t, s) = F_n(0, s-t)$ in such a way that the representation of $X_{P'}^c(0)$ suffices to determine the coefficients, which depend only on $s - t$. By reason of stationarity, indeed, for each $t$,

$$\begin{aligned} X_{P'}^c(t) &= X_{P'}^c(0) \circ \Theta_t^Z \\ &= \sum_{n=1}^{\Phi_{P'}} \int_{-\infty}^0 F_n(0, s) \, dY_n(s) \circ \Theta_t^Z \\ &= \sum_{n=1}^{\Phi_{P'}} \int_{-\infty}^0 F_n(0, s) \, dY_n(s+t) \\ &= \sum_{n=1}^{\Phi_{P'}} \int_{-\infty}^t F_n(0, s-t) \, dY_n(s), \end{aligned}$$

$\Phi_{P'}$ being the (constant) multiplicity. We now write $F_n(0, s-t) = F_n(t-s)$, and using an elegant device due to Hanner (1949, Proposition D), we show that (if $\Phi_{P'} \neq 0$) $\Phi_{P'} = 1$, so that only $F_1$ is required. Supposing the contrary, then $c_2 Y_2$ is a Brownian motion, and we have

$$X_{P'}^c(t) = c_1^{-1} \int_{-\infty}^t F_1(t-s) \, dB_1(s) + c_2^{-1} \int_{-\infty}^t F_2(t-s) \, dB_2(s) + z(t),$$

where $B_1$ and $B_2$ are independent Brownian motions, independent of $z(t)$, and for $s < t$, $B_i(t) - B_i(s)$, $i = 1$ or 2, and $z(s)$ are in $H_{P'}(t)$. Setting for convenience $t = 0$, we can choose a number $c < 0$ such that

$$c_2^{-1} \int_c^0 F_2(s-c) \, dB_1(s) \quad \text{and} \quad c_1^{-1} \int_c^0 F_1(s-c) \, dB_2(s)$$

are both non-vanishing Gaussian random variables. To reach a contradiction it suffices to show that their difference is orthogonal to $X_{P'}^c(t)$ for every $t < 0$, because then it cannot be in $H_{P'}(0)$ are required. But for $t \leq c$ this orthogonality is obvious since $dB_i(s)$, $c < s$, are independent of $H_{P'}(c)$. On the other hand, for $c < t < 0$ we have

$$E^{P'}\left(X_{P'}^c(t)\left(c_2^{-1}\int_c^0 F_2(s-c)\,dB_1(s) - c_1^{-1}\int_c^0 F_1(s-c)\,dB_2(s)\right)\right)$$

$$= E^{P'}\left[\left(c_1^{-1}\int_c^t F_1(t-s)\,dB_1(s) + c_2^{-1}\int_c^t F_2(t-s)\,dB_2(s)\right)\right.$$

$$\left.\times\left(c_2^{-1}\int_c^t F_2(s-c)\,dB_1(s) - c_1^{-1}\int_c^t F_1(s-c)\,dB_2(s)\right)\right]$$

$$= (c_1 c_2)^{-1}\int_c^t (F_1(t-s)F_2(s-c) - F_2(t-s)F_1(s-c))\,ds.$$

Now the change of variables $u = t - s + c$ in the second integral on the right shows immediately that the whole expression vanishes. This completes the proof of Equation (3.18), and the wide-sense assertions follow without further difficulty.

Continuing our modification to the stationary case, we can state immediately the following result.

**Theorem 3.37** *Setting $K_\lambda = \int_0^\infty e^{-\lambda t} F(t)\,dt$, for every $\lambda > 0$ we have $M_\lambda(t) = K_\lambda B(t)$, with $B(t)$ from (3.18).*

Conversely, this shows how to obtain $B(t)$ from $M_\lambda(t)$, i.e. to choose the value of $\lambda_1$ for Theorem 3.31 in the stationary case. Indeed, any value for which $M_{\lambda_1}$ is observed not to vanish will suffice. Then, if the other values $M_\lambda(t)$ are known for a single $t$, one can determine $F(t)$ for all $t$ by inversion of the transform, up to a constant factor. Hence in the stationary case observation of $(M_\lambda)$ or even of $M_n$, $n \geq N$, at a single instant $t$ determines the law of the entire process up to a constant factor, and if $M_\lambda(t)$ is also known for some $\lambda$ in an interval of $t$ (and does not identically vanish) then $B(t)$, and hence the constant factor, are also determined. We hasten to emphasize, however, that observation of $M_\lambda$ requires solution of the prediction problem, which probably requires advance knowledge of the law of the process.

The simplest important example of a stationary Gaussian process is the stationary Ornstein–Uhlenbeck process derived from Exercise 3.31 above. The equation (3.16) with Brownian input is

$$dX(t) = -\beta X(t) + dB(t), \qquad \beta > 0, \tag{3.19}$$

whence (3.15) becomes

$$X_z^c(t) = \int_0^t \exp{-\beta(t-u)}\,dB(u). \tag{3.20}$$

Letting $t \to \infty$ suggests the stationary solution of (3.19) for $-\infty < t < \infty$, viz

$$X_z^c(t) = \int_{-\infty}^t \exp(-\beta(t-u))\,dB(u), \tag{3.21}$$

which is already in proper canonical form. The formula (3.13) for $M_\lambda(t)$ now reads

$$M_\lambda(t) - M_\lambda(0) = (\lambda + \beta)^{-1}\left(X_z^c(t) - X_z^c(0) + \beta \int_0^t X_z^c(s)\,ds\right)$$

which becomes easily

$$M_\lambda(t) - M_\lambda(s) = (\lambda + \beta)^{-1}(B(t) - B(s)), \qquad s < t.$$

Thus $K_\lambda = (\lambda + \beta)^{-1}$, and $F(t) = e^{-\beta t}$ in agreement with (3.21).

For a somewhat less routine example, let us consider the 'triangular' covariance

$$\Gamma(t-s) = \begin{cases} 1 - |t-s| & \text{if } |t-s| \leq 1 \\ 0 & \text{elsewhere.} \end{cases}$$

It is left to the reader to compare our method with the usual method based on the spectral representation. But it should be pointed out that the spectral process itself is not in $H_{P'}(t)$ for any $t$, and hence cannot be determined in practice. Our problem is to determine the prediction $E(X^c(t+s)|\mathscr{L}_t^0)$ directly in terms of $X^c(u)$, $u \leq t$. The expression is complicated, and a short-cut is advisable. Let us notice that a process with covariance $\Gamma$ can be constructed in the form $X(t) = B(t) - B(t-1)$, where $B(t)$ is a two-sided Brownian motion. Since $X(t_1)$ and $X(t_2)$ are independent when $|t_2 - t_1| \geq 1$, there is no problem with prediction for $s \geq 1$; it is just 0. For $0 < s < 1$, we have

$E(X(t+s)|X(u), u \leq t)$

$= E(B(t+s) - B(t+s-1)|B(u) - B(u-1), u \leq t)$

$= E(E(B(t+s) - B(t+s-1)|B(u), u \leq t)|B(u) - B(u-1), u \leq t)$

$= E(B(t) - B(t+s-1)|B(u) - B(u-1), u \leq t). \tag{3.22}$

For purposes of visualization, we can assume that $t = 0$ and $B(0) = 0$ (but the answer must only involve the increments of $B$). Then the 'given' includes $B(-n)$, $n \geq 0$, and in each interval $-(n+1) \leq t \leq -n$ we can represent $B(t)$

in the form

$$B(t) = B(-(n+1)) + \{t + (n+1)\}\{B(-n) - B(-(n+1))\}$$
$$+ \beta_n(t + (n+1)), \quad -(n+1) \le t \le -n, \quad (3.23)$$

where $(\beta_n)$ is a sequence of Brownian bridges independent of $\{B(-n)\}$, and independent of each other. We have easily

$$B(0) - B(s-1) = \{B(0) - B(-1)\}(1-s) - \beta_1(s),$$

and if we show that $\beta_1(s)$ is in the Hilbert space $H(B(u) - B(u-1), u \le 0)$ it follows that the prediction is simply $B(0) - B(s-1)$. Now from (3.23) for $-(n+1) \le t \le -n$ it is easy to see that $\beta_n(t+n-1) - \beta_{n+1}(t+n+1) \in H(B(u) - B(u-1), u \le 0)$, and since $0 \le t+n+1 \le 1$ we can write this as $\beta_n(s) - \beta_{n+1}(s) \in H(B(u) - B(u-1), u \le 0)$. Then, by addition, $\beta_1(s) - \beta_{n+1}(s)$ is also in $H$. But from independence of $\beta_n$,

$$L^2\text{-lim}_{n\to\infty} n^{-1} \sum_{k=1}^{n} (\beta_1(s) - \beta_{n+1}(s)) = \beta_1(s)$$

which is also in $H$, as required. Thus the prediction (3.22) becomes

$$E(X(t+s)|X(u), u \le t) = B(t) - B(t+s-1) \quad (3.24)$$

where the expression for the right side in terms of $\{X(u), u \le t\}$ seems unwieldy, and will be left aside. Continuing in terms of the increments of $B(t)$ (defined in terms of $X^c_{P'}(t)$ on prediction space), we have

$$R^Z_\lambda \hat{\rho}(Z(t)) = \int_0^1 e^{-\lambda s}(B(t) - B(t+s-1)) \, ds.$$

Then applying Definition 3.6 for $M_\lambda(t)$ $(=M_\lambda(t) - M_\lambda(0))$ we obtain easily (taking $B(0) = 0$)

$$M_\lambda(t) = \lambda^{-1}(1 - e^{-\lambda})B(t) - \int_0^1 e^{-\lambda s}(B(t+s-1) - B(s-1)) \, ds$$

$$+ \int_0^t B(u) - B(u-1) \, du - \lambda \int_0^t \int_0^1 e^{-\lambda s}(B(u) - B(u+s-1)) \, ds \, du.$$

In the last integral we can interchange the order and then integrate by parts, whereupon all terms cancel except the first, leaving $M_\lambda(t) = \lambda^{-1}(1-e^{-\lambda})B(t)$. Comparison with Theorem 3.37 shows that $F(t) = I_{0,1}(t)$, so that the proper canonical representation is $X_t = \int_{t-1}^{t} dB(s)$. This could have been anticipated on the grounds of uniqueness once it was shown that $dB(s) \in H_{P'}(t)$, $s < t$, but it illustrates the method.

For an example involving the familiar rational spectral densities, see Example 2.3 of Knight (1983a). Even in such a case, which is presumably the citadel of the spectral method, our method has advantages in obtaining the moving-average representation (3.18) from the predictors $E(X(t + s)|X(u)$, $u \leq t)$. We will see in Chapters 5 and 6 that the same method applies in the non-linear, non-Gaussian case, where the $M_\lambda$ are simply martingales (and the stationary multiplicity may exceed 1).

## Appendix

The following result is from Knight (1983a), Meyer (1983), and a private communication of P.-A. Meyer. Since $M_2 \subset M_1$, and there is no Gaussian hypothesis, it is more than enough for use in Theorem 3.18. Incidentally, another proof of Theorem 3.18 by an unrelated method follows in Chapter 5, but it is somewhat less general.

**Theorem 3.38** *For $z \in M_1$, $t > 0$, $N \geq 1$, we have*

$$\sigma\{R_n^Z \hat{\rho}(Z(0)) + M_n(s), 0 \leq s < t, n \geq N\} \stackrel{z}{\equiv} \mathscr{L}_{t-}^\circ$$

*where $\stackrel{z}{\equiv}$ denotes the equality up to $P^z$-null sets.*

**Proof** We first prove the assertion with $t = \infty$. Since $z \in M_1$, we have $\lim_{t \to \infty} \int_t^\infty e^{-\lambda s} |\hat{\rho}(Z(s))| \, ds = 0$, $P^z$-a.s., and therefore by Hunt's lemma, $\lim_{t \to \infty} e^{-\lambda t} R_\lambda^Z(\hat{\rho}(Z(t))) = 0$, $P^z$-a.s. Let $D_\lambda$ denote the space of all r.c.l.l., real-valued paths $y(t)$ such that $\lim_{t \to \infty} e^{-\lambda t} y(t) = 0$. Thus we have $P^z\{R_\lambda^Z \hat{\rho}(Z(\cdot)) \in D_\lambda\} = 1$, and it is easy to see, using Dellacherie and Meyer (1975, Chapter III, Section 19) that $D_\lambda$ is a measurable Lusin space. Next, let $A$ denote the space of all Borel measurable real-valued paths $x(t)$ such that, for all $\lambda > 0$, $\int_0^\infty e^{-\lambda t} |x(t)| \, dt < \infty$, considered as equivalence classes mod Lebesgue-null sets (i.e. as elements of $L^1(e^{-\lambda t} \, dt)$). Clearly $A$ is a Polish space, and hence a measurable Lusin space with the Borel sets. Also, we have $P^z\{\hat{\rho}(Z(\cdot)) \in A\} = 1$. Now let $B = A \times X_{n \geq N} D_n$, which is also a measurable Lusin space. (Use, for example, the Tychonoff product theorem.) We consider the Borel mapping $\Phi: B \to X_{n \geq N} D$ given by

$$\Phi(x, y_N, y_{N+1}, \ldots)_t = \left( y_n(t) + \int_0^t (x(s) - n y_n(s)) \, ds; n \geq N \right),$$

where $D$ is the measurable Lusin space of all r.c.l.l. real-valued paths. The essential point of the proof is that $\Phi$ is one-to-one. Indeed, since $\Phi$ is linear in $(x, y_N, \ldots)$, it is enough to show that if $0 = y_n(t) + \int_0^t (x(s) - n y_n(s)) \, ds$ for all $t$ and $n \geq N$, then $x = 0 = y_n$, $n \geq N$. But for each $n$ the unique solution (vanishing at $t = 0$) of the above equation is $y_n(t) = e^{nt}(\int_0^t e^{-ns} x(s) \, ds)$. By hypothesis, $\lim_{t \to \infty} e^{-nt} y_n(t) = 0$, and hence $0 = \int_0^\infty e^{-ns} x(s) \, ds$, $n \geq N$. It follows by the Stone–Weierstrass approximation that $\int_0^\infty e^{-ns} f(s) x(s) \, ds = 0$ for $n \geq N$ and all continuous $f$ with compact support, which implies that $x = 0$, and hence also $y_n = 0$ for $n \geq N$.

Consequently, by Lusin's theorem the inverse $\psi = \Phi^{-1}$ is also one-to-one and Borel from $\Phi(B)$ onto $B$. This shows that, apart from the $P^z$-null set where $R_n^z \hat{\rho}(Z(\cdot)) \notin D_n$ for some $n \geq N$ or $\hat{\rho}(Z(\cdot)) \notin A$, the sequence

$$R_n^z \hat{\rho}(Z(s)) + \int_0^s \hat{\rho}(Z(u)) - nR_n^z \hat{\rho}(Z(u))\,du, \qquad n \geq N, s < \infty$$

uniquely determines $\int_0^s \hat{\rho}(Z(u))\,du$, $s < \infty$. This completes the proof in the case $t = \infty$.

The case $t < \infty$ is reduced to this by a modification of the future law; namely, let

$$\tilde{X}_s = \begin{cases} \hat{\rho}(Z(s)) & s < t \\ E^{Z(t-)}\hat{\rho}(Z(s-t)) & s \geq t \end{cases}$$

where we put $\tilde{X}_s = 0$ if the last expectation does not exist (that is, $P^z$-a.s. on a Lebesgue-null set of $s$). Clearly $\tilde{X}_s$ is a measurable process, and $\sigma\{\tilde{X}_s, 0 < s < \infty\} \stackrel{z}{=} \mathscr{L}_{t-}^\circ$ for every $z \in M_1$. Next, let $\tilde{z}$ denote the probability measure of the process $\tilde{X}$. As usual, we consider $\tilde{z}$ as the measure on the prediction space $\Omega_z$ induced by the prediction process $Z_t^{\tilde{z}}$ of $\tilde{X}$. We claim that the joint law of

$$\left( \int_0^s \hat{\rho}(Z(u))\,du, E^{Z(s)} \int_0^\infty e^{-\lambda u} \hat{\rho}(Z(u))\,du \right), \qquad 0 < \lambda, \quad 0 \leq s < t,$$

is the same for $\tilde{z}$ as it is for $z$. Obviously the laws of $\hat{\rho}(Z(s))$ are a.e. the same since $\hat{\rho}(Z(s))$ for $\tilde{z}$ has a.e. the same law as $\tilde{X}_s$ for $z$. Also, for $\tilde{z}$ we have, as process in $s$ ($s < t$),

$$E^{Z(s)} \int_0^\infty e^{-\lambda u} \hat{\rho}(Z(u))\,du = E^{\tilde{z}}\left( \int_0^\infty e^{-\lambda u} \hat{\rho}(Z(u+s))\,du \bigg| \mathscr{L}_s^\circ \right)$$

$$\stackrel{(d)}{=} E^{\tilde{z}}\left( \int_0^\infty e^{-\lambda u} \tilde{X}_{u+s}\,du \bigg| \mathscr{F}_{s+}' \right)$$

$$\stackrel{(d)}{=} E^z\left( \int_0^{t-s} e^{-\lambda u} \hat{\rho}(Z(u+s))\,du \bigg| \mathscr{L}_s^\circ \right)$$

$$+ E^z\left( \int_{t-s}^\infty e^{-\lambda u} E^z(\hat{\rho}(Z(u+s)) | \mathscr{L}_{t-}^\circ)\,du \bigg| \mathscr{L}_s^\circ \right)$$

$$= E^z\left( \int_0^\infty e^{-\lambda u} \hat{\rho}(Z(u+s))\,du \bigg| \mathscr{L}_s^\circ \right),$$

where $\stackrel{(d)}{=}$ denotes equality in joint distribution, $0 < \lambda$, $0 \leq s < t$. The last expression is the same as the second except that $z$ replaces $\tilde{z}$. Obviously the term $\int_0^s \hat{\rho}(Z(u))\,du$ can also be included in the joint equivalence, since the same reasoning applies to it. Thus our claim (transparent enough already) can be proved.

To give a detailed proof of this point, one introduces the binary process $(\hat{\rho}(Z(s)), \tilde{X}(s))$ on $\Omega_Z$ for $P^z$, and considers its prediction process, which includes that of each component separately. Then the previous reasoning shows that the equivalence in law just indicated becomes an a.s. identity between the two corresponding pairs (the first components being identical for $s < t$ by definition of $\tilde{X}_s$). This method of introducing a process with several related components, and defining the combined prediction process, is fundamental whenever several processes must be dealt with simultaneously.

To complete the reduction of $t < \infty$ to $t = \infty$, it now suffices to prove the $t < \infty$ assertion with $\tilde{z}$ in place of $z$. Indeed, the theorem follows if the indicator of every set in $\sigma(\int_0^s \hat{\rho}(Z(u))\,du, s < t)$ has an expression, up to a $P^z$-null set, as a Borel combination of indicators of sets in $\sigma(R_n^Z \hat{\rho}(Z(0)) + M_n(s), s < t, n \geq N)$. But all of these quantities have the same joint law under $P^{\tilde{z}}$ as under $P^z$. Hence it suffices to consider $P^{\tilde{z}}$. But for $P^{\tilde{z}}$, we have $\mathcal{L}_{t-}^\circ \equiv \mathcal{L}_\infty^\circ$, hence, for all $s < \infty$, $M_n(s)$ is $\mathcal{L}_{t-}^\circ$-measurable (up to $P^{\tilde{z}}$-null sets), and since it is a right-continuous martingale, this implies that $M_n(s) \equiv M_n(s \wedge t-)$. Therefore, for $P^{\tilde{z}}$ we have

$$\sigma\{R_n^Z \hat{\rho}(Z(0)) + M_n(s), s < t\} \equiv \sigma\{R_n^Z \hat{\rho}(Z(0)) + M_n(s), s < \infty\}$$

and the case $t < \infty$ is equivalent to the assertion already shown.

# 4

# A classification of measurable processes

> Enfin, le fait même d'avoir proposé une classification, et mis un peu d'ordre dans un domaine assez complexe, n'etait sans doute pas inutile.
>
> Paul Lèvy, 1970, Quelques Aspects de la Pensèe d'un Mathematicien, Premiere partie, 55.

## 4.1 Introduction

When we seek to extend the methods and results of Chapter 3 to non-Gaussian processes, we encounter two separate difficulties. First of all, the variety of processes is greatly increased, and in particular the types of discontinuity that may arise are more varied. This in itself has no necessary connection with prediction, and does not even presume a measure of distance in the state space. It is important, however, in giving insight into the nature of the processes involved. The most basic classification of discontinuities is in terms of the prediction processes rather than the processes themselves. Thus, unlike Chapter 3 where 'Gaussian' is a property of the law of a given process, we are concerned here with classifying processes according to properties of the law of their associated prediction processes. In particular, this includes the discontinuity types of the prediction process. Formally, of course, the two approaches can be made equivalent, since either law determines the other through the equivalence $\tilde{X}_t \equiv \hat{\rho}(Z_t^z)$ (Theorem 2.7). But in fact there is a rather clear and substantial difference in meaning between placing conditions directly on $X_t$ and placing them on $Z_t^z$.

The second difficulty that arises is how to extend the prediction, for which (since we continue to work within the classical least squares framework) it is necessary to have real-valued, or Hilbert-space-valued, processes. Since our methods are somewhat new, we confine attention to the real-valued case. (The real-vector case may be done componentwise.) But even in the real case the Gaussian restriction of Chapter 3 made the problem very special, as is well known. In effect, because of Lemma 3.9 the linear and non-linear Gaussian predictors coincide, so in the Gaussian case one need only work within the Hilbert spaces $H_z(t)$ of the process itself (Notation 3.5). In the general case $z \in M_2$, the linear theory is still available without any change, since it depends only on the mean and covariance structure. However, it need not coincide with the non-linear theory based on the Hilbert spaces $L^2\{\Omega, \mathscr{F}_t, P^z\}$ of the original process (or on $L^2(\Omega_Z, \mathscr{L}_t^0, P^z)$) after translation to the prediction space of Definition 2.13), which leads to the conditional expectations as predictors (in view of Theorem 0.4). Therefore, in terms of prediction errors the Gaussian case is the worst possible for a given covariance structure. One can always improve on its result (in the least squares sense) whenever it does not agree with the non-linear predictor, although it is then necessary to obtain the latter.

In the next chapter we will look into how the prediction methods of Chapter 3 extend to the non-linear case through what is called the analogy between wide-sense and strict-sense results (Doob 1953). It emerges that there are several levels of extension to the general case, and which level will be most useful in arriving at the actual prediction will depend on the problem at hand. This will be illustrated by examples in Chapter 6. In any event, we do not discuss any methods, new or otherwise, for finding a single predictor of a single random variable. Our focus is on the role of certain processes (martingales) in facilitating the prediction of an entire process, or of its value at a time which is not specified in advance.

For the prediction problem, we need only the martingales $M_\lambda(t)$ of Definition 3.6. However, the martingale significance of our classification in the present chapter becomes apparent only when we enlarge the family $M_\lambda(t)$ by including their analogues with the coordinate process $\hat{\rho}(Z_t)$ replaced by $f(Z_t)$ for other functions $f$. Here it would suffice to use $f$ in some dense subset of $C(\bar{M})$ for the Ray compactification $\bar{M}$, but a more natural (non-topological) family is $b(\mathcal{M})$. To be sure, this does not include $\hat{\rho}$ (which may be unbounded), but for purposes of classification we can just as well replace the original processes $X_t$ by $\hat{\varphi}(X_t)$ for a Kuratowski isomorphism $\hat{\varphi}$ (Theorem 1.1) and then $\hat{\rho}$ becomes bounded (see Definition 1.8; note that this is consistent with the more general Definition 2.7 when we specialize to $E = [0, 1]$). In fact, the behaviour of $Z_t$ is relatively immune to replacement of $X_t$ by $\hat{\varphi}(X_t)$, since this merely induces a measurable isomorphism of the state spaces $M$ as in Corollary 1.10, and the definition of $Z_t^z$ is purely measure-theoretic. Our classification is based on the times of discontinuity of $Z_t^z$, which are $P^z$-a.s. the same whether we consider $z$ as a measure on $E$-valued paths or on $\hat{\varphi}(E)(=[0, 1])$-valued paths. More general modifications of the process are deferred to Chapter 6, where it is seen that $Z_t$ has additional stability properties. What is needed below is

**Definition 4.1** The *Kunita–Watanabe martingales* are the family

$$M_\lambda^f(t) = R_\lambda^z f(Z_t) - R_\lambda^z f(Z_0) + \int_0^t f(Z_s) - \lambda R_\lambda^z f(Z_s) \, ds, \qquad 0 < \lambda, f \in b(\mathcal{M}).$$

(4.1)

Since clearly the process $f(Z_t)$ for $P^z$ is in $M_2$, Theorem 3.15(1) says immediately that each $M_\lambda^f$ is a square-integrable martingale additive functional of $Z(t)$ on $M_2$. A basic result of Kunita and Watanabe (1967) and Meyer (1967) now asserts that the family $\{M_\lambda^f\}$ generates, in the sense of the previsible stochastic integral, the class of all square-integrable $(\mathscr{L}_{t+}^0, P^z)$-martingales, for *every* $z \in M$. Note that we use 'square-integrable' to apply for each $t < \infty$, not at $t = \infty$ which would exclude even $M_\lambda^f$. Similarly, the approximation is in the sense of the $L^2$-norm of $P^z$ for each $t < \infty$. Since we are classifying $z$, there is no point in stating the theorem for an arbitrary intial distribution $\mu$ (which, however, is true). In any case, the square-integrable martingales of $(P^z, \mathscr{L}_{t+}^0)$ correspond to the square-integrable martingales of $(P^z, \mathscr{F}_{t+}')$, for any $P^z$ $(=z)$ on the original measurable path space $\Omega'$, and since $z$ is arbitrary it adds nothing at this level to consider an intial distribution $\mu$.

Our limitation to square-integrable martingales may also be avoided. With a suitable reinterpretation of 'generates' the result even applies to all local martingales (see Dellacherie and Meyer (1987, XV, 19)). The extension here is trivial in the continuous case (described in Section 4.2 below), and in the general case it is beyond our scope. But the main innovation in our context should not be overlooked. Namely, starting with an arbitrary $P^z$ on the (measurable) path space $\Omega'$, the theorem reduces the study of all $(P^z, \mathscr{F}_{t+}')$-martingales to that of the special class $\{M_\lambda^f\}$ when we replace $Z(t)$ by its counterpart $Z^z(t)$. Thus $Z^z(t)$ provides a link between the Kunita–Watanabe theorem and an arbitrary almost countably generated probability filtration.

Before proving the theorem of Kunita and Watanabe, we establish a useful auxiliary result about the Kunita–Watanabe martingales.

**Theorem 4.1** (*The resolvent equation for Kunita–Watanabe martingales*). *For fixed $f \in b(\mathscr{M})$,*

$$M_{\lambda_2}^f(t) - M_{\lambda_1}^f(t) = (\lambda_1 - \lambda_2) M_{\lambda_1}^{R_{\lambda_2}^Z f}(t).$$

**Remark** This result also holds for unbounded $f \in \mathscr{M}$ whenever the following proof applies to $|f|$. In particular, it holds for $z \in M_2$ and $f = \hat{\rho}$, $P^z$-a.s. for all $t$.

**Proof** By the usual resolvent equation, we have

$$\lambda_1 R_{\lambda_1}^Z f(Z_s) - \lambda_2 R_{\lambda_2}^Z f(Z_s) = (\lambda_1 - \lambda_2)[R_{\lambda_2}^Z f(Z_s) - \lambda_1 R_{\lambda_1}^Z R_{\lambda_2}^Z f(Z_s)].$$

Substituting this into the expression

$$M_{\lambda_2}^f(t) - M_{\lambda_1}^f(t) = (\lambda_1 - \lambda_2)[R_{\lambda_1}^Z R_{\lambda_2}^Z f(Z_t) - R_{\lambda_1}^Z R_{\lambda_2}^Z f(Z_0)]$$
$$+ \int_0^t \lambda_1 R_{\lambda_1}^Z f(Z_s) - \lambda_2 R_{\lambda_2}^Z f(Z_s) \, ds,$$

the assertion follows.

Effective application of the Kunita–Watanabe martingales depends on using the theory of previsible stochastic integration. At the level which we require, this is very well known and treated in many texts. We shall refer to the encyclopedic work of Dellacherie and Meyer (1987), especially Chapter VIII, but readers will no doubt have their own preferred sources for the results we need. We will treat $\mathcal{L}^0_{t+}$-martingales and therefore our previsible integrands are meant in the sense of $\mathcal{L}^0_{t+}$-previsibility. This partly avoids the usual completions $\mathcal{L}^z_t$ (depending on $z$) and their further extensions (as in Sharpe (1988, VI)). No doubt the completed theory is more general, but it is not required for our purposes. In any case, it is easy to see that every stochastic integral in the completed sense is $P^z$-equivalent, for every $z$, to one in the $\mathcal{L}^0_{t+}$ sense (since this is true for the elementary stochastic integrals) and hence the loss of generality may not be serious.

**Theorem 4.2** (Kunita and Watanabe) *For $z \in M$, and any r.c.l.l. square-integrable martingale $N(t)$ of $(\mathcal{L}^0_{t+}, P^z)$, $N(0) = 0$, there is, for each $\lambda > 0$, a triangular sequence $(M^{f_{n,k}}_\lambda, h_{n,k}; 1 \leq k \leq n)$ of Kunita–Watanabe martingales and $\mathcal{L}^0_{t+}$-previsible processes $h_{n,k}$ such that, for every $t$,*

$$\lim_{n \to \infty} E^z |N(t) - \sum_{k=1}^n \int_0^t h_{n,k}(s) \, dM^{f_{n,k}}_\lambda(s)|^2 = 0.$$

**Proof** If such a sequence exists for a fixed $t$, then by orthogonality of increments the same sequence suffices for any smaller $t$. Hence we obtain a single sequence for all $t$ by choosing the diagonal sequence corresponding to some $t_n \to \infty$, and it suffices to prove the assertion for each $t$. Then if the assertion fails, by the theory of orthogonal decomposition of stochastic integrals (Dellacherie and Meyer (1987), VIII, 51) there would be a non-trivial martingale $N(s)$, $N(0) = 0$, orthogonal on $[0, t]$ to every $M^f_\lambda$. In fact, it is easy to see from Theorem 4.1 that the total class $\{M^f_\lambda\}$ is already obtained for any fixed $\lambda$ so we could have set $\lambda = 1$. But for the present proof it is essential to use $\lambda$ as a variable.

It is enough to show $N(t) = 0$, and to this end we introduce the martingales $J^f_\lambda(t) \doteq \int_0^t e^{-\lambda s} \, dM^f_\lambda(s)$. Then clearly $N(s)$ is orthogonal on $[0, t]$ to all $J^f_\lambda$, and by a familiar integration by parts (Dellacherie and Meyer 1987, VIII, 19) we have

$$J_\lambda(t) = e^{-\lambda t} M^f_\lambda(t) + \lambda \int_0^t e^{-\lambda s} M^f_\lambda(s) \, ds.$$

Now expanding this by routine operations gives

148 | Foundations of the prediction process

$$J_\lambda^f(t) = e^{-\lambda t}(R_\lambda^Z f(Z_t) - R_\lambda^Z f(Z_0)) + e^{-\lambda t}\int_0^t f(Z_s) - \lambda R_\lambda^Z f(Z_s)\,ds$$

$$+ \int_0^t \lambda e^{-\lambda s}(R_\lambda^Z f(Z_s) - R_\lambda^Z f(Z_0))\,ds \qquad (4.2)$$

$$+ \int_0^t \lambda e^{-\lambda s}\int_0^s f(Z_v) - \lambda R_\lambda^Z f(Z_v)\,dv\,ds.$$

Interchanging the order of integration in the last term, it becomes

$$\int_0^t (f(Z_v) - \lambda R_\lambda^Z f(Z_v))(e^{-\lambda v} - e^{-\lambda t})\,dv,$$

and cancelling the second term on the right in (4.2) we are left with

$$J_\lambda^f(t) = e^{-\lambda t}(R_\lambda^Z f(Z_t) - R_\lambda^Z f(Z_0))$$

$$+ \int_0^t e^{-\lambda v} f(Z_v)\,dv - \left(\int_0^t \lambda e^{-\lambda s}\,ds\right) R_\lambda^Z f(Z_0)$$

$$= e^{-\lambda t} R_\lambda^Z f(Z_t) - R_\lambda^Z f(Z_0) + \int_0^t e^{-\lambda v} f(Z_v)\,dv.$$

Therefore, we have by orthogonality

$$0 = E^z(\{N(t) - N(s)\}\{J_\lambda^f(t) - J_\lambda^f(s)\}|\mathscr{L}_{s+}^0)$$
$$= E^z(\{N(t) - N(s)\}\{J_\lambda^f(\infty) - J_\lambda^f(t)\}|\mathscr{L}_{s+}^0),$$

and by addition we get

$$E^z(\{N(t) - N(s)\}\int_s^\infty e^{-\lambda v} f(Z_v)\,dv|\mathscr{L}_{s+}^0)$$
$$= E^z(\{N(t) - N(s)\}\,e^{-\lambda s} R_\lambda^Z f(Z_s)|\mathscr{L}_{s+}^0).$$

The right side is 0 since $N(t)$ is a martingale, so we obtain

$$0 = E^z\left(\int_0^\infty e^{-\lambda u} f(Z_{s+u})\{N(t) - N(s)\}\,du|\mathscr{L}_{s+}^0\right).$$

Now we specialize $f$ to be bounded and continuous, and define the conditional expectation on the right as the integral of regular conditional probability over $\mathscr{L}_\infty^0$. Then we have

$$0 = \int_0^\infty e^{-\lambda u} E^z(f(Z_{s+u})\{N(t) - N(s)\}|\mathscr{L}_{s+}^0)\,du,$$

which may be inverted in $\lambda$ to give

$$0 = E^z(f(Z_{s+u})\{N(t) - N(s)\}|\mathscr{L}^0_{s+}).$$

Now by monotone extensions after separating positive and negative parts of $N(t) - N(s)$, this holds again for $f \in b(\mathscr{M})$, and in particular

$$0 = E^z(f(Z_t)\{N(t) - N(s)\}|\mathscr{L}^0_{s+}). \tag{4.3}$$

With $s = 0$, this is the start of the following induction. For

$$0 < t_1 < \cdots < t_{n+1} = t,$$

and $f_1, \ldots, f_{n+1} \in b(\mathscr{M})$, we have

$$E^z\left(\prod_{k=1}^{n+1} f(Z_{t_k})N(t)|\mathscr{L}^0_{0+}\right)$$

$$= E^z\left(\prod_{k=1}^{n} f_k(Z_{t_k})E^z(f_{n+1}(Z_{t_{n+1}})\{N(t) - N(t_n)\}|\mathscr{L}^0_{t_n+})|\mathscr{L}^0_{0+}\right)$$

$$+ E^z\left(\prod_{k=1}^{n} f_k(Z_{t_k})N(t_n)E^{Z_{t_n}}f_{n+1}(Z(t - t_n))|\mathscr{L}^0_{0+}\right).$$

The first term on the right is 0 by (4.3), and the second is 0 by the induction hypothesis. It follows that $0 = E^z(\prod_{k=1}^{n} f_k(Z_{t_k})N(t))$ for every $n$, and since the products generate $\mathscr{L}^0_t$, and $N(t) \in \mathscr{L}^0_{t+} \equiv \mathscr{L}^0_t$ for $P^z$, we have

$$P^z\{N(t) = 0\} = 1$$

as required.

## 4.2 The continuous prediction processes: intrinsic diffusion

In this section, we consider only $z \in M_c$, where

$$M_c = \{z \in M_E : P^z\{Z_0 = z \text{ and } Z_t \text{ is continuous in } 0 \leq t\} = 1\}.$$

For reasons that will be perhaps clearer later on, we call $M_c$ the *intrinsic diffusions* on $(E, \mathscr{E})$. Note that $E$ must be uncountable in order to have $M_c$ of any interest. But we suppress $E$ in the notation, and our results apply equally to any $E$. It is clear that $M_c \subset D$, and letting $T_\varepsilon = \inf\{t : d(Z_{t-}, Z_t) > \varepsilon\}$ where $d$ is any metric generating the topology of $M$, we have (see also (4.4) ff).

$$\{T_\varepsilon \leq t\} = \bigcap_{\delta > 0} \bigcup_{\substack{r_2 - r_1 < \delta, \\ t + \delta > r_i \in \mathbb{Q}}} \{d\{(Z_{r_2}, Z_{r_1}) > \varepsilon\} \in \mathscr{L}^0_{t+}\}.$$

Thus it follows that

$$M_c = \bigcap_\varepsilon \{z\colon P^z\{T_\varepsilon = \infty\} = 1\} \in \mathcal{M}.$$

We will see below that $M_c$ is a complete Borel demesne in the sense of Definition 2.14. Before turning to this, however, it is important to show that for $z \in M_c$ all martingales of $\mathcal{L}^0_{t+}$ are $P^z$-a.s. continuous. In fact, this last together with $z \in D$ is equivalent to $z \in M_c$.

**Theorem 4.3** $z \in M_c$ is equivalent to $z \in D$ together with the $P^z$-a.s. continuity in $t$ of each (right-continuous) $\mathcal{L}^0_{t+}$-martingale (or only each square-integrable martingale, or even all local martingales).

**Proof** To prove the sufficiency, by Theorem 4.2 and the $L^2$-martingale maximal inequality it is enough to show that $P^z\{M^f_\lambda \text{ is continuous}\} = 1$ for $\lambda > 0$ and all $f \in b(\bar{\mathcal{M}})$. But this follows from continuity ($P^z$-a.s.) of $Z_t$ in the Ray topology (Theorem 2.25(3)) first for $f \in C(\bar{M})$, and then for $f \in b(\bar{\mathcal{M}})$ by the above maximal inequality. Conversely, if all martingales are continuous then so is $R^Z_\lambda f(Z_t)$ for $f \in C(\bar{M})$, since the other term of (4.1) is obviously continuous. By definition this implies that $Z^z_t$ is continuous in the Ray topology. But the Ray topology is stronger than the prediction topology (Definition 2.18), which completes the proof.

**Theorem 4.4** $M_c$ is a complete Borel prediction demesne.

**Proof** We showed earlier that $M_c \in \mathcal{M}$. Therefore $R^Z_\lambda I_{M_c}(z)$ is well defined, and we claim that, for any $\lambda > 0$, $M_c = \{z \in D\colon \lambda R^Z_\lambda I_{M_c}(z) = 1\}$. Indeed, for $z \in M_c$ and each $t$,

$$1 = P^z\{Z_s \text{ is continuous}\}$$
$$\geq P^z\{Z_{t+s} \text{ is continuous in } s\}$$
$$= E^z P^{Z_t}\{Z_s \text{ is continuous}\}.$$

Therefore $P^{Z_t}\{Z_s \text{ is continuous}\} = 1$, $P^z$-a.s., and so $\lambda R^Z_\lambda I_{M_c}(z) = 1$. Conversely, if $\lambda R^Z_\lambda I_{M_c}(z) = 1$, then $P^z\{Z_t \in M_c\} = 1$ for a.e. $t$. But if $P^z\{Z_t \in M_c\} = 1$, then clearly $P^z\{Z_{t+s} \text{ is continuous in } s\} = 1$, and letting $t \to 0$ through a sequence we have $z \in M_c$, as asserted. Therefore, we have $M_c \in \mathcal{M}$, so $M_c$ is Borel.

As we saw in the last proof, $P^z$-a.s. $\lambda R^Z_\lambda I_{M_c}(Z_t)$ is continuous in $t$. Since for $z \in M_c$ it equals 1 for a.e. $t$, it is identically 1, $P^z$-a.s. Hence $P^z\{Z_t \in M_c \text{ for all } t\} = 1$, and since $P^z\{Z_{t-} = Z_t \text{ for all } t\} = 1$, completeness follows immediately.

A classification of measurable processes | 151

From now on, in treating intrinsic diffusions we replace

$$(\Omega', \mathcal{F}'_{t+}, P^z; z \in M_c)$$

by $(\Omega_{M_c}, \mathcal{L}^0_t, P^z; z \in M_c)$, where as in the Remarks to Definition 2.14, $\Omega_{M_c}$ is the space of all continuous, $M_c$-valued paths (for the prediction topology), $\mathcal{L}^0_t$ is the coordinate filtration, and $P^z(S) = P^z(Z^z_{(\cdot)} \in S)$ for $S \in \mathcal{L}^0_\infty$. The next result follows from Theorem 2.24.

**Corollary 4.5** $Z_t$ defines a diffusion process on $(\Omega_{M_c}, \mathcal{L}^0_t, P^z; z \in M_c)$, that is, a Borel right process with continuous paths.

**Proof** Immediate.

As with the other elements of our classification, examples and applications of intrinsic diffusion are deferred to Chapter 6. Here we only note that, in view of Theorem 3.36 *every* real-valued stationary Gaussian process is an intrinsic diffusion.

## 4.3 Classification of discontinuities

Since our classification is formulated in terms of the prediction processes, we work here on the prediction space $(\Omega_Z, \mathcal{L}^0_{t+}, \theta^z_t, P^z)$ of Definition 2.13. As before, a classification of the prediction measures $z \in M_E$ implies a corresponding classification of the measurable processes with state space $E$. We continue to suppress $E$ in the notation.

**Definition 4.2** Let $d(z_1, z_2)$ be a metric generating the topology of $M$, and for $\varepsilon > 0$ let $T^{(1)}_\varepsilon = \inf\{t > 0: d(Z_{t-}, Z_t) \geq \varepsilon\}$, and inductively on $n$ let $T^{(n+1)}_\varepsilon = T^{(1)}_\varepsilon \circ \theta^Z_{T^{(n)}_\varepsilon}$. The product set

$$J = \{(T^{(n)}_\varepsilon(w_z), w_z): T^{(n)}_\varepsilon(w_z) < \infty; 1 \leq n, \varepsilon > 0\}$$

is called the set of *times of discontinuity* of $Z_t$. (Note that $t = 0$ is never a time of discontinuity, even if $P^z$ is a branch point.)

It is easy to see that the times of discontinuity do not depend on the metric $d$. Moreover, for fixed $d$ each $T^{(n)}_\varepsilon$ is a strictly positive stopping time of $\mathcal{L}^0_t$ (and hence also of $\mathcal{L}^0_{t+}$). Indeed,

$$\{T^{(1)}_\varepsilon \leq t\} = \{d(Z_{t-}, Z_t) \geq \varepsilon\}$$

$$\cup \bigcap_n \bigcap_{r \in \mathbb{Q}^+} \left\{\exists r_1 < r_2 < t: d(Z_{r_1}, Z_{r_2}) > \varepsilon - \frac{1}{n}, r_2 - r_1 < r\right\}. \quad (4.4)$$

To see this in detail, note that, given $n$, if $0 < r_2(k) - r_1(k) \to 0$ is a sequence satisfying the last condition, then a subsequence converges to some $t_0 \leq t$, and the absence of discontinuities of the second kind shows that $t_0 < t$ and $d(Z_{t_0-}, Z_{t_0}) \geq \varepsilon - (1/n)$, as required. Now both sets on the right of (4.4) are in $\mathscr{L}_t^0$. The argument for $T_\varepsilon^{(n)}$ can be done similarly, or else by writing $\{T_\varepsilon^{(n+1)} \leq t\} = \bigcup_{r \in \mathbb{Q}} \{T_\varepsilon^{(n)} < r < t; T_\varepsilon^{(1)} \circ \theta_r^Z \leq t - r\}$ and using induction.

Recalling Definition 1.7 for the non-branch points $D$, we have a basic decomposition of the times of discontinuity.

**Definition 4.3** We call the sets

$$J_{D^c} = \{(t, w_z) \in J : Z_{t-} \in D^c\} \quad \text{and} \quad J_D = \{(t, w_z) \in J : Z_{t-} \in D\},$$

respectively, the set of *foreseeable times of discontinuity* and the set of *totally unforeseeable times of discontinuity*.

It follows, as in Definition 4.2, that both $J_{D^c}$ and $J_D$ are given by the graphs of sequences of $\mathscr{L}_t^0$-stopping times restricted to be finite. Thus for the former we can start with

$$T = \inf\{t > 0 : d(Z_{t-}, Z_t) \geq \varepsilon, Z_{t-} \in D^c\},$$

and so forth. However, these are not the only stopping times contained in $J_{D^c}$, and in order to justify the language of foreseeability it is preferable to permit more general stopping times. Let us introduce a convenient definition which really contains little or no novelty.

**Definition 4.4** A $\mathscr{L}_{t+}^0$-stopping time $T$ will be called foreseeable if, for every initial distribution $\mu$ on $\mathscr{M}$, there is a sequence $T_n \leq T$ of $\mathscr{L}_{t+}^0$-stopping times with

$$P^\mu\{T > 0\} = P^\mu\{T_n < T\} = P^\mu\{T_1 < \cdots < T_n \to T \text{ as } n \to \infty\}. \quad (4.5)$$

**Theorem 4.6** *A $\mathscr{L}_{t+}^0$-stopping time $T$ is foreseeable if and only if it is $\mathscr{L}_t^\mu$-previsible for every $\mu$.*

**Remark** The assumption $T_n \leq T$ is really superfluous when (4.5) holds, since we can replace $T_n$ by $T_n \wedge T$. It is an open question whether Definition 4.4 is equivalent to the predictability of the graph $[T, \infty)$ in the sense of Sharpe (1988, I, (5.2), (v)).

**Proof** Now if the conditions holds, then for each $\mu$ define

$$T_n^\mu = \begin{cases} T_n & \text{if } T_1 < T_2 < \cdots < T_k \to T; \\ \dfrac{n-1}{n} T & \text{elsewhere.} \end{cases}$$

Clearly $T_n^\mu \leq T$ is a $\mathscr{L}_t^\mu$-stopping time, and we have $T_1^\mu < \cdots < T_n^\mu \to T$ on $\{T > 0\}$. According to Dellacherie and Meyer (1975, IV, Theorem 71 (a)) it follows that $T$ is $\mathscr{L}_t^\mu$-previsible. Conversely, if $T$ is $\mathscr{L}_t^\mu$-previsible, there is a sequence $T_n^\mu$ having the above properties (Dellacherie and Meyer 1975, IV, Theorem 77 (b)). For each $n$ we can construct an $\mathscr{L}_{t+}^0$-stopping time $T_n$ with $P^\mu\{T_n = T_n^\mu\} = 1$ (by the usual diadic right approximation) and clearly $(T_n, 1 \leq n)$ has the asserted properties.

The main result concerning the foreseeable jumps is

**Theorem 4.7**

(1) *For every $\mu$,*

$$P^\mu\{J_{D^c} = \{(t, w_Z): Z_{t-} \in D^c\}\} = 1.$$

(2) *There is a sequence $T_n$ of foreseeable $\mathscr{L}_{t+}^0$-stopping times such that, for every $\mu$,*

$$P^\mu\left\{J_{D^c} = \bigcup_n (T_n, w_Z)\right\} = 1.$$

**Remarks** Since the set $\{(t, w_Z): Z_{t-} \in D^c\}$ is obviously $P^\mu$-previsible for every $\mu$, the existence of $(T_n)$ for each $\mu$ follows from (1) and a theorem of a the general theory of processes (Dellacherie and Meyer, 1975, Appendix to Chapter IV, 117). However, our construction has the advantage of providing an explicit expression for the $T_n$. For a more sophisticated analysis in the context of right processes, see Sharpe (1988, (44.5)).

**Proof** We remark first that, since $Z_t \in D$ for all $t$ (because of Corollary 2.27) it follows by right-continuity that, for closed $K \subset D^c$, $\{t: Z_{t-} \in K\}$ is *right-isolated*, i.e. it contains no limit points from the right for any $w_Z$. Moreover, $Z_t$ is, with probability 1, r.c.l.l. in the Ray topology (Theorem 2.25(3)) and in the Ray topology (relativized to $M_E$), $D^c$ is easily shown to be an $F_\sigma$-set (Sharpe 1988, Chapter 1, (9.11) (ii)). Let $K_1 \subset K_2 \subset \cdots$ be a sequence of Ray closed sets with $(\cup K_n) \cap M = D^c$. Then, for each $n$, $\{t: Z_{t-} \in K_n\}$ is right isolated, and so its debut $T_n = \inf\{t: Z_{t-} \in K_n\}$ is a strictly positive, $P^\mu$-previsible stopping time for each $\mu$ (Sharpe 1988, (A5.5), (v)). Now the iterates of each $T_n$ have the same properties, but it is not clear that they tend to $\infty$. However, for $\varepsilon > 0$ let

$$K_{n,\varepsilon} = K_n \cap \{z \in D^c: q(0, z, B_\varepsilon^c(z)) \geq \varepsilon\},$$

where $B_\varepsilon(z)$ is the ball of centre $z$, radius $\varepsilon$, in a fixed Ray metric. Since $q(0, z, \{z\}) = 0$ for $z \in D^c$, $\lim_{\varepsilon \to 0} K_{n,\varepsilon} = K_n$. Let us choose $\varepsilon = m^{-1}$, and let $T_{n,m}^{(1)} = \inf\{t: Z_{t-} \in K_{n,m^{-1}}\}$, and $T_{n,m}^{(k+1)} = T_{n,m}^{(k)} + T_{n,m}^{(1)} \circ \theta_{T_{n,m}^{(k)}}^Z$ for all $k \geq 1$. Now $T_{n,m}^{(1)}$ is $P^\mu$-previsible and strictly positive, so there is an announcing sequence $R_k \uparrow T_{n,m}^{(1)}$ satisfying the assumptions of the moderate Markov property (Theorems 1.9(2) and 2.14). It follows that

$$P^\mu\{Z_{T_{n,m}^{(1)}} \in B \mid \mathscr{L}_{T_{n,m}^{(1)}-}^0\} = q(0, Z_{T_{n,m}^{(1)}-}, B), \qquad B \in \mathscr{M},$$

and since $Z_{T_{n,m}^{(1)}-} \in K_{n,m-1} \subset D^c$, we have $(T_{n,m}^{(1)}, w_Z) \in J_{D^c}$, $P^\lambda$-a.s.

It follows by the preservation of stopping times under iteration (Sharpe 1988, Chapter I, Exercise 6.11) that, for every integer $j \geq 1$, $R_k \circ \theta_{T_{n,m}^{(j)}}^Z$ is an announcing sequence for $T_{n,m}^{(j+1)}$, which is therefore foreseeable (by Dellacherie and Meyer (1975), IV, 71). Moreover, by construction we have

$$P^\mu\{Z_{T_{n,m}^{(j)}} \notin B_\varepsilon(Z_{T_{n,m}^{(j)}-}) \mid \mathscr{L}_{T_{n,m}^{(j)}-}^0\} = q(0, Z_{T_{n,m}^{(j)}-}, B_\varepsilon^c(Z_{T_{n,m}^{(j)}-})),$$

which exceeds $\varepsilon$ on $\{T_{n,m}^{(j)} < \infty\}$, from which it is easy to see that $(T_{n,m}^{(j)}, w_Z) \in J_{D^c}$, $P^\mu$-a.s., and $P^\mu\{\lim_{j\to\infty} T_{n,m}^{(j)} = \infty\} = 1$. Hence $\{(T_{n,m}^{(j)}, w_Z); 1 \leq j\}$ comprises all of the times of discontinuity at which $Z_{t-} \in K_{n,m-1}$, with probability 1, and since $\cup_{n,m} K_{n,m-1} = D^c$, the combined sequence $T_{n,m}^{(j)}$ satisfies the assertion (2).

The counterpart of Theorem 4.7 for the totally unforeseeable jumps is easy.

**Theorem 4.8** *For every initial $\mu$, and $\mathscr{L}_{t+}^0$-foreseeable stopping time $T$, $P^\mu\{(T, w_Z) \in J_D\} = 0$.*

**Proof** It suffices to prove the last equality for any $P^\mu$-previsible $T$. Then there exist $\mathscr{L}_t^\mu$-stopping times, $T_n \leq T$, with $T_1 < T_2 < \cdots < T_n \to T$ on $\{T > 0\}$, and so by Theorem 2.14 (via Theorem 1.9(4))

$$P^\mu\{(T, w_Z) \in J \mid \mathscr{L}_{T-}^\mu\} = P^{Z_{T-}}\{Z_T \neq Z_{T-}\} \qquad \text{on } \{T > 0\}.$$

It follows that

$$P^\mu\{(T, w_Z) \in J_D\} = E^\mu\{I_{Z_{T-} \in D} P^{Z_{T-}}\{Z_T \neq Z_{T-}\}\} = 0,$$

as asserted.

Only a few remarks are needed in order to translate the above classification of discontinuities to the prediction process with parameter space $-\infty < t < \infty$, discussed before Theorem 3.33. For matters involving stopping times, the simplest approach is to change the parameter space to $(0, \infty)$ by a fixed homeomorphism, such as $\tau = e^t$. It is easy to see that this maps $\mathscr{L}_t^0 \to \mathscr{L}_{\tau(t)}^0$, $-\infty < t < \infty$, and then all definitions and properties of stopping times $T$ on $[-\infty, \infty)$ can be translated from the corresponding properties on $[0, \infty)$, and $\{T = -\infty\}$ corresponds to $\{T = 0\}$. However, to treat the times of discontinuity, since there is no analogue of $T_\varepsilon^{(1)}$ on $(-\infty, \infty)$, it is necessary to consider the family of prediction processes $\{Z_{t+s}, 0 \leq s\}$, for all $-\infty < t < \infty$, and combine the times of discontinuity of them all to form

J. The definition of the foreseeable and the totally unforeseeable times of discontinuity (Definition 4.3) then extends without change, but their characterization in terms of foreseeable stopping times depended on assertions made 'for every initial $\mu$.' To see how to extend these to $-\infty < t < \infty$, recall that by Theorem 2.29 the distributions of $Z_t$, $t > 0$ for all $P^\mu$ constitute precisely those of $Z_t$, $t > 0$, for all probability entrance laws into $0 < t < \infty$. We therefore *define* the extension of foreseeability to $-\infty < t < \infty$ by replacing 'every initial $\mu$' by 'every entrance law for $q(s, z, B)$ into $-\infty < t < \infty$,' defined in obvious analogy to Definition 2.19. We can then extend Theorem 4.6 simply by applying the change of time $\tau = e^t$ for every entrance law, which preserves stopping times and filtrations (but not the homogeneous transition function).

The same devices serve to extend Theorems 4.7 and 4.8. Thus we prove 4.7(1) for each process $Z_{t+s}$, $s > 0$, where every entrance law on $-\infty < t < \infty$ determines an initial $\mu$ for $Z_{t+s}$, namely the marginal distribution of $Z_t$. Similarly, if $T$ is a foreseeable stopping time of $Z_{t+s}$, then $t + T$ is a foreseeable stopping time of $Z_u$, $-\infty < u < \infty$, and, choosing a sequence $t_k \to -\infty$, we can combine the countable collections of Theorem 4.7(2) applied to each $Z_{t_k+s}$, $1 \le k$, to obtain a single countable collection $T_n$ as required. We note that, since the construction on $\{0 < s < \infty\}$ was valid for every initial $\mu$, it is certainly also valid for all marginals of $Z_{t_k}$ obtained from entrance laws on $\{-\infty < t < \infty\}$, although the latter may be a proper subset. The proof of Theorem 4.8, after replacing $P^\mu$ by any entrance law, carries over without change, since, of course, the moderate Markov property of $Z_t$ continues to hold for $-\infty < t < \infty$. (In proving it, we can confine our attention to the sets $\{t_0 < T\}$ for each fixed $t_0$, where it follows as in the discussion before Theorem 3.33.)

Let us examine the connection between Definition 4.3 and the discontinuities of martingales, and in particular those of $M_\lambda^f(t)$ from Definition 3.6 and (4.1). As we see in Theorem 4.3, the discontinuities of $Z_t$, $P^\mu$-a.s. for any initial distribution $\mu$, coincide with those of the entire set $\{M_\lambda^{f_n}(t)\}$ for any $\lambda > 0$ and $(f_n)$ dense in $C(\bar M)$ in the Ray topology. In fact, since $M_\lambda^f(t)$ is uniquely defined there is no need to restrict here to a countable set, or to write $P^\mu$-a.s. The same holds true identically for all $\lambda > 0$ and $f \in C(\bar M)$. However, when we pass to $f \in b(\mathcal{M})$ by appealing to a maximal inequality, an exceptional $P^\mu$-null set might be introduced where $M_\lambda^f(t)$ does not have the necessary left limits.

For any function $h(t)$ which r.c.l.l., we denote the jump of $h$ at time $t$ by $\Delta_t h = h(t) - h(t-)$. Then the jump of $M_\lambda^f$ at $t$, which is of course a random variable equal to 0 unless $(t, w_z) \in J$, is $P^\mu$-a.s. given by the expression

$$\Delta_t M_\lambda^f = R_\lambda^Z f(Z_t) - R_\lambda^Z f(Z_{t-}), \tag{4.6}$$

so there is no distinction here as to whether $t \in J_{D^c}$ or $t \in J_D$. Quite a different

156 | Foundations of the prediction process

picture prevails for the martingales $Q_\lambda^f(t) \doteq (M_\lambda^f(t))^2 - \langle M_\lambda^f \rangle_t$ as the next theorem shows.

**Theorem 4.9** *For any initial $\mu$, and $f \in b(\mathcal{M})$, and any stopping time $0 < T < \infty$ of $\mathcal{L}_{t+}^0$, $P^\mu$-a.s.,*

(1) *on $\{Z_{T-} \in D\}$ we have $\langle M_\lambda^f \rangle_{T-} = \langle M_\lambda^f \rangle_T$ and $\Delta_T Q_\lambda^f = \Delta_T(M_\lambda^f)^2$;*

(2) *on $\{Z_{T-} \in D^c\}$ we have*

$$\Delta_T \langle M_\lambda^f \rangle = \int q(0, Z_{T-}, dz)(R_\lambda^z f(z))^2 - (R_\lambda^z f(Z_{T-}))^2$$

*and*

$$\Delta_T Q_\lambda^f = (R_\lambda^z f(Z_T))^2 - \int q(0, Z_{T-}, dz)(R_\lambda^z f(z))^2.$$

**Proof** We recall first that the (compensated) quadratic variation $\langle M_\lambda^f \rangle_t$ has, or may be defined up to a $P^\mu$-null set by, the property of being a $\mathcal{L}_t^\mu$-previsible increasing process such that $Q_\lambda^f(t)$ is a $P^\mu$-martingale. In particular, this implies that the times of discontinuity of $\langle M_\lambda^f \rangle_t$, defined by analogy with Definition 4.2, are $\mathcal{L}_t^\mu$-previsible. Also, they are contained in $J$ along with the discontinuity times of both $(M_\lambda^f(t))^2$ and $Q_\lambda^f(t)$. Hence they are contained in $J_{D^c}$, so that $\langle M_\lambda^f \rangle_{T-} = \langle M_\lambda^f \rangle_T$ on $\{Z_{T-} \in D\}$, $P^\mu$-a.s. Conversely, we have $\{Z_{T-} \in D^c\} = \cup_n \{T = T_n\}$ with $T_n$ from Theorem 4.7(2). Letting $R_k \uparrow T_n$ be an announcing sequence, we have, by the moderate Markov property,

$$E\{(R_\lambda^z f(Z_T))^2 | \mathcal{L}_{T-}^\mu\} = \int q(0, Z_{T-}, dz)(R_\lambda^z f(z))^2 \quad \text{on } \{Z_{T-} \in D^c\}.$$

We note that since $M_\lambda^f(t)$ is a martingale, we have on $\{Z_{T-} \in D^c\}$ by similar reasoning $E(R_\lambda^z f(Z_T)|\mathcal{L}_{T-}^\mu) = R_\lambda^z f(Z_{T-})$ and unless $R_\lambda^z f(z) = R_\lambda^z f(Z_T)$ for $q(0, Z_{T-}, dz)$-a.e. $z$, the Schwartz inequality gives

$$(R_\lambda^z f(Z_{T-}))^2 < \int q(0, Z_{T-}, dz)(R_\lambda^z f(z))^2.$$

Since we have easily

$$\Delta_T(M_\lambda^f)^2 = \Delta_T(R_\lambda^z f)^2 - 2\Delta_T(R_\lambda f)\int_0^T (f(Z_s) - \lambda R_\lambda^z f(Z_s)) \, ds, \quad (4.7)$$

it follows that on $\{Z_{T-} \in D^c\}$

$$E(\Delta_T(M_\lambda^f)^2 | \mathcal{L}_{T-}^\mu) = \int q(0, Z_{T-}, dz)(R_\lambda^z f(z))^2 - (R_\lambda^z f(Z_{T-}))^2 = \Delta_T \langle M_\lambda^f \rangle,$$

and so

$$\Delta_T Q_\lambda^f = \Delta_T (M_\lambda^f)^2 - \Delta_T \langle M_\lambda^f \rangle = (R_\lambda^Z f(Z_T))^2 - \int q(0, Z_{T-}, dz)(R_\lambda^Z f(z))^2,$$

as asserted.

We now introduce the two prediction demesnes corresponding to processes whose discontinuity times are all of the same type.

**Definition 4.5** Let $M(J_{D^c})$ (resp. $M(J_D)$) consist of all $z \in M$ for which $P^z\{J = J_{D^c}\} = 1$ (resp. all $z \in D$ for which $P^z\{J = J_D\} = 1$).

We call $M(J_{D^c})$ the processes with foreseeable jumps, and $M(J_D)$ the processes with totally unforeseeable jumps (where $z \in D^c$ is considered to introduce a foreseeable jump at $t = 0$). But it must be borne in mind that it is the jump times, not the jumps themselves, which are or are not foreseeable. Indeed, for any foreseeable stopping time $T$ we have

$$P^z(Z_T \in A | \mathscr{L}_{T-}^z) = q(0, Z_{T-}, A),$$

and on $\{(T, w_z) \in J_{D^c}\}$ the jump itself is certainly not foreseeable. On the other hand, if $T$ is an unforeseeable jump time then the jump itself may or may not be determined by the process $\{Z_{t \wedge T}; 0 < t\}$,* but in any case one can hardly call the jumps foreseeable. More picturesque names for $M(J_{D^c})$ and $M(J_D)$ are respectively *intrinsically discrete time jump processes* and *intrinsically Hunt processes*. Indeed, any discrete sequence $X_0, X_1, \ldots$ of random variables on $(E, \mathscr{E})$ determines a unique $z \in J_{D^c}$ corresponding to the process $X_t = X_n$; $n \leq t < n + 1$, $0 \leq n$, which may be thought of as a prototype of the discrete jump processes, while any Hunt process $X_t$ on $E$ (Definition 2.16) determines a $z \in J_D$. Besides this, any $z \in J_D$ obviously defines a Hunt process on $M$, and in fact (by the following theorem) $q(t, z, A)$ on $J_D$ determines a single *Hunt semigroup* of all intrinsically Hunt processes on $(E, \mathscr{E})$, up to equivalence in law.

**Theorem 4.10.** $M(J_{D^c})$ and $M(J_D)$ are complete Borel prediction demesnes (Definition 2.14).

**Proof** We have $M(J_D) = \{z \in D: P^z\{Z_{t-} \in D \text{ for all } t > 0\} = 1\}$, where, as noted in proving Theorem 4.7, $D^c$ is an $F_\sigma$-set in the Ray topology, say $D^c = (\cup K_n) \cap M$. Then, if $T_{K_n}^- = \inf\{t > 0: Z_{t-} \in K_n\}$, we have

$$M(J_D) = \bigcap_n \{z \in D: P^z\{T_{K_n}^- = \infty\} = 1\},$$

---

* Jumps of the former type are called *strict*, and there is an interesting, but not particularly important, theory of corresponding *strict martingales*, for example, the usual compensated Poisson processes. See LeJan (1978).

and to show that $M(J_D) \in \mathcal{M}$ it suffices to show $\{z: P^z\{T^-_{K_n} = \infty\} = 1\} \in \mathcal{M}$. Now let $K_n = \cap_k \mathcal{O}_{n,k}$, where $\mathcal{O}_{n,k}$ are Ray relatively open sets decreasing to $K_n$ as $k \to \infty$, and note that $\{T^-_{\mathcal{O}_{n,k}} \leq t\} = \cup_{\tau \leq t} \{Z_{\tau-} \in \mathcal{O}_{n,k}\} \in \mathcal{L}^0_t$, where the union is over rational $\tau$. Then $\{T^-_{K_n} = \infty\} = \{\lim_{k\to\infty} T^-_{\mathcal{O}_{n,k}} = \infty\}$, because if $\lim_{k\to\infty} T^-_{\mathcal{O}_{n,k}} < \infty$ then there are $t_k < K < \infty$ with $Z_{t_k-} \in \mathcal{O}_{n,k}$, and moreover there is no decreasing subsequence $t_{k_1} > t_{k_2} > \cdots$, since the existence of such would lead to $Z_{\lim_j t_{k_j}} \in K_n \subset D^c$. Therefore, there is a non-decreasing subsequence whose limit $t$ satisfies $Z_{t-} \in K_n$, contrary to $T^-_{K_n} = \infty$. This implies that $\{T^-_{K_n} = \infty\} \in \mathcal{L}^0_\infty$, and so $P^z\{T^-_{K_n} = \infty\} \in \mathcal{M}$, and finally $M(J_D) \in \mathcal{M}$. Finally, for any $\mathcal{L}^0_{t+}$-stopping time $T < \infty$ we have

$$P^z\{Z_{(T+s)-} \in K^c_n \text{ for all } s > 0\} = E^z P^{Z_T}\{T^-_{K_n} = \infty\},$$

and for $z \in M(J_D)$ this is 1 for all $n$. Therefore, $P^{Z_T}\{T^-_{K_n} = \infty\} = 1$ for all $n$, $P^z$-a.s., and so $Z_T \in M_{J_D}$ a.s. The same reasoning applies with $Z_{t-}$ in place of $Z_t$ for any previsible $T < \infty$, and since $I(Z_t \in M(J_D))$ is $\mathcal{L}^0_{t+}$-optional (and previsible with $Z_{t-}$) the proof is completed easily by an appeal to the optional and previsible section theorems (Dellacherie and Meyer 1975, IV, 84).

Turning to $M(J_{D^c})$, in terms of the $\mathcal{L}^0_t$-stopping times $T^{(n)}_\varepsilon$ of Definition 4.2 (the jump times of $Z_t$) we have

$$M(J_{D^c}) = \{z: P^z\{Z_{T^{(n)}_\varepsilon -} \in D^c, \forall n, \varepsilon: T^{(n)}_\varepsilon < \infty\} = 1\}.$$

Since $\{Z_{T^{(n)}_\varepsilon -} \in D^c\} \in \mathcal{L}^0_\infty$, it is clear that $M(J_{D^c}) \in \mathcal{M}$. Also, for $z \in M(J_{D^c})$ we have

$$P^z\{Z_t \in M(J_{D^c})\} = P^z\{P^{Z_t}\{Z_{T^{(n)}_\varepsilon -} \in D^c, \forall n, \varepsilon: T^{(n)}_\varepsilon < \infty\} = 1\}$$
$$\geq P^z\{Z_{T^{(n)}_\varepsilon -} \in D^c, \forall n, \varepsilon: T^{(n)}_\varepsilon < \infty\}$$
$$= 1,$$

and the same equations hold with optional $T < \infty$ in place of $t$, or with previsible $T < \infty$ and $Z_{T-}$ in place of $Z_T$. Therefore, $M(J_{D^c})$ is a complete demesne, as asserted.

Before continuing our classification of processes, it is to be mentioned that we do not attempt a separation of $M$ into disjoint demesnes. This would involve finding a smallest demesne containing a given $z \in M$, but in general such an object does not exist, even for a Brownian motion on $\mathbb{R}^2$. Instead, we are concerned with classifying the pure types having greatest intrinsic interest. Since, of course, a finite (or countable) intersection or union of demesnes is again a demesne (and completeness is also preserved) one can easily obtain a largest demesne contained in a countable collection of demesnes, or a least one containing them. However, the complement of a demesne is not in general a demesne, so one does not obtain a disjoint classification. For example, it is easy to see that $M(J_{D^c}) \cap M(J_D) = M_c$ (and formally this would even suffice as a definition) but $M(J_D) \cap M^c_c$ is not a demesne because it is easy to construct elements $z \in M(J_D)$ which are not in $M_c$ but which, under $P^z$, evolve into elements of $M_c$ at a time $t_0 > 0$.

The antithesis of $M_c$ is not $M_c^c$, but rather the set of $z$ for which all $P^z$ martingales of $\mathscr{L}_{t+}^o$ are *purely discontinuous*. This is a somewhat technical concept which is not needed for the definition, as we may use the following instead.

**Definition 4.6** Let

$M_d = \{z \in M : \text{there are no } P^z\text{-continuous martingales of } \mathscr{L}_{t+}^o$

(other than those with constant paths)$\}$.

We call $M_d$ the *intrinsic random jump* processes (for want of a better name).

**Theorem 4.11**

(1) *$M_d$ is a complete Borel prediction demesne.*

(2) *We have $z \in M_d$ if and only if*

$$E^z(M_\lambda^f(t))^2 = E^z \sum_{s \leq t} \Delta_s(M_\lambda^f)^2, \qquad (4.8)$$

*for all $f \in b(\mathscr{M})$, $\lambda > 0$, and $t > 0$ (the meaning of $\Delta_s$ is the same as in (4.7)).*

**Proof** We show first that if (2) is false then there is a non-constant continuous $\mathscr{L}_{t+}^o$-martingale. This fact invokes the general decomposition theory of square-integrable martingales, and we shall only quote the references. According to Dellacherie and Meyer (1980, VIII, (43)) every such martingale admits a unique decomposition into a continuous martingale and a 'compensated sum of jumps', for which the square satisfies the equality analogous to (4.8), and these components are orthogonal. Thus if (4.8) fails there is a continuous component $M_c$ with

$$E^z M_c^2(t) = E^z(M_\lambda^f(t))^2 - E^z \sum_{s \leq t} \Delta_s(M_\lambda^f)^2 > 0,$$

and since $M_c(0) = 0$ it does not have constant paths. Conversely, if (4.8) holds, then it also holds for any $M(t) = \sum_{k=1}^n c_k M_{\lambda_k}^{f_k}(t)$. Indeed, by the resolvent equation for Kunita–Watanabe martingales (Theorem 4.1), we can reduce each term in $M(t)$ to the form $c_k M_{\lambda_1}^{g_k}(t)$; $g_k = f_k + (\lambda_1 - \lambda_k) R_{\lambda_k}^z f_k$. Since $M_\lambda^f(t)$ is linear in $f$ for fixed $\lambda$, the expression for $M(t)$ becomes $M(t) = M_{\lambda_1}^g(t)$; $g = \sum_{k=1}^n c_k g_k$, and (4.8) applies by hypothesis. According to Dellacheric and Meyer (1980, VIII, (42)), (4.8) means that $M_\lambda^f$ is a compensated sum of jumps, and the family of all such comprises a stable subspace of square-integrable martingales. Therefore, (4.8) remains true for the stable subspace generated by $\{M_\lambda^f; \lambda > 0, f \in b(\mathscr{M})\}$, which, according to the theorem of Kunita and

Watanabe used above (Theorem 4.2), is all square-integrable martingales. Then if there were a non-constant continuous martingale, we could, by subtracting the initial value and stopping it at a suitable passage time, obtain a non-constant square-integrable continuous martingale starting at 0. For this, (4.3) obviously fails, and we have the desired contradiction proving (2).

Assertion (1) is an easy consequence of (2). Indeed, for $0 \leq t$, $0 < u$,

$$E^z\left(\sum_{t<s\leq t+u} \Delta_s(M_\lambda^f)^2 | \mathscr{Z}_{t+}^o\right) = E^{Z_t} \sum_{s\leq u} \Delta_s(M_\lambda^f)^2, \quad (4.9)$$

and an analogous equality holds for $(M_\lambda^f(t+u) - M_\lambda^f(t))^2$. Since $M_\lambda^f$ has orthogonal increments, it is now easy to see, since (4.8) with $\geq$ always holds by Fatou's lemma, that (4.8) at $t$ and $t+u$ implies (4.8) for $Z_t$, $P^z$-a.s. Since (4.8) for all $\lambda$ and $f$ can be deduced using a countable subcollection not depending on $z$ (for example $\{f_n\}$ dense in $C(\bar{M})$, noting again that the subspace is stable for each $P^z$) it follows immediately that $M_d \in \mathscr{M}$, and that $P^z\{Z_t \in M_d\} = 1$ for $z \in M_d$. The same argument applied at a bounded stopping time $T = \inf\{t: Z_t \notin M_d\} \wedge K$ now shows that $M_d$ is demesne (since both sides in (4.9) are right-continuous), and finally we can replace $T$ by $\inf\{t: Z_{t-} \notin M_d\} \wedge K$, and $\mathscr{Z}_{T+}^o$ by $\mathscr{Z}_{T-}^o$, to show the completeness.

Let us call the complete Borel demesnes $M_d \cap M(J_{D^c})$ and $M_d \cap M(J_D)$ respectively the *foreseeable jump processes*, and the *totally unforeseeable jump processes*, (so that only a change of order and emphasis distinguishes these appellations from those of $M(J_{D^c})$ and $M(J_D)$). These pure jump demesnes now have various subdemesnes, of which we will be content to examine only one, which has proved amenable to analysis precisely because it is defined to make transfinite induction apply to the jumps.

Recalling again the jump times $T_\varepsilon^{(n)}$, we note that the time of first jump $T_0 = \inf_n T_{1/n}^{(1)}$ is also a $\mathscr{Z}_{t+}^o$-stopping time, although it may not be a jump time. However, the set $\Omega_0 = \{w_z: T_0 > 0\}$ is in $\mathscr{Z}_\infty^o$, where we permit $T_0 = \infty$. We define inductively

$$T_1 = T_0 + T_0 \circ \theta_{T_0}^Z, \ldots, T_{n+1} = T_n + T_0 \circ \theta_{T_n}^Z, \ldots.$$

It is easily noted that each $T_n$ is an $\mathscr{Z}_{t+}^o$-stopping time, and hence we have $\Omega_n \doteq \{0 < T_0 < T_1 < \cdots < T_n\} \in \mathscr{Z}_\infty^o$, where we adopt here and in the rest of this section the convention that $\infty < \infty$ holds. Suppose, now, that we have defined $\mathscr{Z}_{t+}^o$-stopping times $T_0, T_1, \ldots, T_\alpha$ and sets

$$\Omega_\alpha = \{0 < T_0 < \cdots < T_\alpha\} \in \mathscr{Z}_\infty^o$$

for a countable ordinal $\alpha$. Then we define $T_{\alpha+1} = T_\alpha + T_0 \circ \theta_{T_\alpha}^Z$. Noting that

$$\Omega_{\alpha+1} = \{T_\alpha = \infty\} \cup (\{T_\alpha < \infty\} \cap \Omega_\alpha \cap \{0 < T_0 \circ \theta_{T_\alpha}^Z\}),$$

it is clear that $\Omega_{\alpha+1} \in \mathscr{Z}_\infty^o$ and $T_{\alpha+1}$ is a $\mathscr{Z}_{t+}^o$-stopping time. On the other hand, if $\alpha$ is a limit ordinal we assume that $\Omega_\beta = \{0 < T_0 < \cdots < T_\beta\}$ is defined for each ordinal $\beta < \alpha$, and satisfies the same two properties. Then let $T_\alpha = \lim_{\beta \to \alpha} T_\beta$, and observe that $\Omega_\alpha = (\cup_{\beta < \alpha} \Omega_\beta) \in \mathscr{L}_\infty^\circ$ (where $\alpha$ is countable) and $\{T_\alpha \leq t\} = \cap_{\beta < \alpha} \{T_\beta \leq t\} \in \mathscr{L}_{t+}^\circ$. Hence we have defined $\Omega_\alpha$ and $T_\alpha$ for all countable ordinals $\alpha$ in such a way that on $\Omega_\alpha$ the $T_\beta$ are strictly increasing or infinite for $\beta \leq \alpha$. It follows that, for every $w_z$, either $w_z \notin \Omega_\alpha$ for large countable $\alpha$, or else $T_\alpha = \infty$ for large countable $\alpha$, otherwise, there would be an uncountable set of disjoint intervals $(T_\alpha, T_{\alpha+1})$ contained in $[0, \infty)$. Furthermore, we observe that $T_\alpha = T_{\alpha+1}$ implies $T_\alpha = T_\beta$ for all $\beta > \alpha$.

**Definition 4.7** The set $M_{\text{WOJ}}$ of *well-ordered jump processes* is given by $M_{\text{WOJ}} = \{z: P^z\{\sup_\alpha T_\alpha = \infty\} = 1\}$, where the supremum is over the countable ordinals.

To justify this definition, we first prove the following lemma.

**Lemma 4.12** *For any initial distribution $\mu$ on $\mathscr{M}$, $\sup_\alpha T_\alpha$ is a $\mathscr{L}_t^\mu$-stopping time. Moreover, for each $t$ the function $P^z\{\sup_\alpha T_\alpha \leq t\}$ is universally measurable on $M$.*

**Proof** It obviously suffices to prove the same assertions for $(\sup_\alpha T_\alpha) \wedge c$ $(= \sup_\alpha (T_\alpha \wedge c))$ for each positive $c < \infty$. Now let $\alpha_n(c)$ be an increasing sequence of countable ordinals, depending on $\mu$, such that

$$\lim_{n \to \infty} E^\mu(T_{\alpha_n(c)} \wedge c) = \sup_\alpha E^\mu(T_\alpha \wedge c).$$

Then if $\alpha(\mu, c) = \inf\{\alpha > \alpha_n(c) \text{ for all } n\}$, $\alpha(\mu, c)$ is a countable ordinal and $E^\mu(T_{\alpha(\mu, c)} \wedge c) = \sup_\alpha E^\mu(T_\alpha \wedge c)$. It follows that

$$P^\mu\{T_{\alpha(\mu, c)} \wedge c = T_{\alpha(\mu, c)+1} \wedge c\} = 1,$$

and so $P^\mu\{T_{\alpha(\mu, c)} \wedge c = \sup_\alpha T_\alpha \wedge c\} = 1$. But $T_{\alpha(\mu, c)} \wedge c$ is a $\mathscr{L}_{t+}^\circ$-stopping time, hence easily $\sup_\alpha (T_\alpha \wedge c)$ is a $\mathscr{L}_t^\mu$-stopping time. It also follows that $P^z\{\sup_\alpha (T_\alpha \wedge c) \leq t\} = P^z\{T_{\alpha(\mu, c)} \leq t\}$ for $\mu$-a.e. $z$, and since the right side is $\mathscr{M}$-measurable the left side is universally measurable.

It follows from Lemma 4.12 that $P^z\{\sup_\alpha T_\alpha = \infty\}$ is well defined and universally measurable in $z$. Also, since $\{\sup_\alpha T_\alpha = \infty\} = \{T_\alpha = \infty$ for all sufficiently large (countable) $\alpha\}$, and since we can find a countable ordinal $\alpha(z)$ such that $P^z\{\sup_\alpha (T_\alpha) = T_{\alpha(z)}\} = 1$ (by taking $\alpha(z) > \alpha(z, c)$ for all integers $c$, where $z$ represents the unit $\mu$ at $z$), we see that

$$M_{\text{MOJ}} = \bigcup_\alpha \{z: P^z\{T_\alpha = \infty\} = 1\}.$$

Let us show that $M_{\text{WOJ}} = \bigcap_\alpha \{z: P^z(\Omega_\alpha) = 1\}$. Clearly $P^z(T_\alpha = \infty) = 1$ implies

162 | Foundations of the prediction process

that $P^z(\Omega_\alpha) = 1$, and since $T_\alpha$ are non-decreasing this implies

$$M_{\text{MOJ}} \subset \bigcap_\alpha \{z: P^z(\Omega_\alpha) = 1\}.$$

Conversely, if $P^z\{\sup_\alpha T_\alpha = \infty\} < 1$ then $P^z\{\sup_\alpha T_\alpha \leq t\} > 0$ for some $t$, and so $P^z\{T_{\alpha(z)} = T_{\alpha(z)+1} \leq t\} > 0$. But $\{T_{\alpha(z)} = T_{\alpha(z)+1} < \infty\} \subset \Omega^c_{\alpha(z)+1}$, hence $P^z(\Omega_{\alpha(z)+1}) < 1$ as required.

**Theorem 4.13** $M_{\text{WOJ}}$ *is a complete prediction demesne (see the Remarks to Definition 2.14: $M_{\text{WOJ}}$ is not asserted to be Borel, but since it is universally measurable it is a Radon space in the language of (Sharpe, 1988; A2)).*

**Proof** We have seen that $P^z\{\sup_\alpha T_\alpha = \infty\}$ is well defined and universally measurable in $z$, but to prove the present result seems to require further tools of probabilistic potential theory. Our proof is based on the fact that $1 - P^z\{\sup_\alpha T_\alpha = \infty\}$ is super-mean-valued, and its excessive regularization vanishes on $M_{\text{WOJ}}$.

We begin by noting that, for $t_1 \leq t_2$, $t_1 + T_0 \circ \theta^Z_{t_1} \leq t_2 + T_0 \circ \theta^Z_{t_2}$. From this it follows by repetition that $t_1 + T_1 \circ \theta^Z_{t_1} \leq t_2 + T_1 \circ \theta^Z_{t_2}$, and by an obvious transfinite induction we obtain $t_1 + T_\alpha \circ \theta^Z_{t_1} \leq t_2 + T_\alpha \circ \theta^Z_{t_2}$ for every countable ordinal $\alpha$. In particular, $T_\alpha \leq t + T_\alpha \circ \theta^Z_t$, which implies that $P^z\{T_\alpha = \infty\} \leq P^z\{T_\alpha \circ \theta^Z_t = \infty\}$. By the Markov property, the right side is $E^z P^{Z(t)}\{T_\alpha = \infty\}$, so we see that $1 - P^z\{T_\alpha = \infty\}$ is super-mean-valued (in the language of Sharpe (1988, (4.11))). Now, for any $z$, if $\mu_t$ denotes the $P^z$-distribution of $Z(t)$, then there exists a countable ordinal $\alpha_\infty$ such that both $P^z\{\sup_\alpha T_\alpha = T_{\alpha_\infty}\} = 1$ and $P^{\mu_t}\{\sup_\alpha T_\alpha = T_{\alpha_\infty}\} = 1$. Indeed, we need only take the larger of two ordinals obtained as in the proof of Lemma 4.12. Then we have

$$P^z\left\{\sup_\alpha T_\alpha = \infty\right\} = P^z\{T_{\alpha_\infty} = \infty\}$$

$$\leq P^z\{T_{\alpha_\infty} \circ \theta^Z_t = \infty\}$$

$$= E^z P^{Z(t)}\{T_{\alpha_\infty} = \infty\}$$

$$= E^z P^{Z(t)}\left\{\sup_\alpha T_\alpha = \infty\right\},$$

proving that $1 - P^z\{\sup_\alpha T_\alpha = \infty\}$ is also super-mean-valued. Now letting $P^{*z}\{\sup_\alpha T_\alpha = \infty\} = \lim_{t \to 0+} E^z P^{Z(t)}\{\sup_\alpha T_\alpha = \infty\}$ it is known that $1 - P^{*z}\{\sup_\alpha T_\alpha = \infty\}$ is an excessive function for $Z(t)$.

Next, let us show that $P^{*z}\{\sup_\alpha T_\alpha = \infty\} = 1$ for $z \in M_{\text{WOJ}}$. Indeed, we have $P^z\{T_1 > 0\} = 1$, and since $T_\alpha = t + T_\alpha \circ \theta^Z_t$ holds on $\{t < T_1\}$ it follows that

$P^z\{T_\alpha = \infty\} = \lim_{t \to 0+} P^z\{T_\alpha \circ \theta_t^Z = \infty\}$. Then with $\alpha(z)$ as before we have

$$1 = P^z\left\{\sup_\alpha T_\alpha = \infty\right\}$$
$$= P_z\{T_{\alpha(z)} = \infty\}$$
$$= \lim_{t \to 0+} P^z\{T_{\alpha(z)} \circ \theta_t^Z = \infty\}$$
$$\leq \lim_{t \to 0+} E^z P^{Z(t)}\left\{\sup_\alpha T_\alpha = \infty\right\},$$

as asserted.

Suppose now, for *reductio ad absurdum*, that for some $z \in M_{\text{WOJ}}$,

$$P^z\{\exists t: Z_t \notin M_{\text{WOJ}}\} \neq 0.$$

Since $P^{*z}\{\sup_\alpha T_\alpha = \infty\} \leq P^z\{\sup_\alpha T_\alpha = \infty\}$ we then have

$$0 \neq P^z\left\{\exists t: P^{*Z(t)}\left\{\sup_\alpha T_\alpha = \infty\right\} < 1\right\}.$$

But, according to Theorem (18.5) of Sharpe (1988, II) (generalizing a result of G. A. Hunt), excessive functions are nearly Borel relative to the Ray topology, and we know that $\mathcal{M}$ is also the Borel field of the Ray topology (Theorems 2.25(1) and 2.35). Hence there are Borel functions

$$f(z) \leq P^{*z}\left\{\sup_\alpha T_\alpha = \infty\right\} \leq g(z)$$

with $P^z\{\exists t: f(Z_t) \neq g(Z_t)\} = 0$. Consequently, we would have

$$0 < P^z\{\exists t: f(Z_t) < 1\}.$$

But $f(Z_t)$ is $\mathcal{L}_t^o$-optional, so by Meyer's optional section theorem, there would be an $\mathcal{L}_t^z$-stopping time $T < \infty$ with $P^z\{f(Z_T) < 1\} > 0$, implying that $P^z\{P^{*Z_T}\{\sup_\alpha T_\alpha = \infty\} < 1\} > 0$, and hence $P^z\{Z_T \in M_{\text{WOJ}}\} < 1$. But then there is a countable $\alpha_\infty$ for which, denoting by $\mu(T)$ the $P^z$ distribution of $Z_T$, $P^{\mu(T)}\{\sup_\alpha T_\alpha = T_{\alpha_\infty}\} = 1$ and $P^z\{\sup_\alpha T_\alpha = T_{\alpha_\infty}\} = 1$, and so by the strong Markov property

$$1 > E^z P^{Z(T)}\left\{\sup_\alpha T_\alpha = \infty\right\} = E^z P^{Z(T)}\{T_{\alpha_\infty} = \infty\} = P^z\{T_{\alpha_\infty} \circ \theta_T^Z = \infty\}.$$

However, since $T_{\alpha_\infty} \leq T + T_{\alpha_\infty} \circ \theta_T^Z$, this would contradict $z \in M_{\text{WOJ}}$, thus completing the proof that $M_{\text{WOJ}}$ is a demesne. Finally, to prove completeness we note that if $P^z\{\exists t: f(Z_{t-}) < 1\} > 0$ then, by the previsible section theorem there would be a $\mathcal{L}_t^z$-previsible $T < \infty$ with $P^z\{Z_{T-} \in M_{\text{WOJ}}\} < 1$, and by applying the moderate Markov property we would arrive at a contradiction in the same way as before.

The demesne $M_{\text{WOJ}} \cap M_d$, and its two subdemesnes formed by intersection with $M(J_D)$ and $M(J_{D^c})$, are basic in the study of jump processes. For example, if $E$ is countable it is a consequence of the theory of Markov chains that a standard chain on $E$ without instantaneous states and whose transition matrix derivatives satisfy $\sum_{j \neq i} p'_{ij}(0) = -p'_{ii}(0)$, for $i \in E$, defines an element $z \in M_{\text{WOJ}} \cap M_d \cap M(J_D)$. However, this demesne contains much more than these because it does not presume any Markov property of the underlying measurable process. Nevertheless, the behaviour of such processes may be analysed in terms of their prediction processes, which are of a type appearing in the literature. For example, it is evident that, after discarding a $P^z$-null set, they have *right-deterministic germ fields* in the sense of Knight (1972), namely, $\mathscr{L}^\circ_t = \mathscr{L}^\circ_{t+}$ for all $t$. In fact the analysis there applies to $Z(t)$ precisely for $z \in M_{\text{WOJ}} \cap M_d \cap M(J_D)$. This, in effect, gives a concrete construction of the 'Lévy systems' of such processes, which are elsewhere obtained only in more abstract terms (for example Sharpe (1988, Section 73)). In a similar vein, the conditions under which $Z(t)$ follows a Markov chain with isolated jumps are studied in Goswami (1990).

On the other hand, $M_{\text{WOJ}} \cap M_d \cap M(J_{D^c})$ contains the discrete parameter jump processes $X_t$ with jumps at integer $t$ (i.e. partial sums of sequences of $E$-valued random variables). But some of the jumps, namely those at time $n$ which are measurable over $\mathscr{L}^\circ_{n-}$, become continuity points of $Z(t)$. In general, of course, the jumps need not be at integer times, or even restricted to any countable set of fixed times, and there may not be any natural reduction of the process to a sequence of random variables.

The processes corresponding to $M_{\text{WOJ}} \cap M_d$ as a whole are studied, for example in Lepingle *et al.* (1981). From (2.2) of that work it follows that, after we discard the $P^z$-null set $\{\sup_\alpha T_\alpha < \infty\}$, for any $\mathscr{L}^\circ_{t+}$-stopping time $T$, $\mathscr{L}^\circ_{T+} \cap \{T_\alpha \leq T < T_{\alpha+1}\} = \mathscr{L}^\circ_{T_{\alpha+}} \cap \{T_\alpha \leq T < T_{\alpha+1}\}$. From this it is easy to show (as was done in Knight (1986b, pp. 3, 4)) that the filtration $\mathscr{L}^z_t$ is generated up to $P^z$-null sets by the *step process*

$$W_t = Z_{T_\alpha} \quad \text{on } \{T_\alpha \leq t < T_{\alpha+1}\} \quad \text{all } \alpha.$$

Let us remark, finally, that any of these demesnes may also be used in the case $-\infty < t < \infty$. The meaning in this case is that it is equivalent to require that $Z_t$ fall in the demesne for all $t$, $P$-a.s., or to require it only along a sequence $t_n \to -\infty$. We shall return to our classification briefly in the Appendix to Chapter 6, to point out that it also gives a classification of probability filtrations, without the specification of a particular generating process.

# 5
# Prediction of measurable processes

> If this is so, then time is also twofold. There is a time for heaven and one for earth. The one remains and at the same time proceeds; the other is borne along in motion...
>
> Proclus, 5th Century A.D., In Park, D. (1980). The Image of Eternity. University of Massachusetts Press, Amherst.

## 5.1 Introduction

The prediction theory of Gaussian processes was dealt with somewhat obliquely in the last half of Chapter 3, beginning with Theorem 3.18 (which was only proved in the Appendix). The major point for applications is actually that the martingales $M_\lambda(t)$, which are jointly Gaussian in the Gaussian case, suffice for a linear representation of the given process. (Theorem 3.26, where the $(Y_n)$ sequence is an orthogonalization of $(M_n)$.) Together with the fact that both $(M_n)$ and $(Y_n)$ tend to be quite tractable processes, and usually we have $\Phi_z(t) = 1$ so that $n = 1$ suffices, the representation (3.6) for $z \in D \cap M_{G(c)}$ provides a useful access to the prediction of $X_z^c(t)$. Indeed, we have immediately

$$E(X_z^c(t) \mid \mathscr{H}_u^o) = \sum_{n=1}^{\Phi_z(t)} \int_0^u F_n(t,s)\,dY_n(s), \qquad u < t. \tag{5.1}$$

Moreover, by definition of the multiplicity $\Phi_z(t)$ in Theorem 3.23, together with our particular choice of $Y_n(t)$ in its proof ((3.5)ff.), it is clear that $Y_n(u) = 0$ for $\Phi_z(u) < n \le \Phi_z(t)$, and so the summation in (5.1) reduces to $n \le \Phi_z(u)$.

In the present chapter we are concerned with extending this method of prediction in the non-Gaussian case, by investigating its strict-sense analogues. Let us begin, therefore, by quoting from Doob (1953, II, Section 3, p. 77) the distinction between a wide-sense and a strict-sense concept.

'Many concepts which are used in the theory of stochastic processes can be formulated in two ways, in a "strict sense" or in a "wide sense." The general principle is the following: suppose a $y_t$ process has a certain property $P$ expressed in terms of variances and covariances. Suppose the corresponding (mean 0) Gaussian process ... has the corresponding but stronger property $P'$. Then $P'$ is called a strict-sense property and $P$ a wide-sense property.'

166 | Foundations of the prediction process

It will be noted that no assertion has been made that $P$ determines $P'$ uniquely, even for given $y_t$. In fact, we will see below that, as a property of Gaussian processes, (5.1) has several different strict-sense interpretations. Moreover, there is an implicit assumption in using this distinction that the property $P'$ covers a reasonably large class of processes $y_t$. In our case, the aim is to cover all processes $y_t$ corresponding to $z \in M_2$. The restriction to $M_2$ is inevitable when we are concerned with prediction in the least squares sense. If it is not fulfilled by a process $y_t$ due to the non-existence of second moments, one could instead predict $f(y_t)$ for some appropriate function $f$. On the other hand, if $y_t$ is merely a measurable process, our theory did not in general apply to $y_t$ itself, but only to some coordinate process $X'_t$ definable in terms of integrals of $y_t$. This is the reason that Theorem 3.26 is limited to $z \in M_{G(c)}$ (continuous covariance) which makes possible the definition of $y_t$ in terms of $X'_t$, $P^z$-a.s. In the non-Gaussian case, continuous covariance is a rather strong requirement. Our strict-sense theory goes through with the weaker restriction that the covariance is either right-continuous or left-continuous at each $t$ and $L_z^2$-bounded on compact sets, and there are even weaker obvious possibilities (existence a.e. of right or left limits, for example). Since we do not wish to overburden the notation with generalities which are not likely to be useful, we limit ourselves to a simpler definition.

**Definition 5.1** Let $M_2^+$ denote

$\{z \in M_2$: there exists a covariance structure $(E(t), \Gamma_z(s, t))$, as in the Assumption of Section 3.1 and Theorem 3.2, such that the covariance $\Gamma_z$ is right-continuous and bounded on compact sets$\}$.

**Remark** Right-continuity in both variables follows from

$$\lim_{\varepsilon \to 0+} \Gamma(t + \varepsilon, t + \varepsilon) = \Gamma(t, t) \quad \text{and} \quad \lim_{\varepsilon \to 0+} \Gamma(t + \varepsilon, t) = \Gamma(t, t) \quad \text{for all } t.$$

Indeed, if $y_t$ denotes a corresponding process with mean 0, then by Schwartz's inequality

$$|\Gamma(s, t + \varepsilon) - \Gamma(s, t)| = |E^z(y_s(y_{t+\varepsilon} - y_t))|$$
$$\leq \Gamma^{\frac{1}{2}}(s, s)\{\Gamma(t + \varepsilon, t + \varepsilon) + \Gamma(t, t) - 2\Gamma(t + \varepsilon, t)\}^{\frac{1}{2}},$$

which tends to 0 along with $\varepsilon$.

It is rather easy to see, generalizing Example 1 of Chapter 3, that $M_2^+$ is not a prediction demesne. However, our prediction theory (Theorem 3.23 ff.) was carried out for each $z$ separately, not in a manner free of $z$, and this will continue to be the case in the present chapter. Hence there is no need for restricting $z$ to a particular demesne. The random process that we wish to predict is defined, for each $z$, just as in Definition 3.7, as follows.

**Definition 5.2**  For $z \in M_2^+$, let

$$X_z^+(t) = L_z^2\text{-}\lim_{\varepsilon \to 0+} \varepsilon^{-1} \int_t^{t+\varepsilon} (\hat{\rho}(Z(s)) - E_z(s))\,ds.$$

Thus $X_z^+(t)$ is defined up to a $P^z$-null set for each $t$, and it is easy to see that $E^z X_z^+(t) = 0$, $E^z(X_z^+(s) X_z^+(t)) = \Gamma_z(s, t)$, and in particular $\Gamma_z$ is uniquely determined by $z \in M_2^+$. Moreover, since $X_z^+$ is right-continuous and locally bound in the $L^2$-sense, its integrals exist in the $L^2(P^z)$-Riemann sense. Indeed $X_z^+(t)$ is $L^2$-continuous except for a countable set of $t$, and

$$P^z\left\{ \int_0^t X_z^+(s)\,ds = \int_0^t (\hat{\rho}(Z_s) - E_z(s))\,ds \right\} = 1 \qquad \text{for every } t.$$

Therefore, the next lemma is obvious.

**Lemma 5.1**  For $z \in M_2^+$, the generated Hilbert spaces $H_z(t)$ of Notation 3.5 become $H_z(t) = L^2\{X_s^+,\ s < t\}$. In particular, $X_z^+(t) \in H_z(t+)$.

Let us discuss next what to look for in a strict-sense interpretation of (5.1). In the Gaussian case, the processes $Y_n$ are independent with independent increments, but that is too much to require for a strict-sense interpretation which is to hold with much generality. Indeed, we could hope to represent processes $X_z^+(t)$ with continuous path functions, but then the $Y_n$ would also need to be continuous. If they had independent increments they would necessarily be Gaussian, and we would be no further than Chapter 3. An intermediate requirement between orthogonality and independence of increments is that $E^z(Y_n(t) - Y_n(u) \mid \mathscr{Y}_{u+}^\circ) = 0$ for $t > u$, which is simply that $Y_n(t)$ be a martingale. In that case, we can follow the construction in Chapter 3 without any change, because the $M_\lambda$ are martingales (Theorem 3.15(1)), and the $Y_n$ are linear combinations of the $M_n$. Hence the $Y_n$ will again be martingales, but they are only orthogonal rather than independent. This does not interfere with (5.1), which depended only on the linearity of the representation of $X_z^c$. The main open question is therefore whether we continue to have such a representation of $X_z^+$ in the non-Gaussian case, at least in sufficient generality. A check of the proof of Theorem 3.18 shows that, although the family $(M_n)$ continues to generate the filtration $\mathscr{Y}_t^\circ$ in the non-Gaussian case, the proof that it generates $H_z(t)$ in the linear sense breaks down. However, a main result (Theorem 5.10) of the present chapter is that it still generates, for any $z \in M_2^+ \cap D$, a Hilbert space containing $H_z(t)$, so the representation (5.1) continues to apply. At the same time, since the proof does not depend on the appendix to Chapter 3, it provides a new proof of the generation of $\mathscr{Y}_{t-}^\circ$ by $\{M_n(s); s < t, 1 < n\}$ in view of the fact that $\mathscr{Y}_{t-}^\circ$ is generated (up to $P^z$-null sets) by $\{X_s^+, s < t\}$. This gives a new basis for the prediction theory of Chapter 3, but it does not seem to yield the extra

168 | Foundations of the prediction process

dividend of the appendix that identifies the filtrations even for $z \in M_1$. On the other hand, since in the non-Gaussian case we do not have $M_n(t) \in H_z(t)$ (or even $\in H_z(t+)$) it is then clear that the generated Hilbert spaces are in general strictly larger than $H_z(t)$. Thus the linear representation by $\{M_n\}$ (or by $\{Y_n\}$) is more general than in the Gaussian case but it also requires more general components. We will call this representation (5.1), with $X_z^+$ for $X_z^c$, the *linear martingale representation* of $X_z^+$ (of course, $X_z^+ \equiv X_z^c$ if $\Gamma$ is continuous). Note that it provides the strict-sense prediction of $X_z^+$.

The linear martingale representation of $X_z^+$ suggests an alternative wide-sense representation of $X_z^+$ which is, in general, not the same as that of Corollary 3.29. (The wide-sense predictions, i.e. projections onto $H_z(u)$ for $u < t$, are the same in any case, but the method of expressing them may be different.) We therefore introduce some notation.

**Notation 5.3** For fixed $z \in M_2$, we set

$$\mathscr{E}^z(M_\lambda(t) \mid H_z(t+)) = \hat{M}_\lambda(t),$$

$$\mathscr{E}^z(Y_n(t) \mid H_z(t+)) = \hat{Y}_n(t)$$

(for $\mathscr{E}$ see Notation 0.3).

In general, $\hat{M}_\lambda \neq \check{M}_\lambda$ and $\hat{Y}_n \neq \check{Y}_n$ (they are, of course, equal in the Gaussian case $z \in M_G$). Indeed, the $(\hat{Y}_n)$ need not be orthogonal as $n$ varies. However, we do have

**Theorem 5.2** *The processes $\hat{M}_\lambda$ and $\hat{Y}_n$ have orthogonal increments.*

**Remark** If the $L^2$-processes $\hat{Y}_n$ are not orthogonal, they can of course be orthogonalized by the same procedure as the $Y_n$ were from $(M_n)$, but we do not know if they suffice for a linear representation of $X_z^+(t)$. At least, the following proof shows that the increments $\hat{M}_\lambda(t+s) - \hat{M}_\lambda(t)$ are orthogonal to $H_z(t+)$, hence also to $\hat{M}_\mu(t)$ for all $\mu$, and the analogue for $\hat{Y}_n(t+s) - \hat{Y}_n(t)$ of course follows as well.

**Proof** We have

$$\mathscr{E}^z(\hat{M}_\lambda(t+s) - \hat{M}_\lambda(t) \mid H_z(t+))$$
$$= \mathscr{E}^z(\mathscr{E}^z(M_\lambda(t+s) \mid H_z(t+s)+) \mid H_z(t+)) - \mathscr{E}^z(M_\lambda(t) \mid H_z(t+))$$
$$= \mathscr{E}^z(M_\lambda(t+s) \mid H_z(t+)) - \mathscr{E}^z(M_\lambda(t) \mid H_z(t+))$$
$$= \mathscr{E}^z(E^z(M_\lambda(t+s) \mid \mathscr{L}_{t+}^o) \mid H_z(t+)) - \mathscr{E}^z(M_\lambda(t) \mid H_z(t+))$$
$$= 0, \qquad (5.2)$$

where we used $H_z(t+) \subset L^2(\mathscr{L}_{t+}^o, P^z)$ for the last step.

We turn now to a second strict-sense interpretation of (5.1), called the *previsible martingale stochastic integral representation* of $X_z^+$. This arises when we impose on the $Y_n$ a stronger type of orthogonality, but yet do not require complete independence as in the Gaussian case. Namely, we require orthogonality of $Y_m(t+s) - Y_m(t)$ and $Y_n(t+s) - Y(t)$ conditionally on $\mathscr{L}_{t+}^o$, for each $t$, $n \neq m$, and $s > 0$. Then from

$$0 = E((Y_n(t+s) - Y_n(t))(Y_m(t+s) - Y_m(t)) \mid \mathscr{L}_{t+}^o)$$
$$= E(Y_m(t+s)Y_n(t+s) - Y_m(t)Y_n(t) \mid \mathscr{L}_{t+}^o)$$

it follows that $Y_n Y_m$ is a martingale for $n \neq m$, and conversely. This is the well-known definition of martingale orthogonality of Meyer (1966). It is not hard to see that if we refine further the sequence $Y_n$ to obtain this stronger orthogonality, yet wish to preserve a representation (5.1), we must also allow the coefficients of integration to become random processes. Indeed, the prevalent situation is that one has only a single martingale dimension in the sense of stochastic integration, hence

$$X_z^+(t) = \int_0^t h_t(s) \, dY_1(s)$$

for a $\mathscr{L}_t^o$-previsible process $h_t$ with

$$E^z \int_0^t h_t^2(s) \, d\langle Y_1 \rangle_s < \infty.$$

Then by uniqueness of the representation it is impossible (unless $h$ is already non-random) to represent $X_z^+(t)$ in the form (5.1) with additional $Y_n$ such that $Y_n Y_m$ are martingales ($n \neq m$) if the coefficients $F_n$ are non-random.

Accordingly, we need to utilize a theory of stochastic integration to handle this type of orthogonality. Fortunately, the appropriate theory for our purposes already exists in detail, as the *previsible stochastic integral*. We refer to Dellacherie and Meyer (1980, Chapter VIII) for the results we need, as well as for the discussion of priorities, but by now there are, of course, many other comprehensive references. Indeed, we require only the most elementary properties, which are found in any textbook on the subject.

The orthogonalization of $(M_n)$ in the sense of stochastic integration will be carried out analogously to that with constant coefficients, and we obtain a new sequence $(Y_n^\#)$, and a new representation (5.1) with previsible integrands $h_n$. However, the Hilbert space of $\mathscr{L}_t^o$-measurable random variables which are so representable is in general much larger than in the case of constant coefficients. Nevertheless, as examples will show, it still does not equal $L^2(\mathscr{L}_t^o, P^z)$ in general. It would be possible to achieve this by the introduction of additional orthogonal martingales based on $M_\lambda^f$ for $f \neq \hat{\rho}$

(in view of Theorem 4.2) and the general situation was treated by Davis and Varaiya (1974). However, our emphasis is on the given process $X_z^+$ and such an extension would lead us too far afield. Another extension would be to use the 'optional' stochastic integral of Dellacherie and Meyer (1987, Chapter VIII, Section 3), thus permitting the integrands to be $\mathscr{L}_t^o$-optional. This gives nothing new for intrinsic diffusion, while in the case of jumps it seems pointless since the jumps themselves are always $\mathscr{L}_t^o$-optional. We will not mention it further in the present chapter, although it is considered briefly in Section 6.3 in connection with Gaussian processes.

Finally, let us discuss a third strict-sense interpretation of (5.1) of a more specialized type. If the variances $\sigma_n^2(t)$ are continuous in the Gaussian case we can write $Y_n(t) = B_n(\sigma_n(t))$, where $(B_n)$ is a sequence of independent Brownian motions. Now it follows immediately from a familiar theorem (Knight 1971) that this continues to hold in the case of intrinsic diffusion $z \in M_c$ if we use $Y_n^\#$ for $Y_n$ and replace $\sigma_n^2$ by the quadratic variation $\langle Y_n^\# \rangle_t$. Then we can write $Y_n^\#(t) = B_n(\langle Y_n^\# \rangle_t)$, where $B_n$ is a sequence of independent Brownian motions. (We assume for convenience that $\lim_{t \to \infty} \langle Y_n^\# \rangle_t = \infty$, thus evading the need to define $B_n(t)$ for $t > \langle Y_n^\# \rangle_\infty$.) In fact, a similar but more general result holds for certain $z \in M(J_D)$ for which some of the processes $Y_n^\#$ are (modified to become) purely discontinuous with unit jumps, and where we supplement $(B_n)$ by including some compensated Poisson processes $(P_n)$ independent both of $(B_n)$ and of each other. This extension is discussed in Knight (1986a), but it will not concern us further below since the class of processes which it covers seems quite limited. However, the extension becomes very enticing in the (apparently) special case that $\langle Y_n^\# \rangle_t$ are measurable for each $t$ over $\sigma\{B_n, P_n; 1 \le n\}$, or when all of this holds for an appropriate modification $(Z_n)$ of $(Y_n^\#)$ which generate $\mathscr{L}_t^z$. Then, according to Lemma 1.3 of Knight (1986b), each $(\langle Z_n \rangle_t, 1 \le n)$ becomes a vector-valued stopping time of the filtration generated by $(B_n, P_n)$, and we obtain a canonical representation of $\mathscr{L}_t^z$ up to $P^z$-null sets as the $\sigma$-field of the stopped process $(V_n(s \wedge \langle Z_n \rangle_t); 0 < s)$, where $\{V_n\} = \{B_n, P_n\}$ are independent Brownian motions and compensated Poisson processes. Unfortunately, the existence of such a representation, even for intrinsic diffusion with orthogonal stochastic integral dimension 1 (where a single $B_1$ should suffice) is a notorious unsolved problem (see, for example, Stroock and Yor (1980) and Knight (1988)). Despite prolonged effort no such general representation has been achieved, and it seems to be asking too much to expect such a result at present (if it exists at all). Therefore this representation of filtrations problem (up to $P^z$-null sets) is left aside.

On the other hand, we can easily classify probability filtrations according to their types of discontinuity, by simply reinterpreting Chapter 4. This has little bearing on prediction, however, and is deferred to Chapter 6.

## 5.2 Strict-sense prediction of square-integrable processes

The linear martingale representation theory (LMR) will be developed in parallel with the previsible martingale stochastic integral representation theory (PMSIR) discussed above, since they are closely analogous. In both cases, our basic martingales are the $M_\lambda$ of Definition 3.6, but the generated Hilbert spaces are no longer the $H_z(t+)$ of Lemma 5.1. It emerges that in the LMR case the Hilbert spaces can be simply characterized, but in the PMSIR case we must be content with only the definitions.

**Definition 5.4**  For $z \in M_2^+$ (Definition 5.1), let $H_z^*(t)$ denote the $L_z^2$-subspace of $L^2\{\Omega_z, \mathscr{L}_t^z, P^z\}$ generated by $\{E^z(X_z^+(u) \mid \mathscr{L}_v^\circ); 0 \leq u; v \leq t\}$. For $z \in M_2$, let $H_z^\#(t)$ denote the $L_z^2$-subspace generated by

$$\left\{ \int_0^t h(s)\, dM_\lambda(s);\ 0 < \lambda,\ h(s, w) \text{ any } \mathscr{L}_t^z\text{-previsible process such that}\right.$$

$$\left. E^z \int_0^t h^2(s)\, d\langle M_\lambda \rangle_s^z < \infty \right\}.$$

**Remarks**  We adopt from now on the usual right-continuous augmentations $\mathscr{L}_t^z$ of $\mathscr{L}_{t+}^\circ$. This conforms with the literature on stochastic integration (Dellacherie and Meyer 1980, VIII), and the inclusion of all $P^z$-null sets does not increase the Hilbert spaces being represented. The reader who prefers a 'clean' $\mathscr{L}_t^\circ$-measurable theory can obtain it for each $z$ merely by restriction to $\mathscr{L}_t^\circ$-measurable representatives. However, we do not go into the question of whether this can be made free of $z$.

We let $H_z^*$ and $H_z^\#$ denote the $L_z^2$-subspaces generated by $\{H_z^*(t), t < \infty\}$ (respectively, by $\{H_z^\#(t), t < \infty\}$). It is clear that, for $z \in M_2^+$ and any $t \geq 0$, $H_z(t+) \subset H_z^*(t+)$. The inclusion is in general proper. For example, if

$$X_t = \begin{cases} X_1 & t \leq \tfrac{1}{2} \\ X_2 & t > \tfrac{1}{2} \end{cases}$$

for any mean 0, $L^2$-random variables $(X_1, X_2)$, then for the corresponding $z$, $E^z(X_2 | X_1) \in H_z^*(1)$, but of course $H_z(1+) \,(= L^2\{X_1, X_2\})$ does not contain $E^z(X_2 | X_1)$ in general.

To treat the $L^2$-theory of $H_z^*$, the covariance structure $(E(t), \Gamma(s, t))$ does not suffice. Instead, we need the following notation.

**Notation 5.5**  Let

$$\Gamma_z^*(u; v_1, v_2) = E^z[E^z(X_z^+(v_1) \mid \mathscr{L}_u^\circ) E^z(X_z^+(v_2) \mid \mathscr{L}_u^\circ)], \qquad z \in M_2^+.$$

**Remark** There is no need to include different values of $u$, because, for $u_1 < u_2$,

$$E^z[(E^z(X_z^+(v_1) \mid \mathscr{L}_{u_1}^\circ)E^z(X_z^+(v_2) \mid \mathscr{L}_{u_2}^\circ)] = \Gamma^*(u_1; v_1, v_2).$$

We also need a notation corresponding to $(m, n)_z^t$ of Theorem 3.20.

**Notation 5.6** For $z \in M_2^+$, let

$$(m, n)_z^*(t) = \int_0^\infty \int_0^\infty e^{-(ms_1 + ns_2)} \Gamma_z^*(t; t + s_1, t + s_2) \, ds_1 \, ds_2.$$

**Remark** It is easy to see that

$(m, n)_z^*(t)$

$$= E^z(R_m^Z \hat{\rho}(Z_t) R_n^Z \hat{\rho}(Z_t)) - \int_0^\infty \int_0^\infty e^{-(ms_1 + ns_2)} E(t + s_1) E(t + s_2) \, ds_1 \, ds_2.$$

The analogue of Theorem 3.20 is now

**Theorem 5.3** For $z \in M_2^+$,

$$E^z(M_m(t)M_n(t)) = (m, n)_z^*(t) - (m, n)_z^*(0) - (m + n) \int_0^t (m, n)_z^*(u) \, du$$

$$+ \int_0^t \int_0^\infty (e^{-ms} + e^{-ns}) \Gamma_z(u; u + s) \, ds \, du.$$

**Remark** In the last term, we used the fact that $\Gamma_z^*(u; u, v) = \Gamma_z(u, v)$ for (Lebesgue) a.e. $v > u$ to remove the last asterisk. Note, however, that in general we do not have $\hat{\rho}(Z_t) - E(t) = X_z^+(t)$, $P^z$-a.s., for every $t$ (only for a.e. $t$). We note also that if $m = n$ a minor simplification occurs, and one could express the general result by polarization, using the case $(m + n, m + n)$.

**Proof** The proof of Theorem 3.20 applies again here. We have only to note that, in the notation of the Remark following that proof,

$$(m, n)_z^*(t) = E^z\left[\mathscr{E}^z\left(\int_0^\infty e^{-ms} X^+(t + s) \, ds \mid H_z^*(t+)\right)\right.$$

$$\left.\times \mathscr{E}^z\left(\int_0^\infty e^{-ns} X^+(t + s) \, ds \mid H_z^*(t+)\right)\right].$$

In other words, we need only rewrite our conditional expectations as projections on the larger Hilbert space, and the same proof goes through.

For the stochastic integral theory, we need to replace $E^z M_n^2(t)$ by $\langle M_n \rangle_t$ for $P^z$, namely, the previsible quadratic variation.

**Definition 5.7** For $z \in M_2$, $\langle M_\lambda \rangle_t^z$ is the $P^z$-a.s. unique $\mathscr{L}_t^z$-previsible increasing process such that $M_\lambda^2(t) - \langle M_\lambda \rangle_t^z$ is a $P^z$-martingale.

Here the existence and uniqueness follow from the Doob–Meyer decomposition of the submartingale $M_\lambda^2(t)$ (Dellacherie and Meyer 1980, VII, 39). Under further assumptions it may be shown that $\langle M_\lambda \rangle_t^z$ may be defined independently of $z$ to become an additive functional (Dellacherie and Meyer 1980, XV, 16). However, in our case this has no sense because $M_2^+$ is not even a demesne. Consequently in the present chapter, where prediction is done for each $z$ separately, we leave these extensions and perfections aside.

The process $\langle M_\lambda \rangle_t^z$ comprises a sort of stochastic mean square of $M_\lambda$, and it is important to obtain for it a more explicit expression.* The first step (for which the priorities are discussed in Dellacherie and Meyer (1980, VII, 21) is the following theorem.

**Theorem 5.4** *For each $t > 0$,*

$$\langle M_\lambda \rangle_t^z = \lim_{n \to \infty} \sum_{k=0}^{n-1} E^z((M_\lambda(t_{k+1}^n) - M_\lambda(t_k^n))^2 \mid \mathscr{L}_{t_k^n}^z)$$

*in the sense of weak $L^1$-convergence, where $t_k^n = kn^{-1}t$ (or any analogous partition).*

**Proof** With $\mathscr{L}_{t-}^z$ in place of $\mathscr{L}_t^z$ this is (Dellacherie and Meyer 1980, VII, 43 1), which actually suffices since $\mathscr{L}_{t-}^z = \mathscr{L}_t^z$ except for a countable set of $t$ (and we could adjust the partition). In the present form, it was first obtained by Doleans (1967). However, this now requires further work, because the limit was shown to be natural rather than previsible. The extra step can be based on Meyer (1966, VII, T49), and is omitted. We note that Theorem 5.4 also can be used as *definition* of $\langle M_\lambda \rangle_t^z$.

---

* Let us recall here the most basic properties. We define $\langle M, N \rangle_t = \frac{1}{2}(\langle M + N \rangle_t - \langle M \rangle_t - \langle N \rangle_t)$ for any pair of square-integrable martingales. Then it is easy to see that $MN - \langle M, N \rangle$ is a martingale. Granting that $\langle M \rangle$ is the $P$-a.s. unique previsible process with paths of locally bounded variation such that $M^2 - \langle M \rangle$ is a martingale, the same uniqueness follows immediately for $\langle M, N \rangle$. (This more general uniqueness of $\langle M \rangle$ is shown in Dellacherie and Meyer (1980, VI, 80 and VII, 25).) The form $\langle M, N \rangle$ is a sort of previsible stochastic inner product. In particular, it is bilinear with respect to previsible processes as well as scalars in the sense that if $h$ is previsible and $E \int_0^t h^2 \, d\langle M \rangle \, (= E(\int_0^t h \, dM)^2) < \infty$, then we have

$$\left\langle \int_0^s h \, dM, N(s) \right\rangle_t = \int_0^t h \, d\langle M, N \rangle_s, \quad P\text{-a.s..}$$

174 | Foundations of the prediction process

To make further progress toward an explicit expression, extra hypotheses seem to be required. The principal difficulty resides in the fact that $\langle M_\lambda \rangle_t^z$ may not be absolutely continuous in $t$, even for $z \in M_c$. Examples of this situation, however, are rare and complicated, so it now seems worthwhile to consider the simplest one we have found.

**Example 1** We remark first that if the coordinate process $X_t = \hat{\rho}(Z_t)$ is already a martingale, then we have $M_\lambda(t) = \lambda^{-1} X_t$, and $\langle M_\lambda \rangle_t^z = \lambda^{-2} \langle X_t \rangle$. Our example is simply a one-dimensional diffusion of a type due to Feller and McKean (1956). Let $B(t)$ be a Brownian motion, with continuous local time $l(t, x)$, and define $\tau(t)$ by $\int_{-\infty}^{\infty} l(\tau(t), x) m(dx) = t$, where $m(dx)$ is a measure concentrated on the rationals with positive mass on each. Then $B(\tau(t))$ is a diffusion, and letting $X_t = B(\tau(t \wedge T))$, where $T = \inf\{t : |B(\tau(t))| = 1\}$, we have a diffusion which is a square-integrable martingale. Since $B^2(t) - t$ is a martingale, it follows by a routine time-change argument (as, for example, in Knight (1981b)) that $\langle X_t \rangle = \tau(t \wedge T)$. Now the local time of $X_t$ with respect to $m(dx)$ is $l(\tau(t \wedge T), x)$, so that if $\mathbb{Q}$ denotes the rationals,

$$\int_0^t I_\mathbb{Q}(X_s) \, ds = \int_{-1}^1 I_\mathbb{Q}(x) l(\tau(t \wedge T), x) m(dx)$$

$$= \int_{-1}^1 l(\tau(t \wedge T), x) m(dx)$$

$$= t \wedge T.$$

Thus $X_t$ spends all of its time on the rationals $\mathbb{Q}$ (although, since it is continuous, there is a Lebesque-null set of $t$ spent elsewhere). Consequently, the set $N = \{t : B(t) \in \mathbb{Q}\}$, which is a Lebesgue-null set with probability 1 because $P\{B(t) \in \mathbb{Q}\} = 0$ for each $t > 0$, satisfies

$$\int_0^t I_N(\tau(s \wedge T)) \, ds = t \quad \text{for all } t \leq T, P\text{-a.s.}$$

But if $\tau(t \wedge T)$ were absolutely continuous, so that (say)

$$\tau(t \wedge T) = \int_0^t h(s, w) \, ds \quad \text{for all } t \leq T,$$

we would have $d\tau(t \wedge T)/dt = h(t, w)$ for a.e. $t$, and in particular there would be an $\varepsilon > 0$ with $\varepsilon < d\tau(t \wedge T)/dt < \varepsilon^{-1}$ on a set $S$ of positive Lebesgue measure ($|S| > 0$). It would follow that for all Borel sets $B$,

$$\varepsilon |B \cap S| < \left| \int_{B \cap S} d\tau(t \wedge T) \right| < \varepsilon^{-1} |B \cap S|.$$

But, considering the inverse set function $\tau^{-1}(B)$, we also have

$$\int_0^T I_B(\tau(s))\,ds = |\tau^{-1}(B \cap (0, \tau(T)])|,$$

since this is trivially true for intervals $(a, b) = B$, and both sides define measures in $B$. In particular, we have $T = |\tau^{-1}(N \cap (0, \tau(T)])|$, $P$-a.s. Thus, setting $M = \tau^{-1}((0, \tau(T)] - N)$, we have $|M| = 0$, and considering $\tau$ also as a set mapping, $\tau(T) = |\tau(M)| + |N| = |\tau(M)|$. Then

$$\int_S d\tau(t \wedge T) = \int_{S \cap M} d\tau(t \wedge T) < \varepsilon^{-1}|S \cap M| = 0.$$

But this contradicts $0 < \varepsilon|S| < \int_S d\tau(t \wedge T)$, and completes the proof that $\langle X \rangle_t$ is not absolutely continuous, concluding Example 1.

On the other hand, if $\langle M_\lambda^f \rangle_t^z$ is absolutely continuous for all $z$ in some appropriate demesne, for a measurable $f$ with $R_\lambda^z|f|$ bounded, then it is known (Dellacherie and Meyer 1987, XV, Theorem 22) that not only $R_\lambda^z f$, but also $(R_\lambda^z f)^2$, is in the domain of the *extended infinitesimal operator* $A$, and

$$\langle M_\lambda^f \rangle_t^z = \int_0^t A(R_\lambda^z f)^2(Z_s) - 2(R_\lambda^z f)A(R_\lambda^z f)(Z_s)\,ds. \tag{5.3}$$

The integrand here is a function of $Z_s$ and is called the square field operator (*carré du champs*). This result does not quite serve our purposes, however, for three reasons: first, we want to express $\langle M_\lambda \rangle_t^z$, which involves $f = \hat{\rho}$ for which $R_\lambda^z|\hat{\rho}|$ may be unbounded; second, we are only concerned with a single $z$, so that the absolute continuity hypothesis for all $z$ in a demesne seems too restrictive; and third (which is the main reason) the above expression is complemented by the one which we obtain, which reduces to it when the extra hypotheses are met and seems to give further insight.

**Notation 5.8** For $z \in M_2^+$, let

$$V_\lambda(z) = E^z\left(\int_0^\infty e^{-\lambda s}\hat{\rho}(Z_s)\,ds\right)^2 - (R_\lambda^z\hat{\rho}(z))^2.$$

Thus, $V_\lambda(z)$ is the variance of $\int_0^\infty e^{-\lambda s}\hat{\rho}(Z_s)\,ds$ for $P^z$. Instead of the absolute continuity hypothesis, we need to assume that $V_\lambda$ is in the range of $R_{2\lambda}^z$ along the paths of $Z_t$. (Generally speaking, this is the same as its being in the domain of the infinitesimal generator, but when $V_\lambda$ is unbounded that becomes undefined.)

The following theorem is tedious, and not required for the sequel.

# 176 | Foundations of the prediction process

**Theorem 5.5** *For $z \in M_2^+$, suppose that there is an $\mathscr{M}$-measurable $f_\lambda$ with $R_{2\lambda}^z|f_\lambda|(z) < \infty$ and such that*

(1) $P^z\{V_\lambda(Z_t) = R_{2\lambda}^Z f_\lambda(Z_t) \text{ for all } t\} = 1$;

(2) $f_\lambda(Z_t)$ *is r.c.l.l. in quadratic mean for $P^z$ (at $t = 0$ it need only be $L^2$-bounded).*

*Then we have*

$$\langle M_\lambda \rangle_t^z = \int_0^t f_\lambda(Z_s) \, ds \quad \text{for all } t, \, P^z\text{-a.s.} \tag{5.4}$$

**Remarks** Intuitively, since $(R_{2\lambda}^Z)^{(-1)} = 2\lambda - A$, we see that

$$-e^{-2\lambda t} \frac{d}{dt} \langle M_\lambda \rangle_t^z$$

is the expected time rate of change of $e^{-2\lambda t} V_\lambda$ along paths of $Z_t$. Assumption (2) is *ad hoc* to some extent, and we do not know even whether $E^z(f_\lambda(Z_t))^2 < \infty$ (which is part of the assumption) is really natural. It seems clear, however, that this type of restriction is considerably less stringent than assuming $f_\lambda(Z_t)$ to be r.c.l.l. in the $P^z$-a.s. sense. The assumption (2) itself can probably be relaxed by using the approximate generators of Meyer (1966, VII, T29) in place of Theorem 5.4 in our proof, but we leave this aside. Let us work out the result in the simplest non-trivial case—that of $\hat{\rho}(Z_t) = B_t$, a standard Brownian motion. Then

$$E^x\left(\int_0^\infty e^{-\lambda s} B(s) \, ds\right)^2 = \left(\frac{x}{\lambda}\right)^2 + E^x\left(\lambda^{-1} \int_0^\infty e^{-\lambda s} \, dB(s)\right)^2 = \left(\frac{x}{\lambda}\right)^2 + (2\lambda^3)^{-1},$$

and so $V_\lambda(z) = (2\lambda^3)^{-1}$. This is free of $z$, and we have easily $f_\lambda = \lambda^{-2}$, in agreement with $\langle M_\lambda \rangle_t^z = \lambda^{-2} \langle B \rangle_t = \lambda^{-2} t$.

**Proof** The first step is to reduce the approximation of Theorem 5.4 to one involving $(R_\lambda^Z \hat{\rho}(Z_t))^2$ instead of $(M_\lambda(t))^2$.

**Lemma 5.6** *For any $P^z$-integrable process $Y_t \in \mathscr{L}_{t+}^o$, we define*

$$\sum_n^Y (t) = \sum_{k=0}^{n-1} E^z(Y(t_{k+1}^n) - Y(t_k^n) \mid \mathscr{L}_{t_k^n}^z).$$

*Then we have*

$$\langle M_\lambda \rangle_t^z = \lim_{n \to \infty} \sum_n^{(R_\lambda^Z \hat{\rho}(Z))^2}(t) + 2 \int_0^t R_\lambda^Z \hat{\rho}(Z_s) \hat{\rho}(Z_s) - \lambda (R_\lambda^Z \hat{\rho}(Z_s))^2 \, ds \tag{5.5}$$

*in the sense of (strong) $L^1$-convergence.*

**Proof** According to Theorem 5.4,

$$\langle M_\lambda \rangle_t^z = \lim_{n \to \infty} \sum_n^{M_\lambda^2}(t).$$

Applying Ito's lemma (Dellacherie and Meyer 1980, VIII, 27) to the semimartingale

$$(R_\lambda^Z \hat{\rho}(Z_t))^2 = \left( M_\lambda \hat{\rho}(Z_t) - \int_0^t \{\hat{\rho}(Z_s) - \lambda R_\lambda^Z \hat{\rho}(Z_s)\} \, ds \right)^2,$$

and denoting the last integral by $\int_0^t$, we have

$$(R_\lambda^Z \hat{\rho}(Z_t))^2 = 2 \int_0^t \left( M_\lambda \hat{\rho}(Z_s) - \int_0^s \right) dM_\lambda(s)$$

$$- 2 \int_0^t R_\lambda^Z \hat{\rho}(Z_s)(\hat{\rho}(Z_s) - \lambda R_\lambda^Z \hat{\rho}(Z_s)) \, ds + \langle M_\lambda \rangle_t^z. \quad (5.6)$$

The first term on the right is a martingale, so that our $\sum_n^Y$ operation applied to it yields 0. It remains only to show that the same operation applied to the second term on the right leaves it unchanged in the limit $n \to \infty$. By hypothesis on $M_2^+$, $\hat{\rho}(Z_t)$ is r.c.l.l. in quadratic mean for $P^z$, and so is $R_\lambda^Z \hat{\rho}(Z_t)$ along with $M_\lambda(t)$. Hence the term in question is the $L^1$-limit of Riemann sums along the partitions $t_k^n$, by Schwartz's inequality. Since $Z(t_k^n) \in \mathcal{L}_{t_k^n}^\circ$, these sums are unchanged by conditioning term by term on $\mathcal{L}_{t_k^n}^\circ$. An application of Jensen's inequality shows that the conditioning decreases the absolute differences in $L^1$, which completes the proof.

**Lemma 5.7** *In the notation of Lemma 5.6,*

$$\lim_{n \to \infty} \sum_n^{V_\lambda(Z)}(t) = 2\lambda \int_0^t V_\lambda(Z_s) \, ds - \int_0^t f_\lambda(Z_s) \, ds,$$

*in the same $L^1$ sense.*

**Proof** Since $V_\lambda(z) = R_{2\lambda}^Z f_\lambda(z)$, writing $f_\lambda = f_\lambda^+ - f_\lambda^-$ we see that $V_\lambda(Z_s)$ is r.c.l.l. in $L^1$, and hence the integrals on the right are $L^1$-limits of Riemann sums along the partitions $t_k^n$ (for example, $R_{2\lambda} f_\lambda^+(Z_s)$ is a $2\lambda$-supermartingale). Now we have

$$\sum_n^{V_\lambda(Z)}(t) = \sum_{k=1}^n E^{Z(t_k^n)} \int_0^\infty e^{-2\lambda s} \left( f_\lambda\left( Z\left( \frac{1}{n} + s \right) \right) - f_\lambda(Z(s)) \right) ds$$

$$= \sum_{k=1}^n E^{Z(t_k^n)} \left[ (e^{2\lambda/n} - 1) \int_{1/n}^\infty e^{-2\lambda s} f_\lambda(Z_s) \, ds - \int_0^{1/n} e^{-2\lambda s} f_\lambda(Z_s) \, ds \right].$$

178 | Foundations of the prediction process

But since $f_\lambda(Z_t)$ is r.c.l.l. in quadratic mean, it follows by the Riemann sum approximation that

$$\lim_{n \to \infty} \sum_{k=1}^{n} E^{Z(t_k^n)} \int_0^{1/n} e^{-2\lambda s} f_\lambda(Z_s) \, ds = \int_0^t f_\lambda(Z_s) \, ds$$

in the sense of $L^2$, as in the previous proof; this is the only point where assumption (2) will be used. Therefore, we shall have

$$\lim_{n \to \infty} \sum_{n}^{V_\lambda(Z)} (t) = \lim_{n \to \infty} 2\lambda n^{-1} \sum_{k=1}^{n} E^{Z(t_k^n)} \int_0^\infty e^{-2\lambda s} f_\lambda(Z_s) \, ds - \int_0^t f_\lambda(Z_s) \, ds,$$

provided that the first limit on the right exists in $L^1$. But this is just $(2\lambda)$ times the Riemann sum for $\int_0^t V_\lambda(Z_s) \, ds$, which completes the proof.

By combining Lemmas 5.6 and 5.7, we see that Theorem 5.5 will be proved if we show that, in an appropriate sense,

$$\lim_{n \to \infty} \sum_{n}^{V_\lambda(Z) + (R_\lambda^Z \hat\rho(Z))^2} (t) = 2\lambda \int_0^t E^{Z(s)} \left( \int_0^\infty e^{-\lambda u} \hat\rho(Z(u)) \, du \right)^2 ds$$

$$- 2 \int_0^t (R_\lambda^Z \hat\rho(Z_s)) \hat\rho(Z_s) \, ds. \quad (5.7)$$

Indeed, the desired result follows when we solve this for $\lim_{n \to \infty} \sum_{n}^{(R_\lambda^Z \hat\rho(Z))^2} (t)$ and substitute into (5.5). Now the left side of (5.7), for fixed $n$, may be written

$$\sum_k E^{Z(t_k^n)} \Bigg[ \int_0^\infty e^{-\lambda s} \left( \hat\rho\left(Z\left(\frac{1}{n} + s\right)\right) - \hat\rho(Z(s)) \right) ds$$

$$\times \int_0^\infty e^{-\lambda s} \left( \hat\rho\left(Z\left(\frac{1}{n} + s\right)\right) + \hat\rho(Z(s)) \right) ds \Bigg]$$

$$= \sum_k E^{Z(t_k^n)} \Bigg[ \left( (e^{\lambda/n} - 1) \int_{1/n}^\infty e^{-\lambda s} \hat\rho(Z_s) \, ds - \int_0^{1/n} e^{-\lambda s} \hat\rho(Z_s) \, ds \right)$$

$$\times \left( (e^{\lambda/n} + 1) \int_{1/n}^\infty e^{-\lambda s} \hat\rho(Z_s) \, ds + \int_0^{1/n} e^{-\lambda s} \hat\rho(Z_s) \, ds \right) \Bigg]$$

$$= \sum_k E^{Z(t_k^n)} \Bigg[ (e^{2\lambda/n} - 1) \left( \int_{1/n}^\infty e^{-\lambda s} \hat\rho(Z_s) \, ds \right)^2$$

$$- 2 \int_{1/n}^\infty e^{-\lambda s} \hat\rho(Z_s) \, ds \int_0^{1/n} e^{-\lambda s} \hat\rho(Z_s) \, ds - \left( \int_0^{1/n} e^{-\lambda s} \hat\rho(Z_s) \, ds \right)^2 \Bigg].$$

$$(5.8)$$

As to the last term, its sum has $P^z$-expectation

$$E^z \sum_k \left( \int_{t_k^n}^{t_{k+1}^n} e^{-\lambda s} \hat\rho(Z_s)\, ds \right)^2 \le \frac{1}{n} \sum_k E^z \left( \int_{t_k^n}^{t_{k+1}^n} \hat\rho(Z_s)^2\, ds \right)$$

$$= \frac{1}{n} \int_0^t E^z(\hat\rho(Z_s))^2\, ds, \tag{5.9}$$

which tends to 0. Since

$$E^{Z(t)} \left( \int_{1/n}^{\infty} e^{-\lambda s} \hat\rho(Z_s)\, ds \right)^2 \le \lambda^{-1} E^{Z(t)} \int_0^{\infty} e^{-\lambda s} (\hat\rho(Z_s))^2\, ds$$

which is bounded in finite time intervals $P^z$-a.s. for $z \in M_2$, as $n \to \infty$ (5.8) becomes

$$\lim_{n \to \infty} \left[ 2\lambda n^{-1} \sum_{k=1}^n E^{Z(t_k^n)} \left( \int_0^{\infty} e^{-\lambda s} \hat\rho(Z_s)\, ds \right)^2 \right.$$
$$\left. - 2 \sum_k E^{Z(t_k^n)} \left( \int_0^{1/n} e^{-\lambda s} \hat\rho(Z_s)\, ds \int_{1/n}^{\infty} e^{-\lambda s} \hat\rho(Z_s) \right) \right].$$

It remains to show that these two sums approach the corresponding terms of (5.7), in a suitable sense. As to the first, we saw in the proof of Lemma 5.7 that $V_\lambda(Z(t))$ is r.c.l.l. in $L^2$, and since $R_\lambda^Z \hat\rho(Z_t)$ is r.c.l.l. in $L^2$, its square is r.c.l.l. in $L^1$. Thus by addition

$$E^{Z(t)} \left( \int_0^{\infty} e^{-\lambda s} \hat\rho(Z_s)\, ds \right)^2$$

is r.c.l.l. in $L^1$, and so the first convergence follows in $L^1$ by the Riemann sum approximation. The second sum may be replaced by

$$\sum_k E^{Z(t_k^n)} \int_0^{1/n} \hat\rho(Z_u) \int_u^{\infty} e^{-\lambda s} \hat\rho(Z_s)\, ds\, du$$

with $L^1$-error tending to 0, by the estimate (5.9) and the $L^1$-boundedness of the sum (using Jensen's inequality to remove the conditional expectations $E^{Z(t_k^n)}$). Next let us subtract the desired limit

$$\sum_k \int_{t_k^n}^{t_{k+1}^n} \hat\rho(Z_u) R_\lambda^Z(\hat\rho(Z_u))\, du.$$

It is easy to see that the partial sums of the difference are now a martingale, but they need not be in $L^2$. However, for $K > 0$, let $T_K = \inf\{t: |R_\lambda^Z \hat\rho(Z_t)| \ge K\}$. We now repeat the above argument with the martingale $M_\lambda(t \wedge T_K)$. Note that the terms of the present sum become

$$E^{Z(t_k^n \wedge T_K)} \int_0^{(1/n) \wedge T_K} \hat{\rho}(Z_u) R_\lambda^Z \hat{\rho}(Z_u) \, du - \int_{t_k^n \wedge T_K}^{t_{k+1}^n \wedge T_K} \hat{\rho}(Z_u) R_\lambda^Z \hat{\rho}(Z_u) \, du,$$

which reduce to 0 for $T_K \le t_k^n$. The partial sums still form a martingale, and now it is in $L^2$ by reason of the bound

$$E^z \left( \int_0^{t \wedge T_K} \hat{\rho}(Z_u) R_\lambda^Z \hat{\rho}(Z_u) \, du \right)^2 \le K^2 E^z \left( \int_0^t |\hat{\rho}(Z_u)| \, du \right)^2$$

$$\le K^2 t E^z \int_0^t (\hat{\rho}(Z_u))^2 \, du.$$

Consequently the mean square reduces to

$$\sum_{k=1}^n E^z \left( E^{Z(t_k^n \wedge T_K)} \int_0^{(1/n) \wedge T_K} \hat{\rho}(Z_u) R_\lambda^Z \hat{\rho}(Z_u) \, du - \int_{t_k^n \wedge T_K}^{t_{k+1}^n \wedge T_K} \hat{\rho}(Z_u) R_\lambda^Z \hat{\rho}(Z_u) \, du \right)^2.$$

But here, the sum of the second terms squared is bounded by

$$K^2 \sum_{k=1}^n E^z \left( \int_{t_k^n \wedge T_K}^{t_{k+1}^n \wedge T_K} |\hat{\rho}(Z_u)| \, du \right)^2 \le K^2 \sum_{k=1}^n \left( \frac{1}{n} E^z \int_{t_k^n \wedge T_K}^{t_{k+1}^n \wedge T_K} (\hat{\rho}(Z_u))^2 \, du \right)$$

$$\le K^2 n^{-1} \int_0^t (\hat{\rho}(Z_u))^2 \, du,$$

and the same bound applies to the first sum by Jensen's inequality. Thus the limit is 0 in $L^2$. The proof of Theorem 5.5 is now completed by letting $K \to \infty$, because $T_K \to \infty$ $P^z$-a.s., which gives convergence in probability to the limit.

**Remarks** The expression for $\langle M_{\lambda_1} - M_{\lambda_2} \rangle_t^z$ now follows from the resolvent equation for $M_\lambda$ (Theorem 4.1), provided that $R_{\lambda_1} \hat{\rho}$ satisfies the hypotheses which Theorem 5.5 assumes for $\hat{\rho}$. Since

$$M_{\lambda_1}^2 - 2M_{\lambda_1} M_{\lambda_2} + M_{\lambda_2}^2 - \langle M_{\lambda_1} - M_{\lambda_2} \rangle_t^z$$

is a $P^z$-martingale, we can thus obtain

$$\langle M_{\lambda_1}, M_{\lambda_2} \rangle_t^z \ (= \tfrac{1}{2}(\langle M_{\lambda_1} - M_{\lambda_2} \rangle_t^z - \langle M_{\lambda_1} \rangle_t^z - \langle M_{\lambda_2} \rangle_t^z)).$$

**Exercise 5.5** (for Theorem 5.5)

(1) Show that when $\hat{\rho}$ is bounded and $(R_\lambda^Z \hat{\rho})^2$ is in the domain of the generator of $Z_t$ on a demesne containing $z$, then (5.3) is equivalent to (5.4). Hint: show that $V_\lambda(z) = 2R_{2\lambda}^Z(\hat{\rho} R_\lambda \hat{\rho})(z) - (R_\lambda^Z \hat{\rho})^2(z)$.

(2) Show that, by the same reasoning, we may write a formula for $\langle M_{\lambda_1} - M_{\lambda_2} \rangle_t^z$ provided that $(R_{\lambda_1}^Z \hat{\rho})^2$, $(R_{\lambda_2}^Z \hat{\rho})^2$, and $(R_{\lambda_1}^Z \hat{\rho}) \cdot (R_{\lambda_2}^Z \hat{\rho})$ are all in the domain of the generator.

**Solution (sketch)** The integrand of (5.3) may be written in the form $A(R_\lambda^z f)^2 - 2\lambda(R_\lambda^z f)^2 + 2\hat{\rho}(R_\lambda^z \hat{\rho})$, whereas that of (5.4) becomes

$$(2\lambda - A)V_\lambda(z) = 2\hat{\rho}(R_\lambda^z \hat{\rho}) - (2\lambda - A)(R_\lambda^z \hat{\rho})^2,$$

which is the same. Furthermore,

$$\langle M_{\lambda_1} - M_{\lambda_2} \rangle_t^z = (\lambda_1 - \lambda_2)^2 \langle M^{R_{\lambda_1}^z \hat{\rho}}_{\lambda_2} \rangle_t^z,$$

which may be expressed as in (5.3) provided that $(R_{\lambda_2}^z R_{\lambda_1}^z \hat{\rho})^2$ is in the domain of the generator. But this is equivalent to $(R_{\lambda_1}^z \hat{\rho} - R_{\lambda_2}^z \hat{\rho})^2$ being in the domain, and the assertion follows.

**Remark** We may observe that the main impediment to absolute continuity of $\langle M_\lambda \rangle_t^z$ is that $(R_\lambda^z \hat{\rho})^2$ may not satisfy the requirements of the domain of the generator. This is the case in Example 1, where $R_\lambda^z \hat{\rho}(x) = \lambda^{-1} x$ ($X_t$ being a martingale), but $x^2$ is not in the domain. Indeed, in this case to be in the domain of $\dfrac{d}{dm}\dfrac{d^+}{dx^+}$ the derivative must have a jump at every rational, whereas $x^2$ is continuously differentiable. On the other hand, if $dm$ has a unit jump at 0, and reduces to $2dx$ elsewhere, one can show by comparison to Brownian motion that $\langle M_\lambda \rangle_t^z$ is absolutely continuous (the jump of $m$ retards the process at 0, thus reducing $d\langle M_\lambda \rangle_t^z$). In this case $\hat{\rho}$ is unbounded, so Theorem 22 of Dellacherie and Meyer (1987, XV) does not apply. But the hypotheses of Theorem 5.5 are satisfied, and after considerable calculation we obtain $f_\lambda(x) = \lambda^{-2}(1 - I_{\{0\}}(x))$.

We turn now to the strict-sense analogues of the Cramér–Hida representation theorem (Theorem 3.23). In the LMR case, however, we only show now that the Hilbert space generated by the $(M_n, 1 \le n)$ is so represented, which is a trivial extension of Theorem 3.23. In Theorem 5.10 we will show that this is $H_z^*(t+)$ if $z \in M_2^+$, which will complete the representation

**Theorem 5.8**

(1) (*The LMR analogue*). For $z \in M_2$, let $H_z'(t)$ denote the $L_z^2$-linear closure of $\{M_n(s), s \le t, 1 \le n\}$. Then there exists a sequence $(Y_n, 1 \le n)$ of square-integrable $\mathcal{L}_t^z$-martingales, $E^z(Y_n Y_m(t)) = 0$ for $n \ne m$, $t > 0$, and $dE^z Y_n^2(t) \ll dE^z Y_m^2(t)$, $m < n$ (in the sense of absolute continuity of measures), such that $H_z'(t)$ is the $L_z^2$-linear closure, of $\{Y_n(s), s \le t, 1 \le n\}$. The equivalence classes of $dE^z Y_n^2$ are uniquely determined. We denote the LMR multiplicity by

$$\phi_z^*(t) = \inf\{k : E^z Y_{k+1}^2(t) = 0\}; \qquad \phi_z^*(t) \le \infty. \tag{5.10}$$

(2) (*The PMSIR analogue*). For $z \in M_2$, there exists a sequence of square-integrable $\mathcal{L}_t^z$-martingales $(Y_n^\#, 1 \le n)$, which are orthogonal ($Y_n Y_m$ is a

## 182 | Foundations of the prediction process

*martingale for $m \neq n$), such that the previsible random measures*

$$d\langle Y_n^\# \rangle_t^z : d\langle Y_n^\# \rangle_t^z(S) = E^z \int_0^\infty I_S(t, w_z)\, d\langle Y_n^\# \rangle_t^z; \qquad S \in \mathscr{P}$$

*satisfy (over the $\mathscr{L}_t^z$-previsible sets $\mathscr{P}$) $d\langle Y_n^\# \rangle_t^z \ll d\langle Y_m^\# \rangle_t^z$ for $m < n$, and such that $H_z^\#(t)$ is the $L_z^2$-linear closure of*

$$\left\{ \int_0^t h(s)\, dY_n^\#(s); h(s, w_z) \text{ any } \mathscr{L}_t^z\text{-previsible process with} \right.$$

$$\left. E^z \int_0^t h^2(s)\, d\langle Y_n^\# \rangle_s^z < \infty \right\}.$$

*The equivalence classes of the previsible measures $d\langle Y_n^\# \rangle_t^z$ are $P^z$-a.s. uniquely determined. We denote the PMSIR multiplicity by*

$$\phi_z^\#(t) = \inf\{k : E^z \langle Y_{k+1}^\# \rangle_t^z = 0\}; \qquad \phi_z^\#(t) \leq \infty. \tag{5.11}$$

**Remark**  As was the case in Theorem 3.23, the novelty and importance of Theorem 5.8 lies not so much in the existence of $Y_n$ and $Y_n^\#$ as in the particular expressions that we give to them in terms of the $M_n$. This feature will be illustrated by examples which are deferred to Chapter 6.

**Proof**  The proof of (1) is the same as that of Theorem 3.23, and need not be repeated. One has only to recognize that the orthogonalization procedure depends only on the inner product formulas $(m, n)_z^t$ of Theorem 3.20. The present result is then obtained by substituting $(m, m)_z^*(t)$ from Notation 5.6. It is, of course, essential to observe that this procedure preserves the martingale property. However, the uniqueness part of the assertion depends only on the orthogonality of increments, and follows just as in the Gaussian case.

Turning to (2), we need to make some further use of the orthogonalization theory of the previsible stochastic integral for which we refer again to Dellacherie and Meyer (1980, VIII, 46–55). We remark that since $z$ is fixed we do not require the corresponding Markovian additive functional theory of Dellacherie and Meyer (1987, XV, 44 ff.) and Sharpe (1988, VI). Our method of proof is more in the spirit of Davis and Varaiya (1974), and applies equally to any sequence of square-integrable martingales, including one which (unlike our $M_n$) generates all of $L_z^2$. We begin by orthogonalizing the $(M_n)$ in the sense of stochastic integration. Let $N_1 = M_1$, and for induction suppose that $N_1, N_2, \ldots, N_n$ have been defined. It is known that we may project $M_{n+1}$ onto the stable subspace of $N_k$ in the form $\int_0^t h_k(s)\, dN_k(s)$, so that $h_k(s)$ is $\mathscr{L}_t^z$-previsible and $M_{n+1}(t) - \int_0^t h_k(s)\, dN_k(s)$ is orthogonal to $N_k$ in the martingale sense. The process $h_k$ is unique ($d\langle N_k \rangle_s$-a.s., $P^z$-a.s.).

It follows that $\langle M_{n+1}, N_k \rangle_t^z = \int_0^t h_k(s) \, d\langle N_k \rangle_s^z$, from which we have

$$h_k(s) = \frac{d\langle M_{n+1}, N_k \rangle_s^z}{d\langle N_k \rangle_s^z}, \qquad d\langle N_k \rangle_s^z\text{-a.e.}, \tag{5.12}$$

in the sense of a Radon–Nikodym derivative. We set

$$N_{n+1}(t) = M_{n+1}(t) - \sum_{k=1}^n \int_0^t h_k(s) \, dN_k(s),$$

with $h_k$ as in (5.12). Then clearly the $N_k$ are orthogonal martingales, and, by solving for $M_{n+1}$, we see that they generate $H_z^\#(t)$ for each $t$, as the least stable subspace containing them.

Next, let $\mathscr{P}$ denote the $\sigma$-field on $[0, \infty) \times \Omega_Z$ of $\mathscr{L}_t^z$-previsible sets ($\mathscr{P}$ is generated by the family of left-continuous, $\mathscr{L}_t^z$-adapted processes). Since $\langle N_n \rangle_t^z$ is previsible, we can consider $d\langle N_n \rangle_t^z$ as a measure on $\mathscr{P}$, namely

$$d\langle N_n \rangle_t^z(S) = E^z \int_0^\infty I_S(t, w_Z) \, d\langle N_n \rangle_t^z, \qquad S \in \mathscr{P}.$$

Now for $n > 1$ there is a Lebesgue decomposition $\Omega_Z = A_n \cup A_n^c$, $A_n \in \mathscr{P}$, such that $I_{A_n^c} d\langle N_n \rangle_t^z \ll d\langle N_1 \rangle_t^z$ (as measures on $\mathscr{P}$, defined by integration) and $d\langle N_1 \rangle_t^z(A_n) = 0$. Setting $B_2 = \bigcup_{k=2}^\infty A_k$, it follows that $I_{B_2^c} d\langle N_n \rangle_t^z \ll d\langle N_1 \rangle_t^z$ for $n > 1$, and $d\langle N_1 \rangle_t^z(B_2) = 0$. Therefore, $I_{B_2^c} d\langle N_n \rangle_t^z \ll I_{B_2^c} d\langle N_1 \rangle_t^z$. We now repeat this construction with the measures $I_{B_2} d\langle N_n \rangle_t^z$, $n > 2$, with $I_{B_2} d\langle N_2 \rangle_t^z$ in place of $d\langle N_1 \rangle_t^z$, to obtain a set $B_3 \in \mathscr{P}$ such that $I_{B_2} d\langle N_2 \rangle_t^z(B_3) = 0$ and for every $n > 2$, $I_{B_3^c \cap B_2} d\langle N_n \rangle_t^z \ll I_{B_3^c \cap B_2} d\langle N_2 \rangle_t^z$. Here we may obviously replace $B_3$ by $B_2 \cap B_3$ and the same relations remain true. Proceeding by induction, we thus obtain a sequence $\Omega_Z = B_1 \supset B_2 \supset B_3 \supset \cdots$ such that $I_{B_k} d\langle N_k \rangle_t^z(B_{k+1}) = 0$, $1 \leq k$, and

$$I_{B_{k+1}^c \cap B_k} d\langle N_n \rangle_t^z \ll I_{B_{k+1}^c \cap B_k} d\langle N_k \rangle_t^z \tag{5.13}$$

for every $n > k$. It follows that the measures $I_{B_{k+1}^c \cap B_k} d\langle N_{k+1} \rangle_t^z$ have disjoint supports and, for every $n$,

$$d\langle N_n \rangle_t^z \ll \sum_{k=1}^n I_{B_{k+1}^c \cap B_k} d\langle N_k \rangle_t^z;$$

in particular, $d\langle N_1 \rangle_t^z \ll I_{B_2^c} d\langle N_1 \rangle_t^z$. We now introduce the first element of $(Y_n^\#)$:

$$Y_1^\#(t) = \sum_{k=1}^\infty 2^{-k}(E^z \langle N_k \rangle_1^z)^{-1} \int_0^t I_{B_{k+1}^c \cap B_k} \, dN_k(s) \tag{5.14}$$

for $0 < t \leq 1$, where $0/0 = 0$. More generally, for $n < t \leq n+1$ we define inductively $Y_1^\#(t) = Y_1^\#(n) + (Y_1^\#(t) - Y_1^\#(n))$, where the process in parentheses is defined as a function of $t - n$ in the same way as for $t \leq 1$, but using $N_k(n + t) - N_k(n)$ in place of $N_k$, and we replace $s$ by $n + s$ in $I_{B_{k+1}^c \cap B_k}$. Clearly we have $E^z(Y_1^\#(t))^2 \leq n + 1$ for $n < t \leq n + 1$, and $Y_1^\#(t)$ is a $\mathcal{L}_t^z$-previsible process. Moreover, $d\langle N_n \rangle_t^z \ll d\langle Y_1^\# \rangle_t^z$ for every $n$ in view of (5.13). Noting that $\int_0^t I_{B_{k+1}^c \cap B_k} dN_k(s)$ is in the stable subspace of $Y_1^\#(t)$, $1 \leq k$, it is easy to see that the sequence $(Y_1^\#, I_{B_2^c \cup B_3} dN_2, \ldots, I_{B_k^c \cup B_{k+1}} dN_k, \ldots)$ generates $H_z^\#(t)$ for every $t$, and that these martingales are orthogonal. Finally, the equivalence class of $d\langle Y_1^\# \rangle_t^z$ on $\mathscr{P}$ is uniquely determined as the smallest with respect to which every $d\langle M_n \rangle_t^z$ is absolutely continuous. Indeed, this is determined by the $H_z^\#(t)$, independently of the choice of generators $M_n$, as the least equivalence class of measures on $\mathscr{P}$ with respect to which $d\langle M \rangle_t^z$ is absolutely continuous for every square-integrable martingale $M(t) \in H_z^\#(t)$ for all $t$.

We now proceed by induction on $n$, in the obvious way. Thus $Y_2^\#$ is defined from $(I_{B_k^c \cup B_{k+1}} dN_k; 2 \leq k)$ in the same way as $Y_1^\#$ was defined from $(dN_k; 1 \leq k)$. Since $I_{B_2^c \cup B_3} dN_2$ is in the stable subspace of $Y_2^\#$, and $I_{B_2^c \cap B_2} dN_2$ is in that of $Y_1^\#$, we see that $N_2$ is in the space of $(Y_1^\#, Y_2^\#)$, and of course $d\langle Y_2^\# \rangle_t^z \ll d\langle Y_1^\# \rangle_t^z$. Assuming the result for $n$, the first element of the next sequence has the form $I_{C_{n+1}^c \cup C_{n+2}} dN_{n+1}$, $C \in \mathscr{P}$, where $I_{C_{n+2}^c \cap C_{n+1}} dN_{n+1}$ is already in the subspace of $(Y_1^\#, \ldots, Y_n^\#)$. Thus $N_{n+1}$ is in the subspace of $(Y_1^\#, \ldots, Y_{n+1}^\#)$, and consequently (by induction) so is $M_{n+1}$.

Finally, the uniqueness of the equivalence classes of $d\langle Y_n^\# \rangle_t^z$, starting with $n = 2$, may be proved also by induction on $n$. As the proof is intricate, and the uniqueness will not be used in this book, we refer the reader to Theorem 2 of Davis and Varaiya (1974) for the proof. (After noting a trivial misprint in formula (12), one can transcribe the rest of the argument letter by letter.) This completes the proof of Theorem 5.8.

**Remark** It should not be assumed from $d\langle Y_{n+1} \rangle_t^z \ll d\langle Y_n \rangle_t^z$ on $\mathscr{P}$ that, for $P^z$-a.e. $w_z$, the Lebesgue–Stieltjes measure $d\langle Y_{n+1}(w_z) \rangle_t^z$ on $\mathbb{R}^+$ is absolutely continuous with respect to $d\langle Y_n(w_z) \rangle_t^z$. For example, suppose that

$$Y_n(t) = 0 \quad \text{for } t < 1;$$

$$Y_n(t) = \begin{cases} +1 \\ 0 \\ -1 \end{cases} \quad \text{for } t \geq 1$$

each with probability $\frac{1}{3}$, independently over $n$. Then $\langle Y_n \rangle_t^z$ has a jump 1 at $t = 1$ with probability $\frac{2}{3}$, and otherwise remains 0. Here $\mathscr{P}$ is degenerate at $t = 1$, so each $d\langle Y_n \rangle_t^z$ has the same integral over $\mathscr{P}$, but they are not mutually absolutely continuous for each $w_z$. This example, however, depends on the foreseeable jump at $t = 1$, i.e. it is in $J_{D^c}$ of Definition 4.3. It is perhaps an open question whether such an example exists for $z \in M(J_D)$.

Prediction of measurable processes | 185

We turn now to the principal fact which enables us to use the family $M_\lambda$ to predict $X_z^+(t)$ for $z \in M_2^+$ (Definition 5.2). This is stated as a lemma, but it is the wide-sense extension of Theorem 3.27. Like the previous Theorem 5.8, it has parts (1) and (2), corresponding to the LMR and the PMSIR analogues of the Gaussian case. (In Chapter 6.3, we will re-examine the PMSIR in the Gaussian case, to see what $H_z^\#(t)$ becomes in that rather special situation.) Since $z$ is fixed, we shall write $X_t^+$ in place of $X_z^+(t)$.

**Lemma 5.9** *Let $z \in M_2^+$, and let $Y(t)$ be any r.c.l.l., square-integrable $\mathscr{L}_t^z$-martingale, $Y(0) = 0$. Then we have the following.*

(1) *Let*

$$F^*(t, s) = \frac{dE^z(X_t^+ Y_s)}{dE^z(Y_s^2)}; \qquad 0 < s \leq t,$$

*(chosen $(\mathscr{B}^+ \times \mathscr{B}^+)$-measurable) and let*

$$G_\lambda^*(s) = \frac{dE^z(M_\lambda(s) Y_s)}{dE^z(Y_s^2)} = \frac{dE^z(M_\lambda(t) Y_s)}{dE^z(Y_s^2)}; \qquad 0 < s < t$$

*in the same sense, suppressing dependence on $z$. Thus we have*

$$X_z^+(t) = \int_0^t F^*(t, s) \, dY_s + V_t^* \quad \text{and} \quad M_\lambda(t) = \int_0^t G_\lambda^*(s) \, dY_s + W_t^*,$$

*where $E^z(Y_s V_t^*) = 0 = E^z(Y_s W_t^*); 0 < s \leq t$.*

*Conclusion:* $G_\lambda^*(t) = \int_0^\infty e^{-\lambda s} F^*(t + s, t) \, ds$ *for all $\lambda > 0$, except for a $dE^z(Y_t^2)$-null set of $t$.*

(2) *Let $X^+(t, s) = E^z(X_t^+ | \mathscr{L}_s^z); 0 < s \leq t$, (so that $X^+(t, s)$ is an r.c.l.l., square-integrable martingale in $s$, starting at 0). Let*

$$F^\#(t, s) = \frac{d\langle X^+(t, s), Y_s \rangle_s^z}{d\langle Y \rangle_s^z}, \qquad 0 < s \leq t.$$

*This may be defined as a $(\mathscr{B}^+ \times \mathscr{P})$-measurable process in $(t, (s, w_z))$, unique for each $t$ up to a $d\langle Y \rangle_s^z$-null set in $\mathscr{P}$. Let*

$$G_\lambda^\#(s) = \frac{d\langle M_\lambda, Y \rangle_s^z}{d\langle Y \rangle_s^z}, \qquad 0 < s \leq t,$$

*as a $(\mathscr{B}^+ \times \mathscr{P})$-measurable process unique up to a $d\langle Y \rangle_s^z$-null set.*

*Conclusion:* $G_\lambda^\#(t) = \int_0^\infty e^{-\lambda v} F^\#(t + v, t) \, dv$ *for all $\lambda > 0$, for $d\langle Y \rangle_t^z$-a.e. $(t, w_z)$, where the integral is in $L^2(d\langle Y \rangle_t^z)$ in finite time intervals, and exists in the pathwise sense for all $\lambda > 0$ (in fact, $F^\#(t + v, t)$ is in $L^2(e^{-\lambda v} dv)$ for all $\lambda > 0$, $d\langle Y \rangle_t^z$-a.e.).*

Note: we recall the definition of $d\langle Y\rangle_t^z$ from Theorem 5.8(2).

**Proof** (1) The proof is essentially that of (3.8), for which we just replace $Y_1(t)$ by the present $Y(t)$, and projection onto $H_z(t+)$ by projection onto $L^2(\mathcal{L}_t^z, P^z)$. However, since we have not yet established the representation of $X_t^+$ in terms of $(Y_n; 1 \le n)$, it is necessary to work directly with the projections onto the $L_z^2$-subspace generated by $(Y_s; 0 < s)$.

As we saw in the former proof, $M_\lambda$ is unchanged if we centre at the mean, replacing $\hat\rho(Z_t)$ by $\hat\rho(Z_t) - E_z(t)$. Then it follows that $X^+(t)$ is $P^z$-equivalent to $\hat\rho(Z_t)$ for a.e. $t$. Letting $\sigma^2(t) = E^z Y^2(t)$, we have

$$\infty > E^z \int_0^\infty e^{-\lambda u}(X_u^+)^2 \, du$$

$$\ge \int_0^\infty e^{-\lambda u} E^z \left(\int_0^u F^*(u,s) \, dY(s)\right)^2 du$$

$$= \int_0^\infty e^{-\lambda u} \int_0^u (F^*(u,s))^2 \, d\sigma^2(s) \, du$$

$$\ge \lambda \int_0^\infty \left(\int_s^\infty e^{-\lambda u} |F^*(u,s)| \, du\right)^2 d\sigma^2(s)$$

$$\ge \lambda e^{2\lambda t} \int_0^t \left(\int_0^\infty e^{-\lambda v} |F^*(t+v, s)| \, dv\right)^2 d\sigma^2(s), \qquad (5.15)$$

so that $\int_0^t (\int_0^\infty e^{-\lambda u} F^*(t+u, s) \, du) \, dY(s)$ exists as an $L_z^2$-integral.

We now write, in the sense of $L_z^2$,

$$R_\lambda^z \hat\rho(Z_t) = E^z \int_0^\infty e^{-\lambda u} E^z(X^+(t+u) \mid \mathcal{L}_t^z) \, du,$$

and project this onto the $L_z^2$-subspace of $(Y_s; 0 < s)$. Since $dY(s)$ is orthogonal to $R_\lambda^z \hat\rho(Z_t)$ for $s > t$, and $R_\lambda^z \hat\rho(Z_t)$ is written as a projection onto $L_z^2(\mathcal{L}_t^z, P^z)$, we can commute the two projections, obtaining

$$E^z\left(\int_0^\infty e^{-\lambda u}\left(\int_0^{t+u} F^*(t+u, s) \, dY(s) \, du \mid \mathcal{L}_t^z\right)\right)$$

$$= \int_0^\infty e^{-\lambda u}\left(\int_0^t F^*(t+u, s) \, dY(s)\right) du,$$

where we also used the fact that projection commutes with $L_z^2$-integration. Now the order of integration may again be reversed, just as in the former proof except that now the last integrand is only $L_z^2$-r.c.l.l. in $u$, along with $X_{t+u}^+$, and so the projection of $R_\lambda^z \hat\rho(Z_t)$ onto the subspace generated by

Prediction of measurable processes | 187

$(Y_s, s \leq t)$ is given by $\int_0^t (\int_0^\infty e^{-\lambda u} F^*(t+u, s) \, du) \, dY(s)$. (This is our replacement for Lemma 3.28; note that since no use is made of a representation of $X^+$ in terms of $(Y_n; 1 \leq n)$ we do not need to assume $z \in D$.)

The balance of the proof is the same as in Theorem 3.26, but we repeat it for the reader's convenience. For the projection of $M_\lambda(t)$ onto the subspace of $(Y_s, s \leq t)$ we have, by orthogonality of the increments of $Y_s$ and the fact that the projection of $X_v^+$ is $\int_0^v F^*(v, s) \, dY(s)$ for $v \leq t$, the expression

$$\mathscr{E}^z(M_\lambda(t) \mid L_z^2(Y_s, s \leq t)) = \int_0^t \left( \int_0^\infty e^{-\lambda u} F^*(t+u, s) \, du \right) dY(s)$$

$$+ \lambda \int_0^t \int_0^v \left( \int_0^\infty e^{-\lambda u}(F^*(v, s) - F^*(v+u, s)) \, du \right) dY(s) \, dv.$$

Now the integrand of the last $dv$-integration is $L_z^2$-r.c.l.l. in $v$, where we can use a 'triangular' argument along with the r.c.l.l. property of

$$\int_0^c \left( \int_0^\infty e^{-\lambda u} F^*(v+u, s) \, du \right) dY(s)$$

in $v$ for fixed $c \leq v$, which follows by that of $\int_0^c F^*(v+u, s) \, dY(s)$ and the bound (5.15). Therefore we see easily that the $dv$ and $dY(s)$ integration may be interchanged and the coefficient $G_\lambda^*(s)$ of $dY(s)$ becomes

$$\int_0^\infty e^{-\lambda u} F^*(t+u, s) \, du + \lambda \int_s^t \left( \int_0^\infty e^{-\lambda u}(F^*(v, s) - F^*(v+u, s)) \, du \right) dv.$$

(5.16)

Again by the finiteness of (5.15), for $dE^z Y^2(s)$-a.e. $s$, we can interchange the order of integration in the last term, whence an integration by parts gives

$$\lambda \int_0^\infty e^{-\lambda u} \int_s^t (F^*(v, s) - F^*(v+u, s)) \, dv \, du$$

$$= -\int_0^\infty e^{-\lambda u} \frac{d}{du} \int_s^t (F^*(v, s) - F^*(v+u, s)) \, dv \, du$$

$$= -\int_0^\infty e^{-\lambda u} \left( \frac{d}{du} \int_{s+u}^{t+u} F^*(w, s) \, dw \right) du$$

$$= \int_0^\infty e^{-\lambda u}(F^*(s+u, s) - F^*(t+u, s)) \, du.$$

The second term of the integrand now cancels the first term of (5.16), leaving the coefficient $G_\lambda^*(s)$ in the form asserted by Lemma 5.9(1).

Turning to (2), we have first to show the existence of a version of $d\langle X^+(t, s), Y\rangle_s^z$ having the asserted measurability in $(t, (s, w_z))$. Now for any choice thereof, say $h(t, s, w_z)$, $\int_0^u h(t, s, w_z) \, dY(s)$, $u \leq t$, is the projection of $X_t^+$ onto the stable subspace generated by $dY$ up to time $u$. This is just an ordinary $L_z^2$-projection (Dellacherie and Meyer 1980, VIII, 51). Writing

$$\int_0^{t+\Delta} h(t+\Delta, s, w_z) \, dY(s) - \int_0^t h(t, s, w_z) \, dY(s)$$

$$= \int_0^t (h(t+\Delta, s, w_z) - h(t, s, w_z)) \, dY_s + \int_t^{t+\Delta} h(t+\Delta, s, w_z) \, dY_s,$$

as $\Delta \to 0+$ the first integral on the right tends to 0 in $L_z^2$ because $X_t^+$ is r.c.l.l. in $L_z^2$. The last integral is approximated in $L_z^2$ by the projection of $X_t^+$ onto the stable subspace of $(dY_s; t < s \leq t + \Delta)$. This Hilbert space can be written as a direct sum of those generated by $dY(s)$, $t + \Delta_{n+1} < s \leq t + \Delta_n$, where $\Delta_1 = \Delta$ and $\Delta_n$ decrease to 0. Thus the projection onto the space generated by $dY(s)$, $t < s \leq t + \Delta_n$ tends to 0 in $L_z^2$. It follows that the projections $\int_0^t h(t, s, w_z) \, dY(s)$ are right-continuous in $L_z^2$. Similarly, if $\int_0^{u-} k(t, s, w_z) \, dY(s)$ defines the projection of $X_{t-}^+$ onto the stable subspace of $(dY_s; 0 < s < u)$, $u \leq t$, then writing

$$\int_0^{t-} k(t, s, w_z) \, dY(s) - \int_0^{t-\Delta} h(t-\Delta, s, w_z) \, dY(s)$$

in triangular form it follows that this difference tends to 0 in $L_z^2$ as $\Delta \to 0+$, so that $\int_0^t h(t, s, w_z) \, dY(s)$ has left limits in $L_z^2$. Now since $dY(s)$ is orthogonal to $X_t$ for $s > t$, if we set $h(t, s, w_z) = 0$ for $s > t$ then, for any fixed $K$, $\int_0^K h(t, s, w_z) \, dY(s)$ defines the projection of $X_t^+$ onto the stable subspace of $dY(s)$, $s \leq K$, and we have

$$E^z \left( \int_0^K h(t_1, s, w_z) \, dY(s) - \int_0^K h(t_2, s, w_z) \, dY(s) \right)^2$$

$$= \int_0^K E^z (h(t_1, s, w_z) - h(t_2, s, w_z))^2 \sigma^2(s).$$

It follows from the r.c.l.l. property of the projections that the functions $h(t, s, w_z)$ considered as functions of $(s, w_z)$ with parameter $t \leq K$ are r.c.l.l. in $L_z^2((0, K] \times \Omega_z; \mathscr{P}; d\langle Y \rangle_s^z; s \leq K)$ for each $K$ (where our notation here means $d\langle Y \rangle_s^z(S) = E^z \int_0^K I_S \, d\langle Y \rangle_s^z)$. For each $K$, this may be normalized to a probability measure, so that $h(t, s, w_z)$ may be regarded as a process in $t$, $t \leq K$. The existence of the measurable version $F^\#(t, s, w_z)$ now follows directly from standard criteria for the existence of measurable modifications of a process (Dellacherie and Meyer 1975, IV, 30). It suffices to apply the result in $K < t \leq K + 1$ for each integer $K \geq 0$ to obtain a definition of $F^\#(t, s, w_z)$ for all $t > 0$.

As in the proof of (1), we may and do assume that $\hat\rho(Z_t)$ has mean 0, so that $\hat\rho(Z_t)$ and $X_t^+$ are $P^z$-equivalent for a.e. $t$. Then we have

$$X_t^+ = \int_0^t F^\#(t,s)\,dY_s + V_t^\# \quad \text{and} \quad M_\lambda(t) = \int_0^t G_\lambda^\#(s)\,dY_s + W_t^\#,$$

where $E^z(V_t^\# | \mathscr{Y}_s^z)$ is a martingale in $s$ orthogonal to $Y_s$, and $W_t^\#$ is orthogonal to $Y_t$. Next, we may replace (5.15) by

$$E^z \int_0^\infty e^{-\lambda u}(X_u^+)^2\,du \geq E^z \int_0^\infty e^{-\lambda u}\left(\int_0^u (F^\#(u,s))^2\,d\langle Y\rangle_s^z\right)du$$

$$= E^z \int_0^\infty \left(\int_s^\infty e^{-\lambda u}(F^\#(u,s))^2\,du\right)d\langle Y\rangle_s^z$$

$$\geq \lambda E^z \int_0^\infty \left(\int_s^\infty e^{-\lambda u}|F^\#(u,s)|\,du\right)^2 d\langle Y\rangle_s^z$$

$$\geq \lambda\, e^{2\lambda t} E^z \int_0^t \left(\int_0^\infty e^{-\lambda v}|F^\#(t+v,s)|\,dv\right)^2 d\langle Y\rangle_s^z.$$

(5.17)

Thus, by Fubini's theorem, $\int_0^\infty e^{-\lambda v} F^\#(t+v,s)\,dv$ exists as a $\mathscr{P}$-measurable function of $(s, w_z)$, $(s \leq t)$, for each $t$. It is easily seen to be continuous in $t$ ($t \geq s$), and the stochastic integral $\int_0^t (\int_0^\infty e^{-\lambda v} F^\#(t+v,s)\,dv)\,dY(s)$ is well defined. To see that this defines the projection of $\int_0^\infty e^{-\lambda v} \hat\rho(Z_{t+v})\,dv$ onto the stable subspace of $(dY(s); 0 < s < t)$ we reason as follows (where, of course, we first replace $\hat\rho(Z_{t+v})$ by $X_{t+v}^+$). Since $X_t^+$ is r.c.l.l. in $L_z^2$, the last integral is a limit of Riemann sums, for which the projection can be done term by term. Thus it follows that the projection commutes with integration, to become, as an $L_z^2$-integral, $\int_0^\infty e^{-\lambda v}(\int_0^t F^\#(t+v,s)\,dY(s))\,dv$. To show that the order of integration may be interchanged, since we are working within the stable subspace of $(dY(s), s \leq t)$, we let $h(s)$ be any bounded, previsible process and show

$$E^z\left(\left(\int_0^t h(s)\,dY(s)\right)\int_0^\infty e^{-\lambda v}\left(\int_0^t F^\#(t+v,s)\,dY(s)\right)dv\right)$$

$$= E^z\left(\left(\int_0^t h(s)\,dY(s)\right)\int_0^t \left(\int_0^\infty e^{-\lambda v} F^\#(t+v,s)\,dv\right)dY(s)\right). \quad (5.18)$$

Now the right side, by the usual orthogonality argument, is

$$E^z \int_0^t h(s)\left(\int_0^\infty e^{-\lambda v} F^\#(t+v,s)\,dv\right)d\langle Y\rangle_s^z.$$

# 190 | Foundations of the prediction process

As to the left side, it is the same as

$$E^z\left(\int_0^t h(s)\,dY(s)\int_0^\infty e^{-\lambda v}X^+_{t+v}\,dv\right).$$

When we again replace the second integral by the approximating Riemann sums, we get

$$\lim_{n\to\infty}\sum_k e^{-\lambda k/n}E^z\left(\left(\int_0^t h(s)\,dY(s)\right)X^+_{t+k/n}\right)n^{-1}$$

$$=\lim_{n\to\infty}\sum_k e^{-\lambda k/n}\left[E^z\int_0^t h(s)F^\#\left(t+\frac{k}{n},s\right)d\langle Y\rangle^z_s\right]n^{-1}$$

$$=\int_0^\infty e^{-\lambda v}\left(E^z\int_0^t h(s)F^\#(t+v,s)\,d\langle Y\rangle^z_s\right)dv,$$

because the item in the last parenthesis is r.c.l.l. and bounded in $v$. There is no difficulty in applying Fubini's theorem to this last expression to complete the proof of (5.18). Since the class of such $\int_0^t h(s)\,dY(s)$ is dense in the stable subspace generated by $(Y(s),\ s\le t)$, we see that the projection of $\int_0^\infty e^{-\lambda v}\hat\rho(Z_{t+v})\,dv$ onto this subspace is indeed given by

$$\int_0^t\left(\int_0^\infty e^{-\lambda v}F^\#(t+v,s)\,dv\right)dY(s),$$

and there is no difficulty in noting that the integrand exists simultaneously in $\lambda>0$ except for a set of $(s,w_z)$ of $d\langle Y\rangle^z_s$-measure 0, where we can define it as 0, preserving the $\mathscr{P}$-measurability.

Finally, we need to consider the projection of the integral term of $M_\lambda$ onto the stable subspace, analogous to (5.16). Interchanging projection with integration in the parameter $v$ (the integrand is $L^2_z$-r.c.l.l.), this becomes, much as above (5.16),

$$\lambda\int_0^t\int_0^v\int_0^\infty e^{-\lambda u}(F^\#(v,s)-F^\#(v+u,s))\,du\,dY(s)\,dv,$$

and our problem again in to interchange $dv$ with $dY(s)$-integration, to obtain

$$\lambda\int_0^t\left(\int_s^t\int_0^\infty e^{-\lambda u}(F^\#(v,s)-F^\#(v+u,s))\,du\,dv\right)dY(s).$$

In the first expression, we replace $\int_0^t$ by an approximating sum in $L^2_z$, then multiply by $\int_0^t h(s)\,dY(s)$ for bounded previsible $h(s)$ as before, and apply $E^z$. The result can be written

$$\frac{\lambda}{n}\sum_{k\le n}\int_0^\infty e^{-\lambda u}\left(E^z\int_0^{kn^{-1}t}(F^\#(kn^{-1}t,s)-F^\#(kn^{-1}t+u,s))h(s)\,d\langle Y\rangle^z_s\right)du,$$

and as $n \to \infty$ this tends to

$$\lambda \int_0^\infty e^{-\lambda u} \left( E^z \int_0^t \int_0^v (F^\#(v, s) - F^\#(v + u, s)) h(s) \, d\langle Y \rangle_s^z \, dv \right) du$$

because, as shown at the start of the proof of (2) $\int_0^v F^\#(v + u, s) \, dY(s)$ is r.c.l.l. in $L_z^2$ as a function of $v$ and $L_z^2$-bounded in $u$. Now applying Fubini's theorem again, this becomes

$$\lambda E^z \int_0^\infty e^{-\lambda u} \int_0^t h(s) \int_s^t (F^\#(v, s) - F^\#(v + u, s)) \, dv \, d\langle Y \rangle_s^z \, du,$$

and one more application reduces it to

$$\lambda E^z \int_0^t h(s) \left( \int_s^t \int_0^\infty e^{-\lambda u} (F^\#(v, s) - F^\#(v + u, s)) \, du \, dv \right) d\langle Y \rangle_s^z,$$

which is the corresponding expression for the desired interchange of integration.

Combining this with the former projection, we have

$$G_\lambda^\#(s) = \int_0^\infty e^{-\lambda u} F^\#(t + v, s) \, dv$$

$$+ \lambda \int_s^t \int_0^\infty e^{-\lambda u} (F^\#(v, s) - F^\#(v + u, s)) \, du \, dv, \quad (5.19)$$

for $d\langle Y \rangle_s^z$-a.e. $(s, w_z)$, $s \leq t$, where the right side is $\mathscr{P}$-measurable. It is clear from (5.17) directly that it is square-integrable with respect to $d\langle Y \rangle_s^z$, $s \leq t$, but at this point it appears to depend on $t$. However, the calculation following (5.16) can now be carried out pathwise, $d\langle Y \rangle_s^z$-a.e., since it involves only the variables $(u, v)$ and the integrand is in $L^1(du, dv)$ for $d\langle Y \rangle_s^z$-a.e. $(s, w_z)$. As before, the dependence on $t$ disappears, and (5.19) reduces to the Conclusion of (2), but with $s$ in place of $t$. The final assertion about the pathwise integration comes from the third line of (5.17), which implies that

$$\infty > E^z \int_0^\infty e^{-\lambda u} (X_u^+)^2 \, du \geq E^z \int_0^\infty e^{-2\lambda t} \left( \int_0^\infty e^{-\lambda v} (F^\#(t + v, t))^2 \, dv \right) d\langle Y \rangle_t^z,$$

showing that, for $d\langle Y \rangle_t^z$-a.e. $(t, w_z)$, $F^\#(t + v, t)$ is $e^{-\lambda v}\, dv$-square-integrable. This completes the proof of Lemma 5.9. The main use of this lemma is to prove the next theorem.

**Theorem 5.10** *Referring to Definition 5.4 and Theorem 5.8, we have for $z \in D \cap M_2^+$*

$$X_z^+(t) \in H_z'(t) = H_z^*(t) = H_z^*(t+) \subset H_z^\#(t), \quad t \geq 0.$$

192 | Foundations of the prediction process

*Moreover, for $z \in M_2^+$ there are representations*

$$X_t^+(t) = \sum_{n=1}^{\phi_z^*(t)} \int_0^t F_n^*(t, s) \, dY_n(s)$$

and

$$X_z^+(t) = \sum_{n=1}^{\phi_z^\#(t)} \int_0^t F_n^\#(t, s) \, dY_n^\#(s),$$

*where $F_n^*(t, s)$ are non-random and $F_n^\#(t, s)$ are $\mathscr{Y}_s^z$-previsible. For each $t$, these integrands are uniquely determined apart from a $dE^z Y_n^2(s)$-null set of $s$ (respectively, a $d\langle Y_n^\# \rangle_s^z$-null set of $(s, w_Z))$.*


**Proof** Let $Y(s)$ be a square-integrable $\mathscr{Y}_t^z$-martingale, orthogonal to $Y_n(s)$ for every $n$ and $s \leq t$, in the linear sense $(E^z(Y_n(s)Y(s)) = 0)$. Then it is orthogonal to each $M_n(s)$, and by Lemma 5.9(1), since $G_n^* \equiv 0$,

$$0 = \int_0^\infty e^{-ns} F^*(t + s, t) \, ds \qquad \text{for all } n > 0$$

and $d\sigma^2$-a.e. $t$, where

$$X_z^+(t) = \int_0^t F^*(t, s) \, dY_s + W_t^*$$

with $0 = E^z(Y_s W_t^*)$, $0 \leq s \leq t$. By (5.15), for $d\sigma^2$-a.e. $t$, $e^{-ns} F^*(t + s, t) \in L^1(ds)$, so by the Stone–Weierstrass theorem we have $F^*(t + s, t) = 0$ for a.e. $s > 0$ and $d\sigma^2$-a.e. $t$. On the other hand, setting $F^*(t, s) = 0$ for $s > t$, $F^*(t, s)$ (as a function of $s$) is right-continuous in $t$ in the sense of $L^2(d\sigma^2(s))$ because $X_z^+$ is r.c.l.l. in $L_z^2$. Since we must have $0 = F^*(t, s)$ for $(dt \times d\sigma^2)$-a.e. $(t, s)$, by Fubini's theorem it also vanishes for $d\sigma^2$-a.e. $s \leq t$, for Lebesgue-a.e. $t$. Thus we have, for $u, v \geq 0$,

$$0 = \int_0^v \int_0^u F^*(t, s) \, d\sigma^2(s) \, dt.$$

Then by right-continuity, $\int_0^u F^*(t, s) \, d\sigma^2(s) = 0$ holds for every $t$ and $u$, which leaves $X_z^+(t) = W_t^*$. Thus, the martingale $E^z(X_z^+(t) \mid \mathscr{Y}_s^z)$ is linearly orthogonal to $Y(s)$. But we can decompose $E^z(X_z^+(t) \mid \mathscr{Y}_s^z)$ into its projection onto the linear martingale space of $(Y_n, 1 \leq n)$ plus a martingale orthogonal to these (the same procedure used for Theorem 5.8(1)). Taking the orthogonal part as $Y(s)$, it follows that $Y(t) = 0$, $P^z$-a.s., because the two components of $E^z(X_z^+(t) \mid \mathscr{Y}_s^z)$ are orthogonal. This completes the proof that $X_z^+(t) \in H_z'(t)$, and that the first representation of $X_z^+$ exists.

Since $R_\lambda^z \hat{\rho}(Z_t) = \int_0^\infty e^{-\lambda s} E^z(X^+(t + s) \mid \mathscr{Y}_t^0) \, ds$, when we again centre at the mean in defining $M_\lambda(t)$, it is clear by the $L_z^2$-right-continuity of $X^+(t)$ that $R_\lambda^z \hat{\rho}(Z_s) \in H_z^*(t)$ for $s \leq t$. A similar reasoning applies to the integral part of $M_\lambda$, so we can infer that $H'_z(t) \subset H^*_z(t)$. Now $H'_z(t) = H'_z(t+)$, because the $Y_n$ are r.c.l.l. in $L^2_z$ with orthogonal increments, so that for $\Delta > 0$, $H'_z(t+\Delta) = H'_z(t) \oplus H'_z(t, t+\Delta]$ where the last Hilbert space tends to $\Phi$ as $\Delta \to 0+$. Thus it is enough to show that $H^*_z(t) \subset H'_z(t)$, and then to take right limits in $t$ to obtain $H^*_z(t+) = H^*_z(t)$. We already have $X^+(s) \in H'_z(t)$, $s \le t$, and from our representation it is clear that $E^z(X^+(u) \mid \mathscr{L}^o_v) \in H'_z(t)$ for $u \le t$. It remains to prove the same fact for $u > t$. We note that, thus far, no use was made of the assumption $z \in D$. We will show that $R^z_\lambda \hat\rho(Z_s) \in H'_z(t)$ for $s \le t$. Indeed, since $X^+_s \in H'_z(t)$ we have $\int_0^s X^+(u)\,du \in H'_z(t)$, and by subtraction from $M_\lambda(s)$ it follows that

$$R^z_\lambda \hat\rho(Z_s) - R^z_\lambda \hat\rho(Z_0) - \lambda \int_0^s R^z_\lambda(\hat\rho(Z_u))\,du \in H'_z(t).$$

Now for $z \in D$, $R^z_\lambda \hat\rho(Z_0)$ is constant $P^z$-a.s., and since we are centring at the mean in defining $M_\lambda$, this constant is 0. Then

$$R^z_\lambda \hat\rho(Z_s) - \lambda \int_0^s R^z_\lambda \hat\rho(Z_u)\,du \in H'_z(t),$$

therefore also

$$\frac{d}{ds}\left(e^{-\lambda s} \int_0^s R^z_\lambda(\hat\rho(Z_u))\,du\right) = e^{-\lambda s}\left(R^z_\lambda \hat\rho(Z_s) - \lambda \int_0^s R^z_\lambda(\hat\rho(Z_u))\,du\right) \in H'_z(t),$$

since the ordinary derivative coincides with the derivative in quadratic mean. But then

$$\int_0^s (e^{\lambda s} - e^{\lambda u}) \frac{d}{du}\left(e^{-\lambda u} \int_0^u R^z_\lambda \hat\rho(Z_v)\,dv\right) du$$

$$- \lambda \int_0^s \left(\int_0^u R^z_\lambda \hat\rho(Z_v)\,dv\right) du \in H'_z(t), \qquad s \le t.$$

Now two differentiations give us $R^z_\lambda \hat\rho(Z_s) \in H'_z(t)$, $s \le t$, as asserted. But

$$R^z_\lambda \hat\rho(Z_s) = \int_0^\infty e^{-\lambda u} E^z(X^+_{s+u} \mid \mathscr{L}^o_s)\,du,$$

where the integrand is r.c.l.l. in $L^2_z(\mathscr{L}^o_s)$. So by inversion of the transform (or by the Stone–Weierstrass theorem) we have $E^z(X^+_{s+u} \mid \mathscr{L}^o_s) \in H'_z(t)$, $s \le t$, $u \ge 0$. This completes the proof of Theorem 5.10.

We have already observed in Section 5.1 that the first representation of Theorem 5.10 provides an independent proof of Theorem 3.38 in the particular case that $z \in M^+_2 \cap D$. Let us conclude with an example to show that the inclusion $H^*_z(t+) \subset H^\#_z(t)$ may be strict. Let $X_t$ be a standard

Brownian motion, $X_0 = 0$. Then since $X_t$ is Gaussian we have $H_z(t) = H_z^*(t)$, and $H_z^*(t) = H_z^*(t+)$ by the 0–1 Law. However, if $T = \inf\{t: X_t \in \pm 1\}$, then

$$X_{T \wedge 1} = \int_0^1 I_{(s<T)} \, dX_s \in H_z^\#(1).$$

But clearly $X_{T \wedge 1} \notin H_z^*(t)$. The central point is that $H_z^\#(t)$ is closed under optional stopping by previsible stopping times, while $H_z^*(t)$ is not.

# 6

# Application to concrete examples, and ramifications of the theory

> Probabilities are as various as the faces to be seen at will in fretwork or paperhangings; every form is there, from Jupiter to Judy, if you only look with creative imagination.
>
> George Eliot, 1872. Middlemarch, 6.

## 6.1 Application to explicit prediction

In order to obtain the coefficients $F_n^*(t, s)$ and $F_n^{\#}(t, s)$ of Theorem 5.10 for a given process $X_z^+(t)$, we can use the same method as discussed in the Gaussian case after Theorem 3.27. However, since $X_z^+(t)$ is not continuous in $L^2$ (only right-continuous) it is first necessary to amplify that discussion slightly to deal with possible discontinuities in $E^z Y_k^2(t)$, and, generally speaking, with the uniqueness of the inversion. The two basic formulas are simply those of Lemma 5.9 with $Y_k(t)$ in place of $Y(t)$, and our starting points for the inversion are the families of $G_\lambda^*(u)$ ($=G_{\lambda,k}^*(u)$) and $G_\lambda^{\#}(u)$ ($=G_{\lambda,k}^{\#}(u)$), respectively, for all $u < t$ (fixed) and $\lambda > 0$. Then we have (suppressing a dependence on $k$) the following result.

**Theorem 6.1** *The identities* $G_\lambda^*(u) = \int_0^\infty e^{-\lambda s} F_k^*(u + s, u)\, ds$ *(respectively, $G_\lambda^{\#}(u) = \int_0^\infty e^{-\lambda s} F_k^{\#}(u + s, u)\, ds$), $u < t$, $\lambda > 0$, determine $F_k^*(v, s)$ uniquely, for each $v < t$, up to a $d\sigma_k^2(s)$ ($\dot{=} dE^z Y_k^2(s)$) null set of $s \leq v$ (respectively, determine $F_k^{\#}(v, s)$ up to a $d\langle Y_k \rangle_s^z$-null set of $(s, w_z)$, $s \leq v$). In other words, they suffice to determine the representation of Theorem 5.10 for $v < t$ by Laplace inversion.*

**Proof** Applying the Laplace inversion formula (Widder 1948, Chapter VII, Section 6) to $G_\lambda^*(u)$ for each $u < t$ (setting the result equal to 0 if it does not converge, which preserves the joint measurability) we determine $F_k^*(u + s, u)$ for $(d\sigma_k^2 \otimes ds)$-a.e. $(u, s)$, $u < t$, $0 < s$ (where we note that here the designation of variables is opposite from that of the Discussion following Theorem 3.27). This implies determination of $F_k^*(v, u)$ for $(dv \otimes d\sigma_k^2)$-a.e. $v \geq u$, $u < t$, and hence by Fubini's theorem, for a.e. $v < t$, $F_k^*(v, u)$ is determined for $d\sigma_k^2$-a.e. $u \leq v$. Thus we determine $\int_{u_1}^t \int_{u_1}^v F_k^*(v, u)\, d\sigma_k^2(u)\, dv$ for $u_1 < t$. But, setting $F_k^*(v, u) = 0$ for $v < u$, by $L^2$-right-continuity of $X_z^+(t)$ we see that $\int_{u_1}^t F_k^*(v, u)\, d\sigma_k^2(u)$ is right-continuous in $v$, and hence is determined for all

$v \in (u_1, t)$. This implies that $F_k^*(v, u)$ is determined for $d\sigma_k^2$-a.e. $u \le v$, for all $v < t$, as asserted. The argument for the second case is similar when we replace $d\sigma_k^2(u)$ by $d\langle Y_k \rangle_u^z$, and, instead of the integral over $(u_1, v]$, we integrate over $(u_1, v] \otimes S$ for an arbitrary set $S \in \mathscr{P}$.

Turning now to explicit examples, suppose that we begin with a real-valued process $X_t$ with a.s. right-continuous paths, with right-continuous $EX_t$, and with $P = z \in M_2^+$. Then it is entirely unnecessary to introduce the prediction space superstructure which we have used to formalize the proofs for measurable processes. We simply proceed directly on the probability space of $X_t$, by replacing $R_\lambda^z(Z(t))$ by $R_\lambda^z(X(t)) \doteq \int_0^\infty e^{-\lambda s} E(X_{s+t} | \mathscr{F}_{t+}^\circ) \, ds$. The existence of a right-continuous version of the former, together with the equivalence in law, assures us of a right-continuous version of $R_\lambda^z(X(t))$, and, using it, we replace $M_\lambda(t)$ of Definition 3.6 by its direct analogue on the probability space of $X$. Then all of the subsequent Hilbert spaces and representation theorems have their analogues on the space of $X$, and we express the predictors in terms of $X_t$ instead of $Z_t$ ($z$ being fixed). Of course, the analogue of $X_t^+$ is $X_t - E^z X_t$ since the $L^2$ and a.s. right limits coincide.

**Convention 1** When no confusion arises, we use the notations $R_\lambda^z(X(t))$, $M_\lambda(t)$, $Y_k$, $G_\lambda^\#$, and $F_\lambda^\#$ for their analogues on the probability space of $X$.

Note that, since $G_\lambda^*$ and $F_k^*$ are non-random, they may of course be used for either space. In actual examples, these functions will have natural continuity properties which determine them uniquely, so the exceptional null sets of Theorem 6.1 are of no concern.

**Example 1** As a first example, let $X_1$ and $X_2$ be any random variables with $EX_1 = EX_2 = 0$ and $0 < EX_1^2, EX_2^2, E(X_1 - X_2)^2$ (so that the joint distribution is non-degenerate), and define

$$X_t = cI(t < 1) + X_1 I(1 \le t < 2) + X_2 I(2 \le t),$$

where $c$ is any constant. Then $X_t$ is a continuous extension of a discrete time process. (Further jumps may be added at times $n > 2$, but would only complicate the discussion.) In the terminology of Definition 4.5, we have $z \in M(J_D^{c)} \cap M_d$, i.e. a pure jump prediction process with all jumps from branch points. Our prediction theory of course yields no new conclusions in this case (where the prediction is trivial) but we can use it to illustrate the theory. First, we write

$$\lambda R_\lambda^z(X(t)) = \begin{cases} c(1 - e^{-\lambda(1-t)}) & 0 \le t < 1 \\ X_1(1 - e^{-\lambda(2-t)}) + E(X_2 | X_1) e^{-\lambda(2-t)} & 1 \le t < 2 \\ X_2 & 2 \le t. \end{cases}$$

Application and ramifications | 197

To calculate $\lambda M_\lambda(t)$, it is easiest to observe that, along with the filtration $\mathscr{F}^\circ_{t+}$, $M_\lambda(t)$ is constant in $[0, 1)$, $[1, 2)$, and $[2, \infty)$, and it is 0 in $[0, 1)$. Since the integral term of $M_\lambda(t)$ is continuous, the jump at times 1 and 2 is just the jump of $R^Z_\lambda(X(t))$ at those times. Thus we obtain

$$\lambda M_\lambda(t) = \begin{cases} 0 & 0 \leq t < 1 \\ (1 - e^{-\lambda})X_1 + e^{-\lambda}E(X_2|X_1) & 1 \leq t < 2 \\ (1 - e^{-\lambda})X_1 - (1 - e^{-\lambda})E(X_2|X_1) + X_2 & 2 \leq t. \end{cases}$$

The orthogonalization procedure of Theorem 5.8(1) need not be followed through when, as happens here and in many other cases, there are simplifications available. Let us introduce formally the next convention.

**Convention 2** When they exist as non-trivial square-integrable martingales, we set $M_\infty(t) = \lim_{\lambda \to \infty} \lambda^k M_\lambda(t)$ and $M_0(t) = \lim_{\lambda \to 0} \lambda^k M_\lambda(t)$, the value of $k$ and the type of limit depending on context.

In the present example, we have $k = 1$ and

$$M_\infty(t) = \begin{cases} 0 & t < 1 \\ X_1 & 1 \leq t < 2 \\ X_1 + X_2 - E(X_2|X_1) & 2 \leq t \end{cases}$$

and

$$M_0(t) = \begin{cases} 0 & t < 1 \\ E(X_2|X_1) & 1 \leq t < 2 \\ X_2 & 2 \leq t \end{cases}$$

Now it is easy to see that $H^*_z(t)$ of Definition 5.4 is $\mathbf{0}$ for $t < 1$, generated by $X_1$ and $E(X_2|X_1)$ for $1 \leq t < 2$, and generated by $\{X_1, E(X_2|X_1), X_2\}$ for $2 \leq t$. Thus we have $\phi^*_z(1) = \phi^*_z(\infty) = 2$, and to get a linear orthogonal representation of $X(t)$ it suffices to orthogonalize $M_\infty$ and $M_0$. Since $\Delta M_\infty(2) = \Delta M_0(2) = X_2 - E(X_2|X_1)$, we need only orthogonalize up to $t = 1$, and then $M_\infty(2)$ will suffice to generate $H^*_z(2)$ along with the past. Thus we are led to the pair $M_\infty(t)$ and

$$N(t) = \begin{cases} 0 & t < 1 \\ X_1 - \dfrac{EX_1^2}{E(X_1 X_2)} E(X_2|X_1); & 1 \leq t. \end{cases}$$

But, since $\mathscr{F}^\circ_{1-}$ is degenerate, it is clear that $M_\infty(t)$ and $N(t)$ are also orthogonal in the stochastic sense, so that we have $\phi^\#_z(t) = \phi^*_z(t)$ (and yet, of course, $H_z^\#(2) \neq H_z^*(2)$, since $f(X_1, E(X_2|X_1))\Delta M_\infty(2)$ is in $H_z^\#(2)$ for nonlinear $f(x_1, x_2)$).

Let us work out the orthogonal representation of $X(t)$ on the basis of $(M_\infty, N)$ by using Lemma 5.9 to calculate the respective coefficients $F_\infty^*(t, s)$ and $F_N^*(t, s)$. Setting

$$G_{\lambda,\infty}^*(t) = \int_0^\infty e^{-\lambda s} F_\infty^*(t+s, t)\, ds \quad \text{and} \quad G_{\lambda,N}^*(t) = \int_0^\infty e^{-\lambda s} F_N^*(t+s, t)\, ds,$$

we need

$$E(M_\lambda(t)M_\infty(t)) = \begin{cases} 0 & t < 1, \\ \lambda^{-1}(1-e^{-\lambda})EX_1^2 + \lambda^{-1}e^{-\lambda}E(X_1X_2) & 1 \leq t < 2, \\ \lambda^{-1}[(1-e^{-\lambda})EX_1^2 + e^{-\lambda}E(X_1X_2) + EX_2^2 - EE^2(X_2|X_1)] & 2 \leq t. \end{cases}$$

Similarly

$$E(M_\lambda(t)N(t)) = \begin{cases} 0 & t < 1, \\ \lambda^{-1}e^{-\lambda}\left[E(X_1X_2) - (EX_1^2)\dfrac{EE^2(X_2|X_1)}{EX_1X_2}\right] & 1 \leq t. \end{cases}$$

It follows (since $F_\infty^*(t, s)$ is constant in $1 \leq t < 2$, and in $2 \leq t$) that we only require

$$G_{\lambda,\infty}^*(1) = \frac{E(M_\lambda(1)M_\infty(1))}{EX_1^2} = \frac{\lambda^{-1}(1-e^{-\lambda})EX_1^2 + \lambda^{-1}e^{-\lambda}E(X_1X_2)}{EX_1^2},$$

and

$$G_{\lambda,\infty}^*(2) = \frac{\Delta E(M_\lambda(2)M_\infty(2))}{\Delta EM_\infty^2(2)} = \lambda^{-1}\frac{EX_2^2 - EE^2(X_2|X_1)}{E(X_2 - E(X_2|X_1))^2} = \lambda^{-1},$$

and similarly

$$G_{\lambda,N}^*(1) = \frac{E(M_\lambda(1)N(1))}{EN^2(1)} = -\lambda^{-1}e^{-\lambda}\frac{E(X_1X_2)}{EX_1^2}.$$

Now inverting these transforms, we have

$$F_\infty^*(1+s, 1) = I_{[0,1)}(s) + I_{[1,\infty)}(s)\frac{E(X_1X_2)}{EX_1^2}$$

$$F_\infty^*(2+s, 2) = I_{[0,\infty)}(s)$$

$$F_N^*(1+s, 1) = -I_{[1,\infty)}(s)\frac{E(X_1X_2)}{EX_1^2},$$

where we use right-continuity of $F^*$ in $t$. To obtain the first representation of Theorem 5.10 (except that $(M_\infty, N)$ replaces $(Y_1, Y_2)$) we need only set $1 + s = t$, to check that

$$X(1) = X_1 = F^*_\infty(1, 1)M_\infty(1) + F^*_N(1, 1)N(1),$$

and

$$X(2) = X_2 = (F^*_\infty(2, 1)M_\infty(1) + F^*_\infty(2, 2)\Delta M_\infty(2)) + F^*_N(2, 1)N(1)$$
$$= \frac{E(X_1 X_2)}{EX_1^2} X_1 + (X_2 - E(X_2|X_1)) - \frac{E(X_1 X_2)}{EX_1^2} X_1 + E(X_2|X_1).$$

We note that the prediction of $X_2$ given $\mathscr{F}^o_{1+}$, namely $E(X_2|X_1)$, is indeed equal to $F^*_\infty(2, 1)M_\infty(1) + F^*_N(2, 1)N(1)$, and these terms are orthogonal. Moreover, since measurable non-random processes are previsible, it follows from uniqueness that the above also gives the stochastic integral representation of $X(t)$ (Theorem 5.10) with $(M_\infty, N)$ in place of $(Y^\#_1, Y^\#_2)$.

**Exercise 6.1** Define the family $\lambda M_\lambda(t)$ for a discrete process with jumps at $t = 1, 2, 3$ having mean 0, finite mean square, and non-degenerate joint distribution. Obtain $M_\infty(t)$ and $M_0(t)$ in this case. (Observe that they no longer suffice for the representation of $H^*_z(1)$.)

**Example 2** As a second example, we shall work out the representation of $X_t = B^3(t)$, where $B(t)$ is Brownian motion starting at 0. Although application of Theorem 5.10 here is not difficult, it affords a chance to compare the result with the wide-sense prediction of Chapter 3, and to illustrate the use of Ito's lemma (McKean 1969, 2.6) which is often a powerful in implementing Theorem 5.10. Since $X_t$ is here a Markov process, we have easily

$$E(X_{t+s}|\mathscr{F}^o_{t+}) - X_t + 3sX_t^{\frac{1}{3}},$$

which yields $R^z_\lambda(X_t) = \lambda^{-1}X_t + 3\lambda^{-2}X_t^{\frac{1}{3}}$. Thus

$$\lambda M_\lambda(t) = X_t + 3\lambda^{-1}X_t^{\frac{1}{3}} - 3\int_0^t X_s^{\frac{1}{3}}\, ds,$$

and, applying Ito's lemma to $B^3(t)$, we get

$$\lambda M_\lambda(t) = 3\int_0^t (B^2(s) + \lambda^{-1})\, dB(s). \tag{6.1}$$

The advantage of (6.1) is that it yields easily

$$\frac{d\langle M_\lambda, M_1 \rangle_t}{d\langle M_1 \rangle_t} = \lambda^{-1}(B^2(t) + \lambda^{-1})/(B^2(t) + 1),$$

and inverting the transform gives us the coefficient of the projection of $X_t$

onto the stochastic integral space of $(M_1(s), s \leq t)$ in the form

$$F^{\#}(t+s, t) = (B^2(t) + 1)^{-1}(B^2(t) + s).$$

It is trivial to see that $M_1$ generates all of $H_z^{\#}$, and the PMSIR becomes

$$X_t = \int_0^t F^{\#}(t, s) \, dM_1(s) = 3 \int_0^t (B^2(s) + (t - s)) \, dB(s). \tag{6.2}$$

Of course, since the projection is simple in this case, there is no practical advantage in using (6.2) to compute it.

Turning to the linear martingale representation (LMR), it is again easiest to look at

$$M_\infty(t) = \lim_{\lambda \to \infty} \lambda M_\lambda(t) = 3 \int_0^t B_s^2 \, dB_s,$$

and

$$M_0(t) = \lim_{\lambda \to 0+} \lambda^2 M_\lambda(t) = 3B_t.$$

Taking $B_t$ as one basis element and orthogonalizing, we have

$$\frac{dEM_0 M_\infty(t)}{dEM_0^2(t)} = t,$$

and so we can use $N(t) = \int_0^t (B^2(s) - s) \, dB(s)$. Comparing with (6.2), we see that the LMR is

$$X_t = 3tB_t + 3N(t). \tag{6.3}$$

Here, of course, the multiplicity is $\phi_z^*(t) = 2$, whereas $\phi_z^{\#}(t) = 1$.

It is natural to suppose that $X_t$ is an intrinsic diffusion (Section 4.2), but this is not obvious. To prove it, we need to show that $M_\lambda^f(t)$ is a.s. continuous for each $f \in b(\mathcal{B})$, and not only for $M_\lambda$ (as in Theorem 4.3). However, it does hold by virtue of the fact that $X_t$ is a diffusion in the customary sense—namely, a strong Markov process with continuous paths.

**Theorem 6.2** *Every diffusion (with a Borel transition function) is also an intrinsic diffusion, i.e. $z = P \in M_c$.*

**Proof** The problem is to connect the natural topology of $\mathbb{R}$ with the topology of $M_R$, and this apparently requires use of a section theorem, as in Dellacherie and Meyer (1987, Chapter XVI). It suffices to show that $R_\lambda f(X_t)$ is a.s. continuous in $t$ for each $f \in b(\mathcal{B})$. Since $X_t$ is continuous, $R_\lambda f(X_t)$ is a previsible process (for the natural filtration, hence also for the right-continuous augmentation). On the other hand, we know from Theorem 2.1 that $R_\lambda f(X_t)$ ($\equiv R_\lambda f \hat{\rho}(Z_t^z)$, by the Remark following Definition 2.7) is r.c.l.l., so that the process $\lim_{s \to t-} R_\lambda f(X_s)$ is well defined ($P^z$-a.s.), and then it is also left-continuous, hence previsible. Thus the Theorem will be proved if we show that these two previsible processes are equivalent. By the previsible section theorem, it is enough to show a.s. equality at any previsible $T < \infty$. Then there exist stopping times $T_n \uparrow T$, $T_n < T$ on $\{T \neq 0\}$. Now we have $R_\lambda f(X_T) = E(\int_0^\infty e^{-\lambda s} f(X_{T+s})\, ds | \mathcal{F}_T^z)$, by the strong Markov property, whereas by Hunt's lemma we have

$$\lim_{s \to T-} R_\lambda f(X_s) = \lim_{n \to \infty} R_\lambda f(X_{T_n})$$

$$= \lim_{n \to \infty} E\left(\int_0^\infty e^{-\lambda s} f(X_{T_n+s})\, ds \Big| \mathcal{F}_{T_n}^z\right)$$

$$= E\left(\int_0^\infty e^{-\lambda s} f(X_{T+s})\, ds \Big| \mathcal{F}_{T-}^z\right).$$

But since $X_T = X_{T-}$, the former conditional expectation given $\mathcal{F}_T^z$ is also measurable over $\mathcal{F}_{T-}^z$. Hence the two are equal, as was to be shown.

Returning to our current example of $X_t = B^3(t)$, we should observe that neither (6.2) nor (6.3) is a linear representation of $X_t$ in terms of the increments of $H_z(s+)$, $s < t$. To obtain these, we must use the wide-sense Cramér–Hida representation of Chapter 3 (and ultimately, of Corollary 3.29), in terms of the wide-sense analogue $\widehat{\mathcal{M}}_\lambda$ of $\mathcal{M}_\lambda$ from Corollary 3.19. This is not as simple to compute as is $\mathcal{M}_\lambda$, but the main difficulty lies in projecting $X_{t+s}$ onto $H_z(t+)$, not in calculating the coefficients of the representation. From the expression $\lambda R_\lambda^z(X_t) = X_t + 3\lambda^{-1} X_t^{\frac{1}{3}}$, we see that the problem reduces to projecting $X_t^{\frac{1}{3}}\ (=B_t)$ onto $H_z(t+)$. Since $X_t^{\frac{1}{3}}$ is continuous in $L^2$, it suffices to project onto $H_z(t)$, and so a reasonable strategy is to guess that the projection has the form $\mathscr{E}(X_t^{\frac{1}{3}}|H_z(t)) = \int_0^t F(t,s) X_s\, ds$, for suitable fixed $F(t, s)$, and to set

$$0 = E\left(\left(\int_0^t F(t,s) X_s\, ds - X_t^{\frac{1}{3}}\right) X_u\right) \qquad (6.4)$$

$$= \int_0^t F(t,s) E(B_s^3 B_u^3)\, ds - E(B_t B_u^3), \qquad 0 < u \leq t.$$

Now a routine calculation based on the Markov property of $B^3(t)$ gives

$$EB^3(s)B^3(t) = 6s^3 + 9s^2 t \qquad s \leq t,$$

and

$$EB(s)B^3(t) = \begin{cases} 3st & s \leq t \\ 3t^2 & s > t, \end{cases}$$

and (6.4) becomes

$$0 = \int_0^u F(t,s)(6s^3 + 9s^2 u)\,ds + \int_u^t F(t,s)(6u^3 + 9u^2 s)\,ds - 3u^2$$
$$0 < u \le t. \quad (6.5)$$

Differentiating twice with respect to $u$ leads to an equation which shows that $F_u(t,u)$ exists, and a third derivative shows that $F_{uu}(t,u)$ exists. Finally, the fourth derivative leads to the Euler equation

$$0 = 3u^2 F_{uu}(t,u) + 18u F_u(t,u) + 16 F(t,u). \quad (6.6)$$

For fixed $t$, the general solution is $u^{-5/2}(c_1 u^a + c_2 u^{-a})$, where $a = \frac{1}{2}(\frac{11}{3})^{\frac{1}{2}}$. Substituting back into (6.5), we must have $c_2 = 0$, but the rest reduces to

$$0 = c_1[6(a-\tfrac{3}{2})^{-1} t^{a-3/2} u^3 + 9(a-\tfrac{1}{2})^{-1} t^{a-1/2} u^2] - 3u^2. \quad (6.7)$$

This is impossible as it stands, so the solution did not have the form (6.4). However, it is easy to see how to remedy the situation. We need to subtract a term in $u^3$, which may be done if we add to $F(t,s)$ a term $c\delta_{(s=t)}$, i.e. a point mass at $t$. This leads to an extra term $c(6u^3 + 9u^2 t)$ on the right of (6.7), and solving simultaneously for $c_1$ and $c$ gives us

$$c = (6t)^{-1}(\sqrt{\tfrac{11}{3}} - 1) \quad \text{and} \quad c_1 = ct^{3/2-a}(\tfrac{3}{2} - a).$$

Hence we have finally

$$\mathscr{E}(X_t^{\frac{1}{2}} | H_z(t)) = (6t)^{-1}(\sqrt{\tfrac{11}{3}} - 1)\left( X_t + t^{3/2-a}(\tfrac{3}{2} - a) \int_0^t s^{-5/2+a} X_s\,ds \right);$$
$$a = \tfrac{1}{2}(\tfrac{11}{3})^{\frac{1}{2}}, \quad (6.8)$$

and the projection of $R_\lambda^z(X_t)$ onto $H_z(t+)$ is $\lambda^{-1} X_t + 3\lambda^{-2}\mathscr{E}(X_t^{\frac{1}{2}} | H_z(t))$. When we substitute this into the definition of $\hat{\mathscr{M}}_\lambda$ (i.e. $M_\lambda$ with $E$ replaced by $\mathscr{E}$) we find after integrating once by parts that

$$\hat{\mathscr{M}}_\lambda(t) = c(t) X_t + d(t) \int_0^t s^{-5/2+a} X_s\,ds,$$

where

$$c(t) = \lambda^{-1} + \lambda^{-2}(2t)^{-1}(\sqrt{\tfrac{11}{3}} - 1),$$
$$a = \tfrac{1}{2}(\tfrac{11}{3})^{\frac{1}{2}}$$
$$d(t) = \lambda^{-1}(\sqrt{\tfrac{11}{3}} - 1)(\sqrt{t})^{1-\sqrt{11/3}}(t + (4\lambda)^{-1}(1 - \sqrt{\tfrac{11}{3}})).$$

This is unwieldy compared to $M_\lambda(t)$ from (6.1), but of course it has the advantage of being linear in $\{X_s, s \leq t\}$. Let us use the same trick as before, setting

$$\hat{\mathcal{M}}_\infty(t) = \lim_{\lambda \to \infty} \lambda \hat{\mathcal{M}}_\lambda(t)$$

$$= X_t + (\sqrt{\tfrac{11}{3}} - 1)(\sqrt{t})^{3-\sqrt{11/3}} \int_0^t (\sqrt{s})^{-5+\sqrt{11/3}} X_s \, ds$$

$$\hat{\mathcal{M}}_0(t) = \lim_{\lambda \to 0} \lambda^2 \hat{\mathcal{M}}_\lambda(t)$$

$$= (2t)^{-1}(\sqrt{\tfrac{11}{3}} - 1)X_t + (\sqrt{\tfrac{11}{3}} - \tfrac{5}{3})(\sqrt{t})^{1-\sqrt{11/3}} \int_0^t (\sqrt{s})^{-5+\sqrt{11/3}} X_s \, ds.$$

It is easy to see that these are linearly independent, whereas $X_t$ is a linear combination of them, so we have $\phi(t) = 2$ and the coefficients of the linear representation do not depend on $s$. We omit the orthogonalization of $\hat{\mathcal{M}}_\infty$ and $\hat{\mathcal{M}}_0$, but it is easy to check that

$$X_t = (3\sqrt{\tfrac{11}{3}} - \tfrac{5}{3})^{-1}((\sqrt{\tfrac{11}{3}} - 3)\hat{\mathcal{M}}_\infty(t) + 4t\hat{\mathcal{M}}_0(t)). \tag{6.10}$$

The corresponding expression from (6.2) is

$$X_t = M_\infty(t) + \int_0^t (t-s) \, dM_0(s), \tag{6.11}$$

for purposes of comparison. The difference in form is to be expected since the $L^2$-norms of $M_\lambda$ and $\hat{\mathcal{M}}_\lambda$ are in general unrelated. For a simpler example of this, we can consider

$$X_t = \begin{cases} 0 & t < 1 \\ \sin \theta & 1 \leq t < 2 \\ \cos \theta & 2 \leq t \end{cases}$$

where $\theta$ is a uniform random variable on $(-\pi/2, \pi/2)$. Then, for $1 \leq t < 2$ and $t + s > 2$, we have $E(X_{t+s}|\mathcal{F}_{t+}^\circ) = \cos \theta = X_{t+s}$, while $\mathscr{E}(X_{t+s}|H_z(t+)) = 0$, since $E(\sin \theta \cos \theta) = 0$. It follows that, for $1 \leq t < 2$, we have

$$\lambda M_\lambda(t) = (1 - e^{-\lambda}) \sin \theta + e^{-\lambda} \cos \theta,$$

while $\lambda \hat{\mathcal{M}}_\lambda(t) = (1 - e^{-\lambda}) \sin \theta$. Obviously they lead to quite different representations of $X_t$.

In this example, the projection of $M_\lambda(t)$ onto $H_z(t+)$ is $\hat{\mathcal{M}}_\lambda(t)$. However, this is not true in general, which is why the projection was denoted by $\hat{M}_\lambda(t)$ for Theorem 5.2. Indeed we have $\hat{\mathcal{M}}_\lambda(t) \neq \hat{M}_\lambda(t)$ in Example 2, but the tedious calculations needed to evaluate $\hat{M}_\lambda$ in this case are omitted.

It may be objected in Example 2 that since the diffusion $B^3(t)$ can be reduced to the martingale $B(t)$ simply by composition with the fixed function $x^{\frac{1}{3}}$ there is no reason to use the $M_\lambda(t)$'s for that purpose. This objection overlooks the fact that composition does not commute with prediction, but in this example it carries some weight. Building on this example, however, let us next give a more serious one, which illustrates the case of strict-sense stationarity.

**Example 3**   Let $X(t) = \int_{t-1}^{t} Y^3(u)\, du$, where $Y(u)$ is a stationary Ornstein–Uhlenbeck process, $-\infty < t < \infty$. We take, for convenience, the parameter $\beta = \frac{1}{2}$, so that by (3.21)

$$Y(t) = \int_{-\infty}^{t} \exp\left(-\frac{t-u}{2}\right) dB(u), \qquad B(0) = 0, \qquad (6.12)$$

where $B(u)$ is a (two-sided) Brownian motion. Clearly $X(t)$ is strictly stationary and non-Markovian. Noting that $dX/dt = Y_t^3 - Y_{t-1}^3 \in \mathcal{F}_t^\circ$, and that $Y_t^3 = L^2\text{-}\lim_{n\to\infty} n^{-1} \sum_{k=1}^{n} (Y_t^3 - Y_{t-k}^3) \in \mathcal{F}_t^z$, we see that $\mathcal{F}_t^\circ$ and $\sigma(B(u), u \leq t)$ are equivalent.

To study the prediction of $X(t)$ it is convenient to express matters in terms of $dB(t)$, which, as just seen, is equivalent to using $X(t)$ itself. Writing

$$Y_{t+s} = e^{-s} Y_t + e^{-s} \int_{t}^{t+s} \exp\left(-\frac{t-u}{2}\right) dB(t),$$

we have easily $E(Y_{t+s}|\mathcal{F}_t^\circ) = e^{-s} Y_t$ and $\text{Var}(Y_{t+s}|\mathcal{F}_t^\circ) = (1 - e^{-s})$. Then, since $Y_t$ is Gaussian, a simple integration gives

$$E\left(\int_0^\infty e^{-\lambda s} Y_{t+s}^3 \, ds \,\Big|\, \mathcal{F}_t^\circ\right) = (\lambda + \tfrac{3}{2})^{-1}(Y_t^3 + 3(\lambda + \tfrac{1}{2})^{-1} Y_t),$$

and consequently, after integrating by parts, we obtain

$$R_\lambda^Z(X_t) = \lambda^{-1} \int_0^1 (1 - e^{-\lambda s}) Y_{t-1+s}^3 \, ds$$
$$+ \lambda^{-1}(1 - e^{-\lambda})(\lambda + \tfrac{3}{2})^{-1}(Y_t^3 + 3(\lambda + \tfrac{1}{2})^{-1} Y_t). \qquad (6.13)$$

To express $M_\lambda(t)$, in accordance with the discussion of the stationary case after Theorem 3.34, it is best to define the increments

$$dM_\lambda(t) = dR_\lambda^Z(X_t) + (X_t - \lambda R_\lambda^Z(X_t))\, dt.$$

Here we can apply Ito's lemma together with (3.19) (i.e. $dY_t = -\tfrac{1}{2} Y_t \, dt + dB(t)$), to write $dY_t^3 = \tfrac{3}{2}(2Y_t - Y_t^3)\, dt + 3Y_t^2 \, dB(t)$, and then, after some routine calculation,

$$dM_\lambda(t) = 3\lambda^{-1}(1 - e^{-\lambda})(\lambda + \tfrac{3}{2})^{-1}(Y_t^2 + (\lambda + \tfrac{1}{2})^{-1}) \, dB_t. \qquad (6.14)$$

It should be observed that, since $M_\lambda(t)$ is a martingale, the terms in $dt$ must vanish in such a way that $dM_\lambda(t)$ is simpler (as a rule) than $dR_\lambda^Z(X_t)$.

In order to derive the representations of $X_t$, we need to observe that the results for the strict-sense representations (in particular, Lemma 5.9 and Theorem 5.10) carry over to the stationary case in the same way as those of Chapter 3. By the same change of variable $\tau(t)$ (preceding Theorem 3.33) we see that the condition $z \in D$ is to be replaced by $\bigcap_t \mathscr{F}_t^\circ \equiv (\phi, \Omega)$, which is satisfied by $X_t$ in view of the 0-1 Law for $B(t)$. Therefore, the LMR and PMSIR of $X_t$ continue to exist, except that the integral lower limit is $-\infty$, and the same formulas relate the coefficients of $M_\lambda$ and $X_t - E(t)$ (where $E(t) = 0$). For the PMSIR, obviously $\phi^\#(t) = 1$ and any $M_\lambda$ suffices, but it is equivalent (and more natural) to use $dB_t$ itself, so we need

$$\frac{d}{dt}\langle M_\lambda, B\rangle_t = 3\lambda^{-1}(1 - e^{-\lambda})(\lambda + \tfrac{3}{2})^{-1}(Y_t^2 + (\lambda + \tfrac{1}{2})^{-1})$$

$$= \int_0^\infty e^{-\lambda s} F^\#(t + s, t)\, ds.$$

The Laplace inversion is not difficult (the factor $\lambda^{-1}(1 - e^{-\lambda})$ indicates the operation $\int_{s-1}^s$) and we obtain

$$F^\#(t + s, t) = 3 \int_{s-1}^s Y_t^2 e^{-3u/2} + (e^{-u/2} - e^{-3u/2})\, du$$

$$= 2(e^{-3s/2}(e^{3/2} - 1)Y_t^2 + 3e^{-s/2}(e^{1/2} - 1) - e^{-3s/2}(e^{3/2} - 1)).$$

Consequently, we have for the PMSIR

$$X_t = \int_{-\infty}^t [2(e^{3/2} - 1)e^{-3(t-s)/2}(Y_s^2 - 1) + 6(e^{1/2} - 1)e^{-(t-s)/2}]\, dB_s. \quad (6.15)$$

Thus, to predict $X_t$ by this method for all $t$, we need only keep track of the two martingales $\int_{-\infty}^t e^{3s/2} Y_s^2\, dB_s$ and $\int_{-\infty}^t e^{s/2}\, dB_s$, where $dB_s$ is obtained from $Y_s$ by (3.19).

The LMR in this case is only a little more involved, but we will leave the wide-sense Cramér–Hida representation aside. It is apparent from the form of $dM_\lambda$ in (6.13) that $\phi^*(t) = 2$, and we need a martingale of the form $Y_t^2\, dB_t$ and another of the form $dB_t$ for a linear representation. Alternatively, one can check that $\lim_{\lambda \to \infty} \lambda^2\, dM_\lambda = 3Y_t^2\, dB_t$ and $\lim_{\lambda \to 0} dM_\lambda = 2(Y_t^2 + 2)\, dB_t$, so these are denoted $M_\infty$ and $M_0$, respectively, and may be used as a basis for the LMR. Noting that $EY_t^2 = 1$, we have an orthogonal basis in the form $(dB_t, (1 - Y_t^2)\, dB_t)$. Let us compute the coefficients by using Lemma 5.9. We have easily

$$\frac{d}{dt} E(M_\lambda B) = 3\lambda^{-1}(1 - e^{-\lambda})(\lambda + \tfrac{1}{2})^{-1},$$

and, noting that $EY_t^4 = 3$, we have also

$$\frac{dE\left(M_\lambda \int_{-\infty}^{t}(1-Y_s^2)\,dB_s\right)}{dE\left(\int_{-\infty}^{t}(1-Y_s^2)\,dB_s\right)^2} = -3\lambda^{-1}(1-e^{-\lambda})(\lambda+\tfrac{3}{2})^{-1}.$$

Inversion of the transforms is elementary (as in the previous case), and we obtain the LMR

$$X_t = 6(e^{1/2}-1)\int_{-\infty}^{t} e^{(s-t)/2}\,dB_s - 2(e^{3/2}-1)\int_{-\infty}^{t} e^{3(s-t)/2}(1-Y_s^2)\,dB_s. \tag{6.16}$$

This, of course, is the same as (6.15) except that now there are two terms (corresponding to $\phi^*(t) = 2$) instead of only one.

**Remarks** This shows that Theorem 3.36 (i.e. $\phi \leq 1$ for real stationary Gaussian processes) does not extend to the LMR of stationary non-Gaussian processes. It also does not extend to the total multiplicity in the sense of stochastic integration. For an example, one has only to replace $dB$ in (6.12) by $dB + dP$, where $P(t)$, $P(0) = 0$, is a (two-sided) bilateral Poisson process (of jump size $\pm 1$ with probability $\tfrac{1}{2}$) independent of $B$. It is clear that a previsible integrand cannot separate $dB$ from $dP$, so we need at least 2 orthogonal generators for $L_0^2(\Omega, \mathscr{F}_t^\circ, P)$.

An underlying difficulty in illustrating the application of Chapter 5 to general processes is that the application depends on how the process is specified, and the ways of specifying processes are too numerous and heterogeneous to be dealt with systematically. It is partly for this reason that we confined Chapter 5 to the theory, leaving specific applications entirely aside. But to develop a sense for how applications may (sometimes) be handled when the situation arises, we conclude this section with one more example and a set of exercises with comments. These make no pretence of covering all situations where the theory may be useful, however.

**Example 4** Let $B(t)$ be Brownian motion, $T(1) = \inf\{t: B(t) = 1\}$, and let us set $X_t = B(t) + I_{\{t \geq T(1)\}}$. Evidently $X_t$ is neither Markovian nor continuous. However, its prediction process $Z_t^z$ is a diffusion, i.e. $z \in M_c$. Indeed, by Theorem 4.3, $z \in M_c$ means *all* martingales of the natural (right-continuous) filtration are continuous. This, however, is a property of the probability filtration itself, not of the process used to generate it. It is easy to see that $X_t$ and $B(t)$ generate the same filtration, hence the assertion.

Now since $B(0)$, for $t < T(1)$ we have $X_t = B(t) < 1$ and $E^{Z(t)} e^{-\lambda T(1)} = \exp(-\sqrt{2\lambda}(1 - B_t))$. It follows that for $t < T(1)$ we have

$$\lambda R_\lambda^Z(X_t) = B_t + \exp(-\sqrt{2\lambda}(1-B_t)),$$

and therefore

$$\lambda M_\lambda(t) = B_t + (\exp - \sqrt{2\lambda})\left[\exp(-\sqrt{2\lambda}\, B_t) - 1) - \lambda \int_0^t \exp(\sqrt{2\lambda}\, B_s)\, ds\right].$$

But $R_\lambda^Z(X_t)$ is continuous at $t = T(1)$, and $\lambda\, dM_\lambda = dB$ for $t > T(1)$, so we can write for all $t$

$$\lambda M_\lambda(t) = B_t + \exp - \sqrt{2\lambda}\left[\exp(\sqrt{2\lambda}\, B_{t \wedge T(1)}) - 1 - \lambda \right.$$
$$\left. \times \int_0^{t \wedge T(1)} \exp(\sqrt{2\lambda}\, B_s)\, ds\right].$$

Since this is a martingale, a natural way to simplify it is to apply Ito's lemma:

$$\exp(\sqrt{2\lambda}\, B(t)) - 1 = \sqrt{2\lambda} \int_0^t \exp(\sqrt{2\lambda}\, B(s))\, dB(s) + \lambda \int_0^t \exp(\sqrt{2\lambda}\, B(s))\, ds.$$

Thus we obtain

$$\lambda M_\lambda(t) = B_t + \sqrt{2\lambda}(\exp - \sqrt{2\lambda}) \int_0^{t \wedge T(1)} \exp(\sqrt{2\lambda}\, B(s))\, dB(s). \qquad (6.17)$$

**Remark**  By a general result (Theorem 6.4) one must have $\lim_{\lambda \to \infty} \lambda M_\lambda(t) = X_t$ in a suitable sense (because $X_t$ is a special semimartingale), but it is not trivial to verify this directly.

We first consider the PMSIR, since it is easier than the LMR. We can write (6.17) in the form

$$\lambda M_\lambda(t) = \int_0^t (1 + \sqrt{2\lambda}\, I_{(s < T(1))} \exp(\sqrt{2\lambda}\{B(s) - 1\}))\, dB_s, \qquad (6.18)$$

from which it is easy to see that

$$M_\mu(t) = \int_0^t \frac{d\langle M_\mu, M_\lambda \rangle_s}{d\langle M_\lambda, M_\lambda \rangle_s}\, dM_\lambda(s).$$

Therefore, we have $\phi^\#(t) = 1$, and any $M_\lambda$ will do as a basis. However, as in Example 3, it is easier to take $B(t)$ as the basis, so that to determine $F^\#(t, s)$ we need to invert from (6.18) the equation

$$\int_0^\infty e^{-\lambda s} F^\#(t+s, t)\, ds = \lambda^{-1}(1 + \sqrt{2\lambda}\, I_{(t < T(1))} \exp(\sqrt{2\lambda}(B(t) - 1))).$$

The result is

$$F^\#(t+s, t) = 1 + 2I_{(t < T(1))}(2\pi s)^{-\frac{1}{2}} \exp(-(B(t) - 1)^2/2s),$$

and thus we obtain the PMSIR

$$X_t - EX_t = \int_0^t 1 + 2I_{(s < T(1))}(2\pi(t-s))^{-\frac{1}{2}}\exp(-(B(s)-1)^2/2(t-s))\,dB(s),$$
(6.19)

and

$$EX_t = \sqrt{\frac{2}{\pi t}}\int_1^\infty \exp(-x^2/2t)\,dx = P\{T(1) \le t\}.$$

Note that it is not trivial to verify such a result directly. The reader is invited to do so for a similar case in Exercise 6.2 below.

We shall only make a brief remark about the LMR in this case. One can see from (6.18) and the linear independence of $\exp \lambda x$ (as $\lambda$ varies) that $\phi^*(t) = \infty$ (there is no finite linear dependence between the integrands of (6.18)). Hence calculation of the LMR is tedious, and probably not useful. Besides, (6.19) is not appreciably more complicated than (6.18), so in this example the stochastic integral representation suffices.

**Exercise 6.2** Consider the process $X_t = |B(t)|$; $B(0) = 0$, $EX(t) = (2t/\pi)^{\frac{1}{2}}$. Show that

$$R^Z_\lambda(X_t) = \lambda^{-1}|B(t)| + (2\lambda^3)^{-\frac{1}{2}}\exp(-\sqrt{2\lambda}|B(t)|),$$

i.e. a linear term plus one which 'compensates' for $X_t$ not being a martingale. By applying Ito's lemma show that

$$M_\lambda(t) = \lambda^{-1}\int_0^t (1 - \exp(-\sqrt{2\lambda}|B_s|))\operatorname{sgn} B_s\,dB_s,$$

where $\operatorname{sgn} B(s) = \pm 1$ according as $B(s) \ge 0$ or $B(s) < 0$. (One can invoke Tanaka's formula from 3.8 of McKean (1969).) Using a Stone–Weierstrass argument, show that

$$H^*_t = \left\{\int_0^t f(s, |B_s|)\operatorname{sgn} B_s\,dB_s : \int_0^t E^z f^2(s, |B(s)|)\,ds < \infty;\ f(s, x)\ \text{measurable}\right\}.$$

Note that $\phi^*(t) = \infty$, but $\phi^\#(t) = 1$ and

$$H^\#_t = \left\{\int_0^t h(s)\operatorname{sgn} B_s\,dB_s : h(s, w)\ \text{is}\ \mathscr{F}^\circ_s\text{-previsible and}\ \int_0^t Eh^2(s)\,ds < \infty\right\}.$$

Choosing $(\operatorname{sgn} B_s)\,dB_s$ as generating martingale, apply Lemma 5.9 to get

$$F^\#(t+s, t) = 1 - |B_t|\int_0^s (2\pi u^3)^{-\frac{1}{2}}\exp\frac{-B^2(t)}{2u}\,du,$$

Applications and ramifications | 209

and deduce the PMSIR

$$X_t - (2t/\pi)^{\frac{1}{2}} = \int_0^t \left(1 - |B_s| \int_0^{t-s} (2\pi u^3)^{-\frac{1}{2}} \exp\frac{-B^2(s)}{2u} du\right) \operatorname{sgn} B_s \, dB_s.$$

**Remarks** It is possible (but not trivial) to verify this directly using Ito's lemma and the identity

$$\frac{\partial}{\partial x} \int_0^{t-s} (2\pi u)^{-\frac{1}{2}} \exp\left(\frac{-x^2}{2u}\right) du = -|x| \operatorname{sgn} x \int_0^{t-s} (2\pi u^3)^{-3/2} \exp\left(\frac{-x^2}{2u}\right) du.$$

One needs to allow for the singular derivative at $x = 0$ by showing that

$$\lim_{x \to 0+} x \int_0^{t-s} u^{-3/3} \exp\left(\frac{-x^2}{2u}\right) du = \sqrt{2\pi}.$$

By Theorem 6.2, we have $z \in M_c$. It is also easy to show that $M_\infty(t) = \lim_{\lambda \to \infty} \lambda M_\lambda(t) = |B(t)| - l(t, 0)$, where $2l(t, 0)$ is the local time at 0 of $|B(t)|$. It follows that $\mathscr{F}^o_{t+} \equiv \sigma(M_\infty(s), s \leq t)$. That this is also true for $M_\lambda$, $\lambda \neq \infty$, is less immediate, but may be verified by using

$$\langle \lambda M_\lambda \rangle_t = \int_0^t (1 - \exp - \sqrt{2\pi}|B_s|)^2 \, ds.$$

**Exercise 6.3** Let $\{a_n; -\infty < n < \infty\}$ be a bilateral sequence of independent random variables with $P\{a_n = 1\} = P\{a_n = -1\} = \frac{1}{2}$, and let $t_0$ be an independent random variable with uniform density 1 on $[-1, 0)$. Let us consider the strictly stationary process $X(t) = a_n$, $t_0 + n \leq t < t_0 + n + 1$, $-\infty < n < \infty$. Then $Z^z_t \in M_d \cap M(J_D)$ (foreseeable jump processes) and it is not hard to check that the covariance of $X(t)$ is the same as in the last example of Chapter 3:

$$\Gamma_z(t) = \begin{cases} 1 - |t| & t \leq 1 \\ 0 & t > 1. \end{cases}$$

Thus the wide-sense linear prediction is the same as in Chapter 3, but since $X(t)$ is non-Gaussian the strict-sense prediction is different. However, since $t_0 \in \mathscr{F}^o_{-\infty} (= \bigcap_t \mathscr{F}^o_t)$, we do not have a LMR or a PMSIR for the unconditional process. It is necessary to take $\mathscr{F}^o_{-\infty}$ as given in order to have (the analogue at $t = -\infty$ of) $z \in D$, and then the process is no longer stationary. However, we can consider the representation of the process $(X_t|\mathscr{F}^o_{0+}); t > 0$. Show that for this process (with $t_0$ fixed) we have $\phi(t) = \phi^*(t) = \phi^\#(t) = 1$, and that $\lambda M_\lambda(t) = (1 - e^{-\lambda}) (\sum_{k=1}^n a_k)$; $t_0 + n \leq t < t_0 + n + 1$, $1 \leq n$, and $M_\lambda(t) = 0$, $0 \leq t < t_0 + 1$. Note that $M_\lambda(t) \in H_z(t)$, whence it follows that the wide-sense representation and the LMR coincide, and since $\phi^* = \phi^\# = 1$ the PMSIR is also the same; this last holds whenever $\phi^* = \phi^\# < \infty$. Taking $M_1(t)$ as basis, derive the LMR for $X_t$ given $\mathscr{F}^\circ_{0+}$, as

$$X_t - EX_t = (1 - e^{-1})^{-1} \int_{0 \vee (t-1)}^t dM_1(s); \qquad 0 < t.$$

Remark, finally, that if we replace $a_{2n}$ by $a_0$ and $a_{2n+1}$ by $a_1$ for all $n$, then $\Gamma(t)$ is unchanged but $X_t$ is strictly deterministic ($X_t \in \mathscr{F}^\circ_{-\infty}$). Then the strict-sense prediction is trivial, while the wide-sense prediction is unchanged.

**Exercise 6.4** Let $(T_n, -\infty < n \neq 0 < \infty)$ be the jump times of a bilateral Poisson process with parameter $\mu > 0$, so that $T_{n+1} - T_n$ are independent, exponential random variables with parameter $\mu$ for $0 \leq n$, and for $n \leq -2$; $T_{-1} < 0 < T_1$. Set $X_t = a_n$ for $T_n \leq t < T_{n+1}$; $n \geq 0$ or $n \leq -2$, and $X_t = a_{-1}$ for $T_{-1} \leq t < T_1$, where $a_n$ are independent, $a_n = \pm 1$ with probability $\frac{1}{2}$, as in Exercise 6.3. Note that $X_t$ is strictly stationary with $Z_t^z \in M(J_{D^c}) \cap M_d$ (totally unforeseeable jump processes), and $\mathscr{F}^\circ_{-\infty} \equiv (\phi, \Omega)$ so the LMR and PMSIR exist over $-\infty < t < \infty$. Derive the expression

$$M_\lambda(t) - M_\lambda(0) = (\lambda + \mu)^{-1}\left(X_t - X_0 + \mu \int_0^t X_s \, ds\right); \qquad 0 < t.$$

It follows that $M_\lambda(t) - M_\lambda(0) \in H_z(t)$, and since $\phi^*(t) = 1$ the three predictive representations of $X_t - X_0$ coincide. Show that, if we take $M_1$ as basis, they are

$$X_t - X_0 = (1 + \mu) \int_0^t \exp(-\mu(t - s)) \, dM_1(s).$$

Applying the same argument to $X_t - X_{t_0}$; $t_0 < 0$, and letting $t_0 \to -\infty$, deduce that the representation on $-\infty < t$ is

$$X_t = (1 + \mu) \int_{-\infty}^t \exp(-\mu(t - s)) \, dM_1(s), \qquad -\infty < t < \infty.$$

**Exercise 6.5** Let $a \neq b$ be positive constants. For $t \geq 0$ set

$$Y_t = aB_t I_{(B_t \geq 0)} + bB_t I_{(B_t < 0)},$$

and define $X_t = |Y_t|$ (reducing to Exercise 6.2 if $a = b$). Note that $X_t$ is not Markovian, but that $Z_t^z$ identifies with $Y_t$. ($X_t$ is a germ-Markov process.) Therefore $Z_t^z$ is a one-dimensional diffusion. Show that

$$\lambda R_\lambda^z(X_t) = X_t + (a + b)(8\lambda)^{-\frac{1}{2}} \exp(-\sqrt{2\lambda}|B_t|),$$

Applications and ramifications | 211

and infer from Ito's lemma that

$$\lambda M_\lambda(t) = X_t - \frac{(a+b)}{2}\int_0^t \exp(-\sqrt{2\lambda}|B_u|)\,d|B_u|$$

$$= \int_0^t \left(\frac{(a-b)}{2} + \frac{(a+b)}{2}\operatorname{sgn} B_u(1-\exp(-\sqrt{2\lambda}|B_u|))\right)dB_u.$$

Taking $B(t)$ as basis, obtain the PMSIR of $X_t - EX_t$, where

$$EX_t = (a+b)(t/2\pi)^{\frac{1}{2}}.$$

**Solution**

$$X_t - EX_t$$

$$= \int_0^t \left[\frac{a-b}{2} + \frac{a+b}{2}\operatorname{sgn} B_s\left(1 - |B_s|\int_0^{t-s}(2\pi v^3)^{-\frac{1}{2}}\exp\left(\frac{-B_s^2}{2v}\right)dv\right)\right]dB_s.$$

**Exercise 6.6**  Consider the process defined as

$$X_t = B_t I_{(t<T(\pm 2))} + (B_t + 1)I_{(t\geq T(\pm 2);\, T(1)<T(-1))}$$

$$+ (B_t - 1)I_{(t\geq T(\pm 2);\, T(1)>T(-1))}, \qquad 0 < t,$$

where $B_t$ is Brownian motion, $B_0 = 0$, and $T(\pm 2) = \inf(t: B(t) \in \{\pm 2\})$, $T(1) = \inf(t: B(t) = 1)$, $T(-1) = \inf(t: B(t) = -1)$. Observe that $X_t$ has jumps, and is not Markovian, and it also has a continuous component. The jumps are previsible, but we do not have $z \in M(J_D)$. Indeed, $Z_t^z$ is continuous, so $z \in M_c$. This process is indicative of what can happen when we permit non-trivial, non-Markovian dependence on the past. We can consider $Z_t^z$ as a one-dimensional diffusion on a graph that can be embedded into $\mathbb{R}^3$ (but not $\mathbb{R}^1$ or $\mathbb{R}^2$). The process $X_t$, and also $Z_t^z$, identifies with $B(t)$ until $t = T(1) \wedge T(-1)$ $(=T(\pm 1))$, but $\pm 1$ are *shunts* sending $Z_t^z$ onto (say) one of two parallel lines in the second dimension (for example, parallel to the y-axis). On whichever line it reaches (i.e. $x = +1$ or $-1$) it next identifies with $B(t)$ on this line (in total arc length coordinates) until $T(\pm 2)$. Then we have four possibilities, which are also shunts, taking $Z_t^z$ onto one of four parallel lines in the third dimension. However, to represent $X_t$ $(= \hat{\rho}(Z_t^z))$ the coordinates on these lines do not start at 0, but at $+3$, $-1$, $+1$, and $-3$, respectively. Then $Z_t^z$ and $X_t$ continue to follow $B(t)$ along whichever line was reached, but in the translated coordinates.

By computing $R_\lambda^z(X_t)$, and then obtaining $M_\lambda(t)$ in a suitable form from Ito's lemma, show that we have $\phi^\#(t) = 1$, and that $B(t)$ may be used as basis. Obtain $\int_0^\infty e^{-\lambda s} F^\#(t+s, t)\,ds$ in this case.

## Solution

$$\lambda R^Z_\lambda(X_t) = \begin{cases} B_t + (\cosh\sqrt{2\lambda}/\cosh 2\sqrt{2\lambda})\sinh(B_t\sqrt{2\lambda}) & 0 < t \leq T(\pm 2) \\ B_t \pm (\cosh(B_t\sqrt{2\lambda})/\cosh 2\sqrt{2\lambda}) & T(\pm 1) \leq t \leq T(\pm 2) \\ & \text{with sign chosen as} \\ & \text{in } B(T(\pm 1)) \\ X_t & T(\pm 2) \leq t. \end{cases}$$

Setting $k(\lambda) = \cosh\sqrt{2\lambda}/\cosh 2\sqrt{2\lambda}$, one has

$$\lambda M_\lambda(t) = \begin{cases} B_t + \sqrt{2\lambda}\, k(\lambda)\int_0^t \cosh(B_s\sqrt{2\lambda}\, dB_s & 0 \leq t \leq T(\pm 1) \\ B_t + \sqrt{2\lambda}k(\lambda)\left[\int_0^{T(\pm)}\cosh(B_s\sqrt{2\lambda})\, dB_s \right. \\ \left. \pm \cosh 2\sqrt{2\lambda}\int_{T(\pm 1)}^t \sinh(B_s\sqrt{2\lambda})\, dB_s\right] & T(\pm 1) \leq t \leq T(\pm 2), \\ & \text{with sign chosen as} \\ & \text{in } B(T(\pm 1)) \\ \lambda M_\lambda(T(\pm 2)) + (B(t) - B(T(\pm 2))) & T(\pm 2) \leq t. \end{cases}$$

Therefore,

$$\int_0^\infty e^{-\lambda s} F^\#(t+s, t)\, ds$$

$$= \begin{cases} \lambda^{-1}(1 + \sqrt{2\lambda}\, k(\lambda)\cosh(B_t\sqrt{2\lambda})) & 0 \leq t \leq T(\pm 1) \\ \lambda^{-1}(1 \pm \sqrt{2\lambda}\,(\cosh\sqrt{2\lambda})\sinh(B_t\sqrt{2\lambda})) & T(\pm 1) \leq t \leq T(\pm 2) \\ \lambda^{-1} & T(\pm 2) \leq t. \end{cases}$$

## 6.2 Adaptation to a general filtration

A characteristic feature of the modern general theory of processes not available thus far in our prediction theory is that it applies to a process measurable with respect to (*adapted to*) a more or less arbitrary probability filtration. Thus, definitions such as those of a Markov process or a martingale, or even of a Brownian motion, are made relative to an unspecified filtration. A related extension is needed in treating filtering problems. There part of the problem is to estimate a process which is unobserved, and hence contributes to the filtration, but not to the processes available for the estimation.

Under the assumption that the filtration generates a $\sigma$-field $\mathscr{F}_\infty (= \bigvee_t \mathscr{F}_t)$ which is almost countably generated (Definition 0.2), there is no serious obstacle to incorporating both of these extensions in the theory of the prediction process. In fact, we refrained from doing so at the beginning only to lighten an already heavy load of concepts. The extension to cover filtering is due to Yor (1977). Here we shall derive it by following the same argument as in Chapter 1. It is plausible, indeed, that it should follow from that result by taking conditional expectations given the 'observed' filtration, but we are unable to derive it by such a short-cut. The former extension is similar, but easier because nothing must be 'forgotten'. Instead, we have only to adjoin an auxiliary process to fill out the generated filtration to the prescribed $\mathscr{F}_t$, and then to predict both processes. It is then easy to restrict attention to prediction of the original process. Since the auxiliary process is of no interest *per se*, its precise definition is largely a matter of convenience. When $\mathscr{F}_\infty$ is almost countably generated the existence of such an auxiliary process in the form of a sequence of measurable martingales is guaranteed by the discussion on page 7 of the Introduction. But ordinarily the extra information is already generated by a process, hence there is no need to look further. This is the case, for example, when it is given that $X_t = \int_{-\infty}^t F(t, s)\, dB(s)$. Then one would adjoin the process $B_t$ immediately. It does not cause difficulty even if it should happen that $B_t$ and $X_t$ generate the same filtration, which may be difficult to ascertain.

Instead of assuming a single process $X_t$, our starting point is thus a pair $(X_t, Y_t)$ of measurable process with values in Lusin spaces $E_1$ and $E_2$. For convenience we again assume them to be uncountable, and we regard $(X_t, Y_t)$ as a measurable vector process with values in the Lusin space $E = E_1 \otimes E_2$. Let $\hat{\rho}_1$, $\hat{\rho}_2$, and $\hat{\rho}$ be isomorphisms of $E_1$, $E_2$, and $E$, respectively, onto ([0, 1], $\mathscr{B}[0, 1]$), as in Theorem 1.1, and define $\rho'_1(e_1, e_2) = \rho_1(e_1)$ on $E$. We have $\Omega'_E = \Omega'_{E_1} \otimes \Omega'_{E_2}$, $\mathscr{F}'_E(t) = \mathscr{F}'_{E_1}(t) \otimes \mathscr{F}'_{E_2}(t)$, and finally

$$P'(S') = P\{(X_{(\cdot)}, Y_{(\cdot)}) \in S'\}, S' \in \mathscr{F}'_E(\infty),$$

just as in Chapter 1. Then using $\hat{\rho}$ we map $\mathscr{F}'_E$ onto $\mathscr{F}'(=\mathscr{F}'_{[0,1]})$ and replace $E$ by [0, 1] to construct the prediction process $Z_t^z$ of $(X_t, Y_t)$ relative to $\mathscr{F}'_{t+}$.

It remains to examine the case of incomplete information, where $\mathscr{F}'_{E_1}(t+)$ is given instead of $\mathscr{F}'_E(t+)$. Letting $g(x) = \rho'_1 \hat{\rho}^{-1}(x), 0 \leq x \leq 1$, the filtration $\mathscr{F}'_{E_1}(t)$ maps under $\hat{\rho}$ onto the filtration $\mathscr{G}'_t = \sigma(\int_0^s g(w(u))\, du, s < t)$, since this determines the (first) component in $\Omega'_{E^1}$. We now proceed to construct a new prediction process as in Definition 1.5, but conditioning by $\mathscr{G}'_{t+}$ instead of by $\mathscr{F}'_{t+}$. This construction needs no change, and we obtain a process with values in $M_E$ having the same continuity and measurability properties as before (Theorem 1.8(1) and (2)) except that $\mathscr{G}'_t$ replaces $\mathscr{F}'_t$. The elementary

Markov property (Theorem 1.8(3)) also carries over. To prove the strong and moderate Markov properties relative to $\mathscr{G}'_T$, as in Theorem, we need to replace arguments involving $\int_0^s w(u)\,du$ by their analogues with $\int_0^s g(w(u))\,du$ instead (as in Exercise 1.9, for example). Apart from this, there is no serious change, and going back to the original state space $E$ we obtain Corollary 1.10, relative to $\mathscr{F}'_{E_1}(t+)$.

**Definition 6.1** Let $G_t^z(S)$, $S \in \mathscr{F}'_E$, denote the prediction process on $(\Omega'_E, \mathscr{F}'_E)$ relative to $\mathscr{F}'_{E_1}(t+)$ just constructed, and let $q_E^*(t, z, A) = P^z(G_t^z \in A);\ A \in \mathscr{M}_E$.

It follows that $q_E^*$ has the same measurability properties as $q_E$ of Definition 1.6, so that it is a Markov transiton kernel on $(M_E, \mathscr{M}_E)$ for each $t$. It remains, however, to show that the Markov properties of $G_t^z$ hold relative to $q_E^*$, so that in particular $q_E^*$ satisfies the Chapman–Kolmogorov equation. This result for $Z_t^z$ and $q_E$ was Theorem 1.14, which depended on the identity $Z_{T+t}^z(w) = Z_t^{Z_T^z}(\theta_T w)$ of Theorem 1.11. Accordingly, we must re-prove Theorem 1.11 for $(G_t^z, \mathscr{G}_t)$, namely, that for $\mathscr{G}'_{t+}$-optional $T < \infty$, we have $G_{T+t}^z(w, \cdot) = G_t^{G_T^z(w)}(\theta_T w, \cdot)$, and for $\mathscr{G}'_t$-previsible $T$ the exponent on the right may be replaced by $G_{T-}^z(w)$. The argument is essentially unchanged. We again reduce to $t = s_k$: $G_{(T+t)-}^z \equiv G_{T+t}^z$, and the (long) argument leading to Corollary 1.13 applies to show that it is enough (for part (1)) to prove that

$$E^z(g(h \circ \theta_T) G_{T+t}^z f) = E^z(g(h \circ \theta_T) G_t^{G_T^z(w)}(f \circ \theta_T)) \tag{6.20}$$

for $g \in b(\mathscr{G}'_{T+})$, $h \in b(\mathscr{G}'_t)$, and $f \in C(\Omega')$. (We recall that the proof need only be done for the space $E = [0, 1]$, where $\Omega'$ is compact.) Just as before, the left side becomes

$$E^z(gE^z(h \circ \theta_T(f \circ \theta_{T+t}) | \mathscr{G}'_{T+t})) = E^z(gE^z(h \circ \theta_T(f \circ \theta_{T+t}) | \mathscr{G}'_T))$$
$$= E^z(gG_T^z h(f \circ \theta_t))$$
$$= E^z(gG_T^z(hG_t^{G_T^z} f)), \tag{6.21}$$

and the problem is again to remove the first $G_T^z$. Now, in place of $h(w') G_t^{G_T^z(w)}(w', f)$, let $K(w, w')$ be any bounded $(\mathscr{G}'_{T+} \otimes \mathscr{G}'_{t+\varepsilon})$-measurable random variable. (The former has this measurability, since Theorem 1.8(2) applies to $G_t^z$ with $\mathscr{G}'_{t+\varepsilon}$ in place of $\mathscr{F}'_{t+\varepsilon}$.) Then if $K(w, w') = K_1(w) K_2(w')$; $K_1 \in b(\mathscr{G}'_{T+})$, $K_2 \in b(\mathscr{G}'_{t+\varepsilon})$, it follows by definition of $G_T^z$ as a conditional expectation that $E^z(gE^{G_T^z(w)} K(w, w')) = E^z(gK(w, \theta_T(w)))$. Since both sides are monotone and linear in $K$, the result extends to general $K$, and hence implies that the last term of (6.21) equals $E^z(g(h \circ \theta_T) G_t^{G_T^z(w)}(f \circ \theta_T))$, as required. The extension to $\mathscr{G}'_t$-previsible $T$ of course also carries over, and since the rest of the proof of Theorem 1.14 depended only on the argument just extended, we may assert (by mapping back to the original state space $E_1 \otimes E_2$) the following result.

## Theorem 6.3

(1) (Strong Markov property) *For every $z$ and $\mathscr{F}'_{E_1}(t+)$-stopping time $T < \infty$,*

$$P^z(G^z_{T+t} \in A | \mathscr{F}'_{E_1}(t+)) = q^*(t, G^z_T, A), \quad A \in \mathscr{M}_E, t \geq 0.$$

(2) (Moderate Markov property) *For $\mathscr{F}'_{E_1}(t)$-stopping times $T > 0$ (automatically $\mathscr{F}'_{E_1}(t)$-previsible, by the extension of Theorem 1.9(3) to $\mathscr{G}^1_t$, using $\int_0^s g(w(u))\,du$ in place of $\int_0^s w(u)\,du$),*

$$P^z(G^z_{T+t} \in A | \mathscr{F}'_{E_1}(T)) = q^*_E(t, G^z_{T-}, A), \quad A \in \mathscr{M}_E, t \geq 0.$$

We again define the non-branch points $D = \{z : q^*_E(0, z, \{z\}) = 1\}$, and Theorem 1.15 remains true for $G^z_t$ ($P^z\{G^z_t \in D$ for all $t \geq 0\} = 1$). However, the meaning of $z \in D$ is quite different from before, because, if we use the same coordinates $\tilde{X}_t$, then $\tilde{X}_t \notin \mathscr{F}'_{E_1}(t+)$ since $\tilde{X}_t$ includes the unobserved process $w_2(t)$. To understand the difference most easily, consider the following example.

**Example 5** Let $(X_t, Y_t) = (|B_t|, B_t)$, where $B_t$ is the usual Brownian motion. Here it is clear that $G^z_t$ assigns probability $\frac{1}{2}$ to the law of $B_{t+s}$ starting at $|B_t|$, and $\frac{1}{2}$ to the law of $B_{t+s}$ starting at $-|B_t|$, so that $G^z_t$ (momentarily) appears to be a branch point. Contradiction with Theorem 1.15 is avoided because the definition of $D$ refers only to the future of $G^z_t$, and this does not include the choice of sign at $t = 0$.

As we see from this example, if we want Theorem 1.16 to extend to $G^z_t$ we must redefine the coordinates to include only the first (observed) component. This was determined by $w_1(t)$ (or by $g(w(t))$ in the case $E = [0, 1]$). Here it is irrelevant to define coordinates for $E = [0, 1]$, since that was only a device for the proofs. We need $E = E_1 \otimes E_2$, and hence we modify Definition 2.6 by replacing

$$\rho : \bar{E} \to \underset{1}{\overset{\infty}{\times}} [0, 1] \quad \text{by} \quad \rho_1 : \bar{E}_1 \to \underset{1}{\overset{\infty}{\times}} [0, 1],$$

and setting $\rho'_1(e_1, e_2) = \rho_1(e_1)$, $(e_1, e_2) \in E$. Then letting $e_{-1}$ be a fixed element of $E_1$, we define

$$\tilde{\rho}_{E_1}(w) = \begin{cases} e_1 & \text{if } \exists e_1 = \rho_1^{-1}(\tilde{\rho}_k) \\ e_{-1} & \text{otherwise,} \end{cases}$$

just as in Definition 2.6, and then we introduce $\tilde{X}_{E_1}(t) = \tilde{\rho}_{E_1}(w \circ \theta_t)$, $0 \leq t$. It is clear that these coordinates have the same role, relative to $\mathscr{F}'_{E_1}(t+)$, as $\tilde{X}_E(t)$ had for $\mathscr{F}'_E(t+)$. The extension of Definition 2.7 to define $\hat{\rho}_{E_1}$ follows the same lines as indicated, and we recover Theorem 2.7 to the effect that either of $\{\hat{\rho}_{E_1}(G^z_s), s < t\}$ or $\{G^z_s, s < t\}$ generates $\mathscr{F}'_{E_1}(t)$ up to $P^z$-null sets.

216 | Foundations of the prediction process

We pass over the extension of Section 2.2 to the treatment of Markov processes relative to a general filtration. This is evidently possible, but routine, and may be left to the reader.

The (slight) modification of $G_t^z$ so as to be r.c.l.l. in the prediction topology of $M_E$ (Theorem 2.4) follows without change, and we are in a position to introduce prediction space (Definition 2.13) for $G_t^z$. Formally, this has the same denotation $(\Omega_Z, \mathscr{L}_{t+}^\circ, \theta_t^z, P^z)$ as for $Z_t^z$, except that now $P^z$ is the law of the Markov process with transition function $q_E^*$ (starting from $z \in M_E$) instead of $q_E$, and for this process $\mathscr{L}_{t-}^\circ$ is generated (up to $P^z$-null sets) by $\{\tilde{\rho}_E(w_Z(s)), s < t\}$ instead of by $\{\hat{\rho}_E(w_Z(s)), s < t\}$. The strong and moderate Markov properties of $w_Z(t)$ for $q_E^*$ follow immediately as before (Theorem 2.14), and we see that on $D$ the coordinate process $w_Z(t)$ is again a Borel right process (Theorem 2.24).

For the purposes of prediction and filtering, it is unnecessary to represent the *unobserved* coordinate on prediction space. Instead, if it is real and suitably bounded (namely, if $EY_t^2 < \infty$ and $E^z \int_0^t |w_2(u)| \, du < \infty$), and if we follow the same prescription as in defining $\hat{\rho}_{E_1}(z)$ but with $E_2$ in place of $E_1$, we obtain a function $\hat{\rho}_{E_2}(z)$ which defines the *conditional expectation* of the unobserved coordinate given $\mathscr{L}_{0+}^\circ$. In more detail, we do not use

$$\rho_2 : \bar{E}_2 \to \underset{1}{\overset{\infty}{\times}} [0, 1]$$

since $E_2 = R$, but rather we simply define

$$\tilde{\rho}_{E_2}(w) = \limsup_{n \to \infty} n \int_0^{1/n} w_2(u) \, du,$$

$\tilde{X}_{E_2}(t) = \tilde{\rho}_{E_2}(w \circ \theta_t)$, and $\hat{\rho}_{E_2}(z) = E^z(\tilde{\rho}_{E_2}(w))$. Then it follows that, for a.e. $t$, $\tilde{\rho}_{E_2}(w_Z(t))$ represents the expected value of the unobserved coordinate given $\mathscr{L}_{t+}^\circ$. For purposes of prediction and filtering, this suffices in place of the unobserved coordinate itself (if the second coordinate is right-continuous, it is easy to see that the representation holds for every $t$). Of course, it is not necessary to go over to prediction space in order to do prediction or filtering for a fixed $P$ ($= z$). In any case, $\tilde{\rho}_{E_2}(G_t^z)$ has the same meaning on $\Omega_E'$ as $\tilde{\rho}_{E_2}(w_Z(t))$ has on $\Omega_Z$.

On the other hand, the concept of a prediction demesne (Definition 2.14) is the same whether one is using prediction relative to $\mathscr{F}_E'(t+)$ or $\mathscr{F}_{E_1}'(t+)$ (i.e. using $q_E$ or $q_E^*$ to define $P^z$ on prediction space). The actual demesnes are quite different, however. For example, if the component processes $(X(t), Y(t))$ are independent, then $\tilde{\rho}_{E_2}(w_2(t))$ is a fixed function when $q_E^*$ defines $P^z$, but it has the law of $Y(t)$ when $q_E$ is used. Moreover, as soon as both components are of interest (assuming them to be real) the demesnes of finite $r$th moment (Definition 3.4) must be defined for each component separately, and in particular we must distinguish $M_{2,E_1}$ and $M_{2,E_2}$ in such a way that their intersection is the demesne of square-integrable processes of two components (when $q_E^*$ is being used, one would write $M_{2,E_i}^*$ to distinguish that situation). To be sure, analogous remarks apply to processes of $n$ components, some observed and some unobserved, etc. and we make no attempt to present all possible cases. It is worth mentioning, however, that in the jointly Gaussian case under $q_E^*$, the second component $\hat{\rho}_{E_2}(w_2(t))$ remains Gaussian in such a way that the general results applying to Gaussian processes (Chapter 3) may be adapted to the filtering problem.

For $z \in M_{2,E_i}$ or $z \in M_{2,E_i}^*$, $i = 1$ or 2, it is at once evident that the Definition 3.6 of the fundamental martingales $M_\lambda(t)$, as well as the corresponding Theorem 3.15 to the effect that they are square-integrable martingale additive functionals of $Z(t)$, immediately carry over to the more general situation. Indeed, that proof applies unchanged to any real function $f(Z(s))$ such that for $P^z$ the process $f(Z(s))$ is in $M_2$. Thus it applies in particular to $\hat{\rho}_{E_i}(Z_s)$ on the corresponding demesne. We denote by $M_{\lambda,1}(t)$, $M_{\lambda,2}(t)$, $M_{\lambda,1}^*(t)$, and $M_{\lambda,2}^*(t)$ the corresponding martingale additive functionals. On the space $\Omega_E'$, where we replace $Z(s)$ $(= w_Z(s))$ by $Z^z(s)$ in the definitions (we should perhaps write $Z^{*z}(s)$ when $\mathscr{F}_E'$ is replaced by $\mathscr{F}_{E_1}'$) these are martingales for the corresponding filtrations $\mathscr{F}_E'(t+)$ or $\mathscr{F}_{E_1}'(t+)$, but the additive functional property is not available. Indeed, they are not additive functionals for the translation operators $\theta_t$ on $\Omega_E'$; in short, the superscript $z$ would have to depend on $t$ and $Z^z(t)$ is not a function of $w(t)$.

In each of our four cases, the prediction theory representation of Corollary 3.29 (the wide-sense linear representation) as well as the LMR and the PMSIR of Theorem 5.10, carry over to the corresponding coordinate process centred at the mean, provided that it is suitably bounded and right-continuous in $L^2(P^z)$ as in Definition 5.1. The set $M_2^+$ of that definition thus gives rise to four sets, according to which coordinate is being represented and which filtration is taken as given. The essential point to realize is that the only feature of the coordinates $\hat{\rho}(Z(t)) - E_n(t)$ which is not available in the present situation is that they generated $\mathscr{L}_t^\circ$. But it is easy to see that this feature was not used in developing their representation in terms of $(M_\lambda(t), 0 < \lambda)$. In particular, the key Lemma 5.9 needs no change. In short, the coefficients $\mathscr{G}_\lambda(s)$ of the projections of $M_{\lambda,i}$ or $M_{\lambda,i}^*$ onto $\{Y(s), s \leq t\}$ are related to the coefficients $F(t, s)$ of the corresponding centred $L^2$-coordinate process (say $X_i^+(t)$ or $X_i^{*+}(t)$, $i = 1$ or 2) by the same Laplace transform relation $\mathscr{G}_\lambda(t) = \int_0^\infty e^{-\lambda s} F(t+s, t)\, ds$ as before (to avoid overburdening the notation, we take this to cover all twelve cases). Indeed, the same proofs apply as well to any other function $f(Z(t))$ which has the same boundedness and right-continuity as assumed for the coordinates $\hat{\rho}(Z(t))$, when we replace $M_\lambda$ by the corresponding Kunita–Watanabe martingale. The end result is formally the same, and we obtain a representation in terms of martingales $M_\lambda$ or $M_\lambda^*$ of $\mathscr{L}_{t+}^\circ$, but in terms of the original situaton on $\Omega_E'$, $M_{\lambda,i}$ represents prediction relative to $\mathscr{F}'_{E_1 \otimes E_2}(t+)$, whereas $M^*_{\lambda,i}$ represents prediction relative to $\mathscr{F}'_{E_1}(t+)$.

We refrain from presentation of more examples which may better be left to others more knowledgeable on the subject of filtering. Let us mention, however, that Yor (1977) and Knight (1981a, Essay I) contain an application of the prediction process to filtering of semi-Markov processes. For an example of the use of the enlarged filtration alone, we refer again to the last example of Chapter 3, where we treated the process $X(t) = B(t) - B(t-1)$, $-\infty < t < \infty$. The most difficult part of that treatment was in showing ((3.23) ff.) that $\{B(t) - B(u), u < t\}$ is in the $L^2$-space of $\{X(s), s \leq t\}$. This is now rendered unnecessary if we are content to predict relative to the filtration $\sigma\{B(u), u < t\}$ instead of $\sigma\{X(u), u < t\}$. We have only to introduce $B(t)$ as a second component, and then use $M_{\lambda,1}(t)$, relative to the first component, in the representation theory.

In practical examples such as this and the kind discussed in Section 6.1, it is naturally unnecessary to change the probability space either to $\Omega'_E$ or to $\Omega_Z$. Instead, one has only to define $R^Z_\lambda(X(t)) = \int_0^\infty e^{-\lambda s} E(X_{s+t}|\mathscr{F}_{t+})\,ds$, much as in Section 6.1 but with the general filtration $\mathscr{F}_t$ instead of $\mathscr{F}^\circ_t$. This leads directly to the martingales $M_\lambda(t)$ relative to $\mathscr{F}_{t+}$, and the remainder of the representation is then carried out on the original probability space. The interested reader may like to consider Example 3 in this light, without taking as known the fact that $\mathscr{F}^\circ_t$ is equivalent to $\sigma(B_s, s \leq t)$. The latter then plays the role of $\mathscr{F}_t$, and the same results follow.

## 6.3 Decomposition of semimartingales, and Gaussian processes

This section contains two completions of earlier work, the first going back to the Introduction, and the second relevant to Chapter 3 as well as to the first. In the Introduction, and several times thereafter, we have used the martingales $M_\infty(t) = \lim_{\lambda \to \infty} \lambda M_\lambda(t)$ when the limit existed in a suitable sense. Here we propose to examine this limit more carefully, with a view to obtaining from it the decomposition of a special semimartingale into a local martingale plus a previsible process of locally finite variation. The decomposition, going back to J. L. Doob and P.-A. Meyer, is well known and well understood (Dellacherie and Meyer (1980, VII. 23), at least in theory. However, as Result 0.1 already illustrates, putting it into practice may pose difficulties. Moreover, the end result may be basic in studying a given process. For example, when the martingale component is tractable (as in Result 0.1) the decomposition makes available a stochastic change of measure (ascribed to Cameron–Martin and Girsanov; see for example Revuz and Yor (1991, Chap. VIII) for more details) for reduction of the probabilities to those of the martingale alone.

It thus becomes of importance to obtain the decomposition explicitly, and the following method may be seen as a supplement to the other known methods. In various situations, especially those involving processes (such as Brownian motion) for which an explicit resolvent operator is known, $\lambda M_\lambda$ is a more tractable object than, for example, the *Laplaciens approches* of Meyer (1966, Chapter VII, T28) which may be directed to a similar purpose (Dellacherie and Meyer 1980, VII. 22). Furthermore, the present method applies without knowing in advance that the process $X_t$ is a special semimartingale. In short, we obtain necessary and sufficient conditions for a special semimartingale in terms of the behaviour of $\lambda M_\lambda$, whereas the existent literature is apparently confined to implications of the semimartingale hypothesis.

We begin with a filtered probability space $(\Omega, \mathscr{F}_t, P)$ such that $\mathscr{F}_t$ is right-continuous containing all $P$-null sets, $\mathscr{F}_\infty$ is almost countably generated, and $P$ is complete, and a real-valued process $X_t$ with right-continuous paths, adapted to $\mathscr{F}_t$; we write $X_t \in \mathscr{F}_t$. Then $X_t$ is called a special semimartingale of $(\mathscr{F}_t, P)$ if there exists a decomposition $X_t = X_0 + M'_t + A'_t$, in which $M'_t$ is a local martingale of $\mathscr{F}_t$, and $A'_t$ is $\mathscr{F}_t$-previsible, with (right-continuous) paths of finite variation in finite time intervals (Dellacherie and Meyer 1980, VII. 23). For our purposes, it is easier however to take a different definition, the equivalence of which is proved in Dellacherie and Meyer (1980, VII. 25).

**Definition 6.2** Under the above conditions on $(\Omega, \mathscr{F}_t, P)$, a process $X_t \in \mathscr{F}_t$ is a special semimartingale if (and only if) there exists a decomposition $X_t = X_0 + M'_t + A'_t$, $(A'_t \in \mathscr{F}_t)$, and a sequence $T_n \uparrow \infty$ of $\mathscr{F}_t$-stopping times, such that, for each $n$, $M'(t \wedge T_n)$ is a uniformly integrable martingale $(M'_0 = 0)$ and $A'(t \wedge T_n)$ is right-continuous, previsible, and of integrable total variation.

The decomposition is unique (up to a $P$-null set), even without the consideration of stopping times (Dellacherie and Meyer 1980, VII. 23). The case in which $T_n = \infty$ for all $n$ seems worth giving a separate designation. We propose to call $X_t$ a *totally integrable special semimartingale* if, in the above, $M'_t$ is a uniformly integrable martingale, and $A'_t$ is a process of integrable total variation. Obviously, then, $X_t$ is a special semimartingale if and only if there is a sequence $T_n \uparrow \infty$ of $\mathscr{F}_t$-stopping times such that, for each $n$, $X(t \wedge T_n)$ is a totally integrable special semimartingale. Thus, to decompose $X_t$ it suffices (by uniqueness) to decompose each $X(t \wedge T_n)$ in the form $X_0 + M'(t \wedge T_n) + A'(t \wedge T_n)$. As we shall not be concerned with the problem of how to determine the $T_n$ (which will be obvious in the cases treated below) it suffices for our purposes to identify, and determine the decomposition of, a totally integrable special semimartingale.

**Theorem 6.4** *Let $X_t$ and $(\Omega, \mathscr{F}_t, P)$ be as above. Then $X_t$ is a totally integrable* *special semimartingale if and only if*:

(1) $X_t$ is $L^1$-right-continuous and uniformly integrable;
(2) *for each t*, $\lim_{\lambda \to \infty} \lambda M_\lambda(t) = M_\infty(t)$ *exists in the weak $L^1$-topology; and*
(3) *setting* $M_\infty(t+) = \limsup_{r \to t+, r \in \mathbb{Q}} M_\infty(r)$, *the process*

$$A(t) = X_t - X_0 - M_\infty(t+)$$

*is previsible and of integrable total variation.*

Furthermore, in this case $M_\infty(t+) = M_\infty(t)$ a.s., $A$ is the finite variation component determining the decomposition of $X_t$, and for every bounded stopping time $T$ we have $A(T) = -\lim_{\lambda \to \infty} \lambda \int_0^T X_u - \lambda R_\lambda^Z(X_u) \, du$ in the weak $L^1$-topology.

**Remark** We refrain from changing the setting to $\Omega'$ or $\Omega_Z$. Thus $M_\lambda(t)$ is defined using $R_\lambda^Z(X_t)$ as in Section 6.1, but relative to the (possibly enlarged) filtration $\mathscr{F}_t$ as in Section 6.2.

**Proof** Suppose, first, that properties (1)–(3) are satisfied. By (1) it is clear that $P \in M_1$, so that $M_\lambda(t)$ exists. Then, for $S \in \mathscr{F}_t$ and $0 < s$,

$$E(I_S M_\infty(t+s)) = \lim_{\lambda \to \infty} E(I_S \lambda M_\lambda(t+s)) = \lim_{\lambda \to \infty} E(I_S \lambda M_\lambda(t)) = E(I_S M_\infty(t)),$$

so that $M_\infty(t)$ is an $\mathscr{F}_t$-martingale. On the other hand, since $X_t$ is right-continuous in $L^1$, $\lim_{\lambda \to \infty} \lambda R_\lambda^Z(X_t) = X_t$ in $L^1$. Therefore, in the weak $L^1$-topology we have

$$M_\infty(t) = X(t) - X(0) + \lim_{\lambda \to \infty} \lambda \int_0^t X_u - \lambda R_\lambda^Z(X_u) \, du.$$

It is well known (Doob 1953) that $M_\infty(t+)$ is a right-continuous martingale, hence $A(t)$ is also right-continuous and adapted to $\mathscr{F}(t)$. Assumption (3) then gives the conditions under which the decomposition

$$X(t) = X(0) + M_\infty(t+) + A(t)$$

satisfies the defining property of a totally integrable special semimartingale.

The converse assertion relies heavily on the proof of the Doob–Meyer decomposition in Meyer, (1966). Suppose that $X_t$ is a totally integrable special semimartingale. Obviously (1) is satisfied, and there is a (unique) decomposition $X_t = X_0 + M'_t + A'_t$. We can write $A'_t = \int_0^t |dA'_s| - \bar{A}_t$, where $\int_0^t |dA'_s|$ is adapted to $\mathscr{F}_t$ (Dellacherie and Meyer 1980, VI. 53), in such a way that clearly $M'_t - \bar{A}_t$ and $-\int_0^t |dA'_s|$ are right-continuous supermartingales of class D. We may decompose each into the sum of a uniformly integrable martingale and a potential of class D (Meyer 1966, VI, Theorem 11) and it follows from Meyer (1966, VII, T29) that for any stopping time $T < \infty$,

$$A'(T) = -\lim_{h \to 0} A_T^h \text{ (weak } L^1\text{-topology),}$$

$$\text{where} \quad A_T^h = h^{-1} \int_0^T (X_u - E(X_{u+h}|\mathscr{F}_u)) \, du, \quad (6.22)$$

the integrand being chosen right-continuous. (We note that our $A'(t)$ is the negative of $A(t)$ in Meyer (1966, VII, T29).) On the other hand, the operations defining $A_T^h$ and $\lambda M_\lambda(t)$ are linear, and leave uniformly integrable martingales invariant in such a way that we have

$$A_T^h = h^{-1} \int_0^T A_u' - E(A_{u+h}'|\mathscr{F}_u) \, du, \quad (6.23)$$

and

$$\lambda M_\lambda(T) = M_T' + \lambda R_\lambda^Z(A'(T)) - \lambda R_\lambda^Z(A'(0)) + \lambda \int_0^T A_u' - \lambda R_\lambda^Z(A_u') \, du. \quad (6.24)$$

It is easy to see that $\lim_{\lambda \to \infty} \lambda R_\lambda^Z A'(T) = A'(T)$ in the $L^1$-norm ($A_t'$ being dominated by its total variation), hence the theorem will follow if we show that, for fixed $t$, $\lim_{\lambda \to \infty} \lambda \int_0^T A_u' - \lambda R_\lambda^Z(A_u') \, du = -A'(T)$ in the weak $L^1$-topology. Indeed, by (6.24) this implies that $\lim_{\lambda \to \infty} \lambda M_\lambda(t) = M'(t)$, and since $M'(t)$ is right-continuous it follows that

$$\lim_{\lambda \to \infty} \lambda M_\lambda(t) = M_\infty(t+) = M'(t),$$

$P$-a.s. Hence the process $A(t)$ of (3) is the same as $A'(t)$ of the assumed decomposition, which will complete the proof.

Let $S \in \mathscr{F}_\infty$, and rewrite Meyer's result (6.22) in the form

$$E(A'(t)I_S) = -\lim_{h \to 0} h^{-1} \left( \int_0^t E(I_S A_u') \, du - \int_0^t E(I_S E(A_{u+h}'|\mathscr{F}_u)) \, du \right)$$

$$= \frac{d}{dh} F(0),$$

where $F(h) = \int_0^t E(I_S E(A_{u+h}'|\mathscr{F}_u)) \, du$. Thus it only remains to observe that, by Fubini's theorem,

$$\lim_{\lambda \to \infty} \lambda \int_0^t E\left[\left(\lambda \int_0^\infty e^{-\lambda s} E(A_{s+u}'|\mathscr{F}_u) \, ds - A_u'\right) I_S\right] du$$

$$= \lim_{\lambda \to \infty} \lambda \left[\left(\lambda \int_0^\infty e^{-\lambda s} F(s) \, ds - F(0)\right)\right]$$

$$= \lim_{\lambda \to \infty} \lambda \int_0^\infty \lambda e^{-\lambda s}(F(s) - F(0)) \, ds$$

222 | Foundations of the prediction process

$$= \lim_{\lambda \to \infty} \lambda \int_0^\varepsilon \lambda\, e^{-\lambda s}\left(s\,\frac{d}{dh}F(0) + o(s)\right)ds$$

$$= \frac{d}{dh}F(0). \tag{6.25}$$

Here we used the fact that $F(s)$ is bounded by $t\cdot$ (the mean total variation of $A$) to eliminate the integral from $\varepsilon$ to $\infty$.

The final assertion concerning $A(T)$ follows similarly from (6.22) and (6.23), if we redefine $F(h) = E(I_S \int_0^T E(A'_{u+h}|\mathscr{F}_u)\,du)$ and apply Fubini's theorem pathwise to the double integral $\int_0^T \int_0^\infty e^{-\lambda s} E(A'_{s+u}|\mathscr{F}_u)\,ds\,du$. This is not difficult to justify. We first choose $E(A'_{u+r}|\mathscr{F}_u)$ to be right-continuous in $u$ for each $0 < r \in \mathbb{Q}$, and then set $E(A'_{s+u}|\mathscr{F}_u) = \lim \sup_{r \to s+, r \in \mathbb{Q}} E(A'_{u+r}|\mathscr{F}_u)$. Since $E(A'_{u+r}|\mathscr{F}_u) = E(\int_0^{u+r}|dA'_v| - \bar{A}_{u+r}|\mathscr{F}_u)$ is a difference of two right-continuous supermartingales in $u$ (as in Meyer (1966, VII. 27)), and these are monotone in $r$, we see by Lemma 2.2 that the effect is to define a version of $E(A'_{s+u}|\mathscr{F}_u)$ which is a.s. right-continuous for each $s$, and jointly integrable with respect to $e^{-\lambda s}\,ds\,du$, $P$-a.s. The balance of the argument follows the lines of (6.25).

Two simple illustrations, which also introduce the topic to follow, are described in the next two exercises.

**Exercise 6.7** Let $X_z^c(t)$ $(=X(t))$ be the Ornstein–Uhlenbeck process of (3.21). Show that $X_z^c(t) - X_z^c(0)$ is an absolutely integrable special semimartingale relative to the Brownian filtration $\mathscr{F}_t$, and at the same time obtain the finite variation component $A(t)$, $t > 0$. Solution: $A(t) = -\beta \int_0^t X_s\,ds$.

**Exercise 6.8** Take $X(t) = B(t) - B(t-1)$, relative to the Brownian filtration $\mathscr{F}_t$; $X(t)$ is the process of (3.22). Show that, in a locally uniform pathwise sense,

$$\lim_{\lambda \to \infty} \lambda \int_0^t (X_u - \lambda R_\lambda^Z X_u)\,du = B(-1) - B(t-1), \quad 0 < t.$$

Conclude, since this is of unbounded variation, that $X(t)$ is not a special semimartingale. (This illustrates the use of Theorem 6.4 in establishing that a given process is not a semimartingale.)

These exercises raise the question of how to determine when a regular stationary Gaussian process $X_t$ is a semimartingale, and, when it is, of determining its decomposition. Indeed, it follows immediately from a familiar dichotomy for separable stationary Gaussian processes (Belayev 1961) that if $X_t = X_0 + M_t + A_t$, where $A_t \in \mathscr{F}_t$ and is of locally finite variation (and $M_t \in \mathscr{F}_t$ is a martingale), then $X_t$ has continuous paths. Since $P \in M_c$ for stationary Gaussian processes (Section 4.2), it follows that $M_t$, and therefore $A_t$, also have continuous paths. Consequently, $A_t$ is automatically previsible, so the semimartingale $X_t$ is always special.

For stationary processes one cannot expect total integrability over $0 < t < \infty$. Instead, the relevant definition is obtained by stopping the process at some fixed $t_0$, and it is enough to set $t_0 = 1$. For Gaussian processes, at least, the condition of total integrability on $(0, 1]$ then also follows automatically for a semimartingale. To see this, we note that since $dA(t)/dt$ exists for a.e. $t$ (by a theorem of Lebesgue) it follows from Fubini's theorem that, in fact, $dA(t)/dt$ exists almost surely for a.e. $t$. But if $dA(t_0)/dt$ exists a.s., then since, by uniqueness, $A(t)$ has stationary increments, $dA(t)/dt$ exists as at $t = 0$. Now if $A(t)$ is Gaussian (and it follows at the end of the proof of Theorem 6.5 below that this is the case) then the a.s. derivative at $t = 0$ is also an $L^2$-derivative, and it is also Gaussian. Then it is quite simple to show, by a polygonal approximation of $A(t)$, that the expected total variation of $A(t)$, $0 < t \leq 1$, is simply $\int_0^1 E|dA(s)/ds|\, ds$, which equals $E|dA(0)/dt|$ by stationarity. Hence total integrability follow from finite variation.

Finally, we have the familiar Wold decomposition

$$X_t = \int_{-\infty}^{t} F(t - s)\, dB(s) + V_t,$$

where $V_t \in \mathscr{F}_{-\infty}$ and $B$ is a Brownian motion (by no means unique, however). The answer to our question for $V_t$ is then contained in Jain and Monrad (1982, Theorem 1.7). Because the martingale component of $V_t$ must obviously have constant paths, $V_t$ is a semimartingale if and only if it has locally finite variation. The necessary and sufficient condition for this is $\int x^2 \mu\, (dx) < \infty$, where $\mu(dx)$ is its spectral measure.

This brings us, finally, to the main result.

**Theorem 6.5** *A regular stationary separable Gaussian process*

$$X_t = \int_{-\infty}^{t} F(t - v)\, dB(v)$$

*is a semimartingale, relative to the filtration of $B$, if and only if $F$ is absolutely continuous and $F' \in L^2(0, \infty)$. In that case, $F(0+)$ exists and we have $A(t) = X(t) - X(0) - F(0+)B(t)$, where $t \geq 0 = B(0)$.*

**Remark** It seems noteworthy that such a simple conclusion was not previously known. Meanwhile, an interesting project for further research might be to test the present methods in the non-stationary Gaussian case.

**Proof** A formal application of Theorem 6.4 for the stopped process $X(t \wedge 1)$

224 | Foundations of the prediction process

would require re-calculation of $R_\lambda^Z X(t)$ to allow for the stopping. However, this is not necessary, and we can simply use

$$R_\lambda^Z(X(t)) = \int_0^\infty e^{-\lambda s}\left(\int_{-\infty}^t F(t+s-v)\,dB(v)\right)ds,$$

where the right side follows easily from independence of $B$-increments. Indeed, a trivial calculation gives, for $0 < t < 1$,

$$R_\lambda^Z(X(t \wedge 1)) - R_\lambda^Z(X(t)) = \int_{-\infty}^t \int_{1-t}^\infty e^{-\lambda s}(F(1-v) - F(t+s-v))\,ds\,dB(v),$$

with mean square bounded by

$$e^{-2\lambda(1-t)} \int_{-\infty}^t \left(\int_0^\infty e^{-\lambda u}(F(1-v) - F(1+u-v))\,du\right)^2 dv$$

$$\le e^{-2\lambda(1-t)}\lambda^{-1} \int_{-\infty}^t \int_0^\infty e^{-\lambda u}(F(1-v) - F(1+u-v))^2\,du\,dv$$

$$< 2e^{-2\lambda(1-t)}\lambda^{-2}\left(\int_{-\infty}^t F^2(1-v)\,dv\right)$$

$$\le 2e^{-2\lambda(1-t)}\lambda^{-2}\left(\int_0^\infty F^2(x)\,dx\right). \tag{6.26}$$

Thus, in calculating $\lim_{\lambda\to\infty} \lambda M_\lambda(t)$ for $t < 1$, we can use $M_\lambda$ for the original process $X(t)$ rather than for $X(t \wedge 1)$, with an error which tends to 0 in $L^2$ and hence does not affect weak $L^1$-convergence.

Since $\lim_{\lambda\to\infty} \lambda R_\lambda^Z(X_t) = X_t$ in $L^1$, as in the proof of Theorem 6.4 it suffices to find conditions for the existence of

$$\lim_{\lambda\to\infty} \lambda \int_0^t (X_u - \lambda R_\lambda^Z X_u)\,du = -A_t \quad \text{(weak } L^1\text{-topology)}, \tag{6.27}$$

and then to study the properties of this $A_t$. Assuming that $X(t \wedge 1)$ is a semimartingale, let $T_n \uparrow 1$ be stopping times such that $X(t \wedge T_n)$ is absolutely integrable. Then the finite variation component of $X(t \wedge T_n)$ is $A_{t \wedge T_n}$. For fixed $t$, substituting $R_\lambda^Z X_u$ into the integral (6.27), we can interchange the integrations and collect the coefficient of $dB(v)$ (which is a special case of the operation used in ((5.16)) and we obtain

$$\lambda \int_0^t (X_u - \lambda R_\lambda^Z X_u)\,du = \int_{-\infty}^t C_\lambda(v)\,dB(v),$$

where

$$C_\lambda(v) = \lambda^2 \int_{v \vee 0}^t \left(\int_0^\infty e^{-\lambda s}(F(u-v) - F(u+s-v))\,ds\right)du$$

and we have suppressed the dependence on $t$. To study the convergence as $\lambda \to \infty$, we take the Fourier transform $\tilde{C}_\lambda(r) = \int_{-\infty}^{t} e^{irv} C_\lambda(v)\, dv$. Separating the cases $v < 0$ and $0 < v < t$, and setting $F(u) = 0$ for $u < 0$, we obtain, after integration by parts,

$$\tilde{C}_\lambda(r) = -\lambda^2 \int_0^\infty e^{-\lambda s}\left[\int_{-\infty}^0 e^{irv}\left(\int_{s-v}^{t+s-v} F\, du - \int_{-v}^{t-v} F\, du\right) dv\right.$$

$$\left. + \int_0^t e^{irv}\left(\int_s^{t+s-v} F\, du - \int_0^{t-v} F\, du\right) dv\right] ds$$

$$= (ir)^{-1}\lambda^2 \int_0^\infty e^{-\lambda s}\left[\int_{-\infty}^0 e^{irv}(-F(t+s-v) + F(s-v) + F(t-v)\right.$$

$$\left. - F(-v))\, dv + \int_0^t e^{irv}(-F(t+s-v) + F(t-v))\, dv\right] ds$$

$$= (ir)^{-1}\lambda^2 \int_0^\infty e^{-\lambda s}\left[\int_{-\infty}^t e^{irv}(F(t-v) - F(t+s-v))\, dv\right.$$

$$\left. + \int_{-\infty}^0 e^{irv}(F(s-v) - F(-v))\, dv\right] ds.$$

Now denoting $\tilde{F}(r) = \int e^{irv} F(-v)\, dv$, this becomes

$$\tilde{C}_\lambda(r) = (ir)^{-1}\lambda^2 \int_0^\infty e^{-\lambda s}\left[(e^{irt} - e^{ir(t+s)})\tilde{F}(r) + \int_t^{t+s} e^{irv} F(t+s-v)\, dv\right.$$

$$\left. + (e^{irs} - 1)\tilde{F}(r) - \int_0^s e^{irv} F(s-v)\, dv\right] ds$$

$$= \tilde{F}(r)(1 - e^{irt})\lambda/(\lambda - ir) + (ir)^{-1}(e^{irt} - 1)\lambda^2$$

$$\times \int_0^\infty e^{-\lambda s + irs} \int_0^s e^{-iru} F(u)\, du\, ds$$

$$= \tilde{F}(r)(1 - e^{irt})\lambda/(\lambda - ir) + (ir)^{-1}(e^{irt} - 1)\lambda/(\lambda - ir)\int_0^\infty \lambda e^{-\lambda s} F(s)\, ds.$$

Inspection of this Fourier transform reveals a very simple alternative. Either $\lim_{\lambda \to \infty} \int_0^\infty \lambda e^{-\lambda s} F(s)\, ds$ exists (call it $L$) and then the Fourier transform converges in $L^2$ to $\tilde{F}(r)(1 - e^{irt}) + (ir)^{-1}(e^{irt} - 1)L$, or else there are subsequences $\lambda_{j,1} \to \infty$ and $\lambda_{j,2} \to \infty$ along which there are two distinct limits, or a limit $+\infty$. In these last cases it is trivial to see, by invoking Plancherel's theorem, that (6.27) has no weak $L^1$ limit, contrary to hypothesis. In the former case, by inversion of the transform it converges in $L^2$ to

$$\int_{-\infty}^{t} \{F(-v) - F(t-v)\} \, dB(v) + \{B(t) - B(0)\}L = -\{(X_t - X_0) - LB(t)\}.$$

This convergence holds, *a fortiori*, over $\{t \le T_n\}$, and letting $n \to \infty$ it follows that the only candidate for $A(t)$ is

$$A_t = (X_t - X_0) - \left(\lim_{\lambda \to \infty} \int_0^\infty \lambda e^{-\lambda s} F(s) \, ds\right) B(t).$$

So it remains only to give the conditions under which this is of finite variation. In particular, we see that $A(t)$ is necessarily Gaussian. (It may be noted, at this point, that this result is consistent with Exercises 6.7 and 6.8 above.)

**Lemma 6.6** *For $F \in L^2(0, \infty)$, there exists $\lim_{h \to 0+} h^{-1}(F(x+h) - F(x))$ in the $L^2$-norm if and only if $F$ is absolutely continuous and $F' \in L^2$. In this case, the limit function is $F'$.*

**Proof** Suppose, first, that an $L^2$ limit $g$ exists. Then for $0 \le x_0 < x$ we have $L^1$-convergence over $(x_0, x)$, and therefore

$$0 = \lim_{h \to 0+} \int_{x_0}^{x} g(y) - h^{-1}(F(y+h) - F(y)) \, dy$$

$$= \int_{x_0}^{x} g(y) \, dy - \lim_{h \to 0+} h^{-1} \left( \int_{x}^{x+h} F(y) \, dy - \int_{x_0}^{x_0+h} F(y) \, dy \right).$$

Choosing $x_0$ so that $\lim_{h \to 0+} h^{-1} \int_{x_0}^{x_0+h} F(y) \, dy = F(x_0)$, we see that

$$\int_{x_0}^{x} g(y) \, dy = F(x) - F(x_0)$$

for a.e. $x$, and therefore $g(x) = F'(x)$ for a.e. $x$. Since $g \in L^2$, this proves necessity.

Conversely, suppose that $F$ is absolutely continuous, with $F' \in L^2(0, \infty)$. Now we have

$$\lim_{h \to 0+} \int_0^\infty (h^{-1}(F(x+h) - F(x)) - F'(x))^2 \, dx$$

$$= \lim_{h \to 0+} \int_0^\infty \left( h^{-1} \int_x^{x+h} F'(y) - F(x) \, dy \right)^2 dx.$$

Since $F'(x)$ is locally in $L^1$, the integrand has limit 0 for a.e. $x$. Hence, by the dominated convergence theorem, it is sufficient to observe that the integrand is bounded by

$$h^{-1}\int_x^{x+h}(F'(y)-F'(x))^2\,dy \le 2h^{-1}\int_x^{x+h}(F'(y))^2\,dy + 2(F'(x))^2,$$

where we have

$$2h^{-1}\int_0^\infty\int_x^{x+h}(F'(y))^2\,dy\,dx = 2h^{-1}\int_0^\infty(F'(y))^2\left(\int_{(y-h)\vee 0}^y dx\right)dy$$

$$\le 2\int_0^\infty (F'(y))^2\,dy$$

$$< \infty.$$

To complete the proof of Theorem 6.5, we recall that, by an argument preceeding the theorem, (Gaussian) $A(t)$ is of finite variation if and only if $dA(0)/dt$ exists in $L^2$. Now

$$\lim_{h\to 0+} h^{-1}(A(h)-A(0))$$

$$= \lim_{h\to 0+}\left(\int_{-\infty}^0 h^{-1}(F(h-v)-F(-v))\,dB(v) + h^{-1}\int_0^h (F(h-v)-L)\,dB(v)\right).$$

By independence of increments, the first term must converge in $L^2$, and by Lemma 6.6 this is equivalent to $F$ being absolutely continuous with $F' \in L^2$. In that case, it is trivial to see that

$$\lim_{\lambda\to\infty}\int_0^\infty \lambda e^{-\lambda s}F(s)\,ds = \lim_{h\to 0+}F(h) = F(0+),$$

and the last term has mean square

$$\lim_{h\to 0+} h^{-2}\int_0^h (F(h-v)-F(0+))^2\,dv$$

$$= \lim_{h\to 0+} h^{-2}\int_0^h \left(\int_0^u F'(w)\,dw\right)^2 du$$

$$\le \lim_{h\to 0+} h^{-2}\int_0^h u\int_0^u (F'(w))^2\,dw\,du$$

$$\le \lim_{h\to 0+} \tfrac{1}{2}\int_0^h (F'(w))^2\,dw$$

$$= 0.$$

Consequently, this term poses no further restriction on $F$, and the proof is complete.

We shall conclude this section with a theoretical result concerning the Gaussian demesne $M_G$ of Theorem 3.11, and its relationship to the Hilbert space $H_z^\#$ of Definition 5.4, generated by all previsible stochastic integrals $\int h(s) \, dM_\lambda(s)$, $0 < \lambda$. In the general case, $z \in M_2$, no effective test of $X \in H_z^\#$ has been given, nor is any available here. It suffices to give an example where $H_z^\# \neq L_0^2(\Omega_Z, \mathscr{L}_\infty^z, P^z)$ to see that characterization of $H_z^\#$ may be problematical. For such an example, we need only consider $X_t = \int_0^t B_2(s) \, dB_1(s)$ for two independent Brownian motions $B_i$. Since $X_t$ is a martingale, we have $\lambda M_\lambda(t) = X_t$ for all $\lambda$, hence the space $H_z^\#$ has only one dimension ($\phi^\# = 1$). On the other hand, both $|B_2|$ and $\int_0^t \operatorname{sgn} B_2 \, dB_1$ are measurable over the generated filtration in such a way that, by a well-known result of K. Ito, the space $L_0^2(\Omega_Z, \mathscr{L}_\infty^z, P^z)$ has dimension 2 in the sense of stochastic integration.

The result we wish to show assets that such an example does not exist for Gaussian processes, except when they have (fixed) times of discontinuity (Knight 1986, Corollary 1.4).

**Theorem 6.7** *For $z \in D \cap M_G$, we have $H_z^\#(t) = L_0^2(\Omega_Z, \mathscr{L}_t^z, P)$ if and only if $P^z\{Z_s \text{ is continuous}, s \leq t\} = 1$. In particular, $H_z^\# = L_0^2(\Omega_Z, \mathscr{L}_\infty^z, P^z)$ if and only if $z \in M_c$, i.e. $z$ an intrinsic diffusion.*

**Proof** We recall from Theorem 3.16 that, for $z \in M_G$, the discontinuities of $Z_t$ (if any) occur with probability 1 at fixed times $t_j$. Then, if such a $t_j$ exists, we have $f(Z_{t_j-}, Z_{t_j}) \in L^2(\Omega_Z, \mathscr{L}_{t_j}^z, P^z)$ for any bounded $f(z_1, z_2) \in \mathscr{M} \times \mathscr{M}$. Since $h(t_j) \in \mathscr{L}_{t_j-}^z$ for previsible $h$, it is clear that functions of the form $f(R_\lambda^Z \hat{\rho}(Z_{t_j}) - R_\lambda^Z \hat{\rho}(Z_{t_j-}))$ can only be represented as elements of $H_z^\#(t_j)$ if they are in $H_z(t+) \cap H_z^\perp(t)$, and in particular they must by Gaussian with mean 0. (Of course, $\Delta R_\lambda^Z \hat{\rho}(Z_{t_j})$ is Gaussian and independent of $\mathscr{L}_{t_j-}^z$.) This proves the necessity.

For the converse, suppose that $Z(s)$ is continuous, $0 \leq s \leq t$, and that $0 \neq X \in L_0^2(\Omega_Z, \mathscr{L}_t^z, P^z)$ cannot be represented as an element of $H_z^\#(t)$, namely, that $X \neq \sum_{n=1}^{\phi_z^\#(t)} \int_0^t h_n(t,s) \, dY_n^\#(s)$ for any previsible $h_n$ with $E \int_0^t h_n^2(t,s) \, d\langle Y_n^\# \rangle_s < \infty$. Then, by the usual orthogonalization procedure, the martingale $E(X|\mathscr{L}_s^z)$ has a non-trivial projection onto the subspace complementary to $H_z^\#(t)$ in the space of square-integrable, mean 0, $\mathscr{L}_s^z$-martingales, $0 \leq s \leq t$. Letting $Y(s)$ denote this martingale, we note that since the $Y_n^\#$ are Gaussian, they must coincide with the $Y_n^*$ of the LMR and also with the $Y_n$ of Theorem 3.23. Indeed, the $Y_n$ are independent in this case, hence they are orthogonal in the sense of stochastic integration. In particular, this implies that $\sigma(Y_n^\#(s), s \leq t, n < \phi_z^\#(t) + 1) \equiv \mathscr{L}_t^z$, where, of course, $\phi_z^\#(t) = \phi_z(t)$. Since the family $(Y; Y_n, n < \phi_z(t) + 1)$ are orthogonal martingales for $s \leq t$, it follows by the theorem of Knight (1971) that there are independent Brownian motions $(B, B_n)$ (which are extended beyond the times $\langle Y \rangle_t$ and $\langle Y_n \rangle_t$ by an independent product space construction) such that $Y(s) = B(\langle Y \rangle_s)$ and $Y_n(s) = B_n(\langle Y_n \rangle_s)$, $0 \leq s \leq t$. But $\langle Y_n \rangle_s = EY_n^2(s)$, being non-random, implies that $\mathcal{L}_t^z \subseteq \sigma(B_n(s), 0 < s, n < \phi_z^\#(t) + 1)$, up to $P^z$-null sets, while $\langle Y \rangle_s$, $s \leq t$, although perhaps random, is in $\mathcal{L}_t^z$ and therefore independent of $B(t)$. Thus if $\mathcal{L}_t^z$ is given, $B(\langle Y \rangle_s)$ is an independent Brownian motion with a given time parameter $\langle Y \rangle_s$. This contradicts $B(\langle Y \rangle_s) \in \mathcal{L}_t^z$ unless $\langle Y \rangle_s \equiv 0$, $s \leq t$, which would be contrary to the non-triviality of $Y_t$, completing the proof.

**Remark**  A means of extending the previsible stochastic integral of Theorem 5.10 to general optional integrands is given in Dellacherie and Meyer (1980, VIII. 2. 37) under the name *compensated stochastic integral of a local martingale*. Whether or not this suffices to represent all of $L_0^2(\Omega_Z, \mathcal{L}_t^z, P^z)$ for $z \in D \cap M_G$ is not clear to us, but in any case an effective method of determining the integrands would be lacking.

## 6.4 Modifications of the future

In this final section we explore briefly the effect and utility of modifying one or both components in the pairing $(\tilde{X}_t, Z_t^z)$ of a process $(\Omega'_E, \mathcal{F}'_t, \tilde{X}_t, P^z)$ with its prediction process $Z_t^z$ on the same space (for the notation $\tilde{X}_t$, see Definition 2.7). One should distinguish between the effect of modifying $\tilde{X}_t$, for example killing $\tilde{X}_t$ at a random time which in general changes $Z_t^z$ for all $t$, and the effect of modifying $Z_t^z$ in a similar way which may be carried out on the canonical space $(\Omega_Z, \mathcal{L}_t^\circ, P^z)$ and does not change $\hat{\rho}(Z_t)$ ($\equiv^d \tilde{X}_t$) prior to the time of modification. These are evidently two different types of operation, and we will consider them separately in this order.

A natural question as to our usage of the martingales $M_\lambda$ is the manner in which they depend on the future of $X_t$, and especially the distant future, whose law may not be known in practice. A simple observation is the following.

**Theorem 6.8**
(1) *Let $Y_t$, $0 < t$, be adjoined to $(\Omega', P^z)$ as a process independent of $\mathcal{F}'_\infty$ with law $z_Y \in M_2$ (by a product space construction), and set*

$$U_t = \begin{cases} X_t & t < t_0; 0 < t_0 \text{ fixed.} \\ Y_{t-t_0} & t \geq t_0 \end{cases}$$

*Thus, denoting $M_\lambda$ for $U_t$ by $M_\lambda^U$, we have for, $t < t_0$,*

$$M_\lambda^U(t) = \int_0^{t_0-t} e^{-\lambda s} E(\tilde{X}_{t+s} | \mathcal{F}'_{t+}) \, ds - \int_0^{t_0} e^{-\lambda s} E(X_s | \mathcal{F}'_{0+}) \, ds$$
$$+ \int_0^t \left( X_u - \lambda \int_0^{t_0-u} e^{-\lambda s} E(\tilde{X}_{u+s} | \mathcal{F}'_{u+}) \, ds \right) du.$$

## 230 | Foundations of the prediction process

*In particular, the result is free of Y. For $t \geq t_0$,*

$$M^U_\lambda(t) = M^U_\lambda(t_0-) + M^Y_\lambda(t-t_0) + \int_0^\infty e^{-\lambda s}(E(Y_s|\mathcal{U}_{t_0+}) - EY_s)\,ds,$$

*where $\mathcal{U}_t$ is the filtration $\sigma(U_s, s \leq t)$.*

(2) *Let $e_\mu$ be an exponential random variable with parameter $\mu > 0$, independent of $\tilde{X}_t$, and set*

$$X^\mu_t = \begin{cases} \tilde{X}_t & t < e_\mu \\ 0 & t \geq e_\mu. \end{cases}$$

*Then, denoting $M_\lambda$ for $X^\mu_t$ by $M^\mu_\lambda$, we have*

$$M^\mu_\lambda(t) = \begin{cases} M_{\lambda+\mu}(t) + \mu \int_0^t R^Z_{\lambda+\mu}(\tilde{X}_u)\,du & t < e_\mu \\ M^\mu_\lambda(e_\mu-) - R^Z_{\lambda+\mu}(\tilde{X}_{e_\mu}) & t \geq e_\mu. \end{cases}$$

(3) *If we combine the operations of (1) and (2) by replacing $t_0$ by $e_\mu$ and setting*

$$V^\mu_t = \begin{cases} X_t & t \leq e_\mu \\ Y_{t-e_\mu} & t > e_\mu, \end{cases}$$

*then, for $t < e_\mu$,*

$$M^{V^\mu}_\lambda(t) = M_{\lambda+\mu}(t) + \mu \int_0^t R^Z_{\lambda+\mu}(\tilde{X}_u)\,du - (\lambda+\mu)^{-1}\lambda\mu t \left(\int_0^\infty e^{-\lambda s}(EY_s)\,ds\right).$$

*For $t \geq e_\mu$, the continuation is given below by (6.29).*

**Proof** For (1) we merely collect the contribution to $M^U_\lambda(t)$ from $s > t_0$, getting easily

$$\int_{t_0-t}^\infty e^{-\lambda s} EY(t-t_0+s)\,ds - \int_{t_0}^\infty e^{-\lambda s} E(Y(s-t_0))\,ds$$

$$- \lambda \int_0^t \left(\int_{t_0-u}^\infty e^{-\lambda s} E(Y(u-t_0+s))\,ds\right) du = 0.$$

The assertion for $t \geq t_0$ then follows directly.

Turning to (2), for $t < e_\mu$ we have

$$R^Z_\lambda(X^\mu_t) = \mu \int_0^\infty e^{-\mu t'} \int_t^{t+t'} e^{-\lambda s} E(\tilde{X}_{t+s}|\mathcal{F}'_{t+})\,ds\,dt'$$

$$= \int_0^\infty e^{-(\mu+\lambda)t'} E(\tilde{X}_{t+t'}|\mathcal{F}'_{t+})\,dt' = R^Z_{\lambda+\mu}(\tilde{X}_t).$$

Thus the result follows by substitution in $M^\mu_\lambda(t)$. For $t \geq e_\mu$, we get no further contribution to the terms of $M^\mu_\lambda(t)$ beyond the left limit $M^\mu_{\lambda+\mu}(t-)$, and at time $t = e_\mu$ we lose the term $R^Z_{\lambda+\mu}(\tilde{X}_{e_\mu})$ due to the killing, as required. Finally, as to (3) we have, for $t < e_\mu$,

$$R^Z_\lambda(V^\mu(t)) = R^Z_\lambda(X^\mu_t) + \mu \int_0^\infty e^{-\mu v} \int_v^\infty e^{-\lambda s} EY(s-v)\,ds\,dv$$

$$= R^Z_{\lambda+\mu}(\tilde{X}_t) + \frac{\mu}{\lambda+\mu} \int_0^\infty e^{-\lambda s} EY(s)\,ds,$$

hence the extra contribution due to $Y$ in $M^{V^\mu}_\lambda$ is

$$\frac{-\mu\lambda}{\lambda+\mu}\left(\int_0^t du\right) \int_0^\infty e^{-\lambda s} EY(s)\,ds,$$

as required.

This result suggests that, at any rate, it is not much more complicated to use $M^\mu_\lambda$ or $M^{V^\mu}_\lambda$ than to use $M_\lambda$ itself, for $t \leq e_\mu$.

In part (2), for example, we may write

$$M^\mu_\lambda(t) = M_{\lambda+\lambda}(t \wedge e_\mu) + \left[\mu \int_0^{t \wedge e_\mu} R^Z_{\lambda+\mu}(\tilde{X}_u)\,du - I_{(t \geq e_\mu)} R^Z_{\lambda+\mu}(\tilde{X}_{e_\mu})\right],$$

$$0 \leq t \quad (6.28)$$

which expresses $M^\mu_\lambda$ as the martingale $M_{\lambda+\mu}(t \wedge e_\mu)$ plus a compensated jump martingale with jump $-R^Z_{\lambda+\mu}(\tilde{X}_{e_\mu})$ at time $e_\mu$. The same reasoning shows how to extend (3) for $t \geq e_\mu$. We have, for $t \geq e_\mu$,

$$M^{V^\mu}_\lambda(t) = M^{V^\mu}_\lambda(t \wedge e_\mu-) - I_{(t \geq e_\mu)}\left[R^Z_{\lambda+\mu}(\tilde{X}_{e_\mu}) - (\lambda+\mu)^{-1}\lambda \int_0^\infty e^{-\lambda s} E(Y_s)\,ds\right]$$

$$+ I_{(t \geq e_\mu)} M^Y_\lambda(t - e_\mu), \quad (6.29)$$

thus including the jump whose compensator is

$$\mu \int_0^t R^Z_{\lambda+\mu}(X_u)\,du - (\lambda+\mu)^{-1}\lambda\mu t \int_0^\infty e^{-\lambda s} EY_s\,ds.$$

By a decomposition result of Meyer (1966, VIII, T31) the above compensated jump martingale terms are orthogonal to $M_{\lambda+\mu}(t \wedge e_\mu)$ in (2), and to $M_{\lambda+\mu}(t \wedge e_\mu) + I_{(t \geq e_\mu)} M^Y_\lambda(t - e_\mu)$ in part (3). Let us see how the representation theories for $X^\mu_t - EX^\mu_t$ are related to those of $\tilde{X}_t - E\tilde{X}_t$ in part (2)—we pass over the representation of $V^\mu_t - EV^\mu_t$ for (3), which of course would require also assumptions about $Y_t$.

Since the two component martingales of $M^\mu_\lambda$ in (6.28) are orthogonal, the

232 | Foundations of the prediction process

projection of $X_t^\mu - EX_t^\mu$ onto the space of $M_\lambda^\mu$ can be written as a corresponding sum. However, the projection onto $M_{\lambda+\mu}(t \wedge e_\mu)$ is not a trivial consequence of that onto $M_{\lambda+\mu}(t)$, so we prefer to proceed directly. For ease of notation, we fix $\lambda = \lambda_0$ and assume the situation of Theorem 5.5 for $\tilde{X}_t$, so that $\langle M_\lambda \rangle_t = \int_0^t f_\lambda(Z_s^z)\,ds$. Letting $F_{\lambda_0}^\mu(t,s)$ denote the integrand of the projection of $X_t^\mu - EX_t^\mu$ onto $M_{\lambda_0}^\mu$ in the stochastic integral representation (PMSIR) and using orthogonality, we have

$$\int_0^\infty e^{-\lambda s} F_{\lambda_0}^\mu(t+s,t)\,ds$$

$$= \frac{d\langle M_{\lambda+\mu}, M_{\lambda_0+\mu}\rangle_{t \wedge e_\mu} + \mu R_{\lambda_0+\mu}^Z(\tilde{X}_t) R_{\lambda+\mu}^Z(\tilde{X}_t)\,d(t \wedge e_\mu)}{d\langle M_{\lambda_0+\mu}\rangle_{t \wedge e_\mu} + \mu (R_{\lambda_0+\mu}^Z(\tilde{X}_t))^2\,d(t \wedge e_\mu)}.$$

Inverting the transform in $\lambda$ in terms of the corresponding unkilled integrands $F_\lambda^\#(t,s)$, we obtain for $t < e_\mu$,

$$F_{\lambda_0}^\mu(t+s,t) = e^{-\mu s} \frac{[f_{\lambda_0+\mu}(Z_t^z) F_{\lambda_0+\mu}^\#(t+s,t) + \mu R_{\lambda_0+\mu}^Z(\tilde{X}_t)\tilde{X}_t]}{f_{\lambda_0+\mu}(Z_t^z) + \mu (R_{\lambda_0+\mu}^Z(\tilde{X}_t))^2}. \quad (6.30)$$

Consequently, if $\phi^{\#,\mu}(t) = 1$, the PMSIR of $X_t^\mu$ is

$$X_t^\mu - EX_t^\mu = \int_0^{t \wedge e_\mu} e^{-\mu(t-s)} \frac{[f_{\lambda_0+\mu}(Z_s^z) F_{\lambda_0+\mu}^\#(t,s) + \mu R_{\lambda_0+\mu}^Z(\tilde{X}_s)\tilde{X}_s]}{f_{\lambda_0+\mu}(Z_s^z) + \mu (R_{\lambda_0+\mu}^Z(\tilde{X}_s))^2}$$

$$\times (dM_{\lambda_0+\mu}(s) + \mu R_{\lambda_0+\mu}^Z(\tilde{X}_s)\,ds - \delta_{\{s=e_\mu\}} \cdot R_{\lambda_0+\mu}^Z(\tilde{X}_s)), \quad (6.31)$$

where the last term includes a unit jump $\delta$ at time $e_\mu$ if $e_\mu \le t$, and we assume that $M_{\lambda_0}^\mu$ suffices for the representation, so that we do not have to orthogonalize and use $Y_1^{\#,\mu}$ instead.

**Remark** Formula (6.30) exhibits a connection between $F_{\lambda_0}^\mu$ and $F_{\lambda_0+\mu}^\#$, and thus between $M_{\lambda_0}^\mu$ and $M_{\lambda_0+\mu}$, which ought to be directly explainable.

**Exercise 6.9** Verify (6.31) in the case $X_t = B_t$, so that $f_{\lambda_0+\mu} = (\lambda_0+\mu)^{-2}$ and $F_{\lambda_0+\mu}^\#(t,s) = \lambda_0+\mu$.

**Exercise 6.10** Obtain an expression analogous to $M_t^\mu(t)$ for the process $\tilde{X}_t$ stopped at time $e_\mu$, say

$$X_t^{-,\mu} = \begin{cases} X_t & t < e_\mu \\ X_{e_\mu} & t \ge e_\mu. \end{cases}$$

Solution:

$$M_\lambda^{-,\mu}(t) = (1+\mu) M_{\lambda+\mu}(t \wedge e_\mu) - \mu \int_0^{t \wedge e_\mu} (X_u - (1+\mu) R_{\lambda+\mu}^Z(X_u))\,du.$$

Another interesting modification of the future is that of the appendix to Chapter 3, namely, if we replace $\tilde{X}_s$, $s \geq t$, by $E^z(\tilde{X}_s|\mathcal{F}'_{t+})$, for fixed $t$ and $z \in M_1$, then $M_\lambda(s)$ will not be affected for $s \leq t$. (In the appendix we used instead $E^{Z(t-)}\hat{\rho}(Z(s-t))$, which is the analogue of $E^z(\tilde{X}_s|\mathcal{F}'_{t-})$ when we represent $X_t$ on $\Omega'$ instead of on the prediction space.) But this is not useful for improving $M_\lambda$. In fact, the considerations of stationarity in Chapter 3 (Theorem 3.34, ff.) and Example 3 of this chapter suggest that, in order to improve $M_\lambda$, a good choice for the future law is one which makes the entire process as nearly stationary as possible, in distinction to the abrupt changes of Theorem 6.8, but this idea seems difficult to make precise.

We turn now to the effect of modifications which are made primarily in the prediction process $Z_t^z$. This implies, of course, corresponding modification of $\hat{\rho}(Z_t^z)$, and hence of $\tilde{X}_t$. Indeed, the primary purpose of such an operation for us would be to use the Markov properties of $Z_t^z$ to get information about the (non-Markovian) $\tilde{X}_t$. In particular, the extensive theory of stopping times, terminal times, and other special random times of Markov processes (see Blumenthal and Getoor (1968) and Sharpe (1988)) may in theory be directed to obtaining information about $\tilde{X}_t$ through $\hat{\rho}(Z_t^z)$.

To expedite such a procedure in terms of stopping times (such as first passage times) which depend on the process at individual $t$, it is necessary to know that $\tilde{X}_t = \hat{\rho}(Z_t^z)$ holds simultaneously in $t$; for penetration times, on the other hand, this is not needed. We shall assume that the original process $X_t$ has r.c.l.l. paths in a Lusin space $E$. Then, according to the Remarks after Theorem 2.7, we have $P^z\{\tilde{X}_t = \hat{\rho}_E(Z_t^z)$ for all $t\} = 1$. Of course, if the paths are left-continuous (l.c.r.l.) one could define $\hat{\rho}_E^-$ like $\hat{\rho}_E$ but 'to the left,' and then $P^z\{\tilde{X}_t = \hat{\rho}_E^-(Z_{t-}^z)$ for all $t > 0\} = 1$.

Next, in order to make use of translation operators (involved in many operations, such as the definition of terminal times), it is necessary to model our process on prediction space $\Omega_Z$ instead of on $\Omega'_E$, since $\theta_t^Z$ is only available on $\Omega_Z$. Then, since $P^z\{\hat{\rho}_E(Z_t^z)$ is r.c.l.l.$\} = 1$ in $\Omega'_E$, it follows by the Remark in the proof of Theorem 2.19(1) that $P^z\{\hat{\rho}_E(Z_t)$ is r.c.l.l.$\} = 1$ in $\Omega_Z$, as a set in the $P^z$-completion $\mathcal{L}_\infty^z$. Denoting this set by $\{\text{r.c.l.l.}\}_Z$, we model the original process $X_t$ as $\hat{\rho}_E(Z_t)$ on $\{\text{r.c.l.l.}\}_Z$, with the same probability $P^z$, filtration $\mathcal{L}_t^z$, and $\theta_t^Z$, all restricted to this subset. (Similar considerations apply to $\{\text{l.c.r.l.}\}$, or even to $\{\text{r.c.}\}_Z$ and $\{\text{l.c.}\}_Z$—see Dellacherie and Meyer (1975, IV, Theorem 34—depending on what is known for $X_t$.) An advantageous fact here is that, in view of Theorem 2.7, the filtration $\sigma\{\hat{\rho}_E(Z_s), s < t\}$ is $P^z$-equivalent to $\mathcal{L}_{t-}^0$ for every $z$. Thus it does not matter which we use in defining the usual right-continuous filtrations $\mathcal{L}_t^z$, or their intersections over all $z$ (or over all initial distributions $\mu$) which are used in the theory of Markov processes.

One of the basic problems about $X_t$ which can be approached advantageously by using $Z_t$ is the calculation of passage time distributions and arrival probabilities. Thus, let $B \in \mathcal{E}$ and define the arrival time $T_B^X = \inf\{t > 0: X_t \in B\}$, and the absorbed process $X_B(t) = X(t \wedge T_B^X)$. To treat such a process in terms of $Z_t$, one need only introduce the set, time, and process $A = \{z: \hat{\rho}_E(z) \in B\}$, $T_A = \inf\{t > 0: Z_t \in A\}$, and $Z_A(t) = Z(t \wedge T_A)$, respectively. Then clearly the processes $X_B(t)$ and $\hat{\rho}_E(Z_A(t))$ are equivalent in law, and the passage time $T_A$ has the same law as $T_B$. Also for $C \in \mathscr{E}$,

$$P\{X_B(T_B) \in C\} = P^z\{Z_A(T_A) \in \hat{\rho}_E^{(-1)}(C)\},$$

i.e. the arrival probabilities of $Z_A$ include those of $X_B$. It should be noted, of course, that the passage times of $Z_t$ are more general than those of $\hat{\rho}_E(Z_t)$, there being many elements of $\mathscr{M}_E$ not of the form $\hat{\rho}_E^{(-1)}(B)$, $B \in \mathscr{E}$.

A general method for calculating passage time distributions and arrival probabilities for the Markov process $Z_t$ is to consider the function $E^z(e^{-\lambda T_A}f(Z(T_A)))$, $f \in b(\mathscr{M}_E)$, $f = 0$ on $A^c$. Generally speaking, this is the unique solution on $D$ (where $D$ are non-branch points) of the equation

$$\mathscr{A}F(z) = +\lambda F(z) \quad \text{on} \quad A^c \cap D; \quad F(z) = f(z) \quad \text{on} \quad A \cap D, \tag{6.32}$$

where $\mathscr{A}$ is the weak infinitesimal generator of the semigroup of $Z_A(t)$ on $b(\mathscr{M}_E)$, and we assume that $A$ is closed and $A = A^r$, where $A^r = \{z: T_A = 0, P^z\text{-a.s.}\}$ is the set of regular points of $A$. Then we have

$$E^z(e^{-\lambda T_A}f(Z(T_A))) = \lambda R_\lambda^{Z_A} f(z), \quad z \in A^c \cap D, \tag{6.33}$$

and (6.32) reduces to the familiar resolvent identity $\lambda R_\lambda^{Z_A} f - \mathscr{A}(R_\lambda^{Z_A} f) = f$. However, there is evidently no hope of solving (6.32) everywhere on $D$, and we refrain from considering uniqueness of the solution in a setting as general as this.

The important point here is that, in many cases, a given process $X_t$ brings with it a natural demesne $U \subset D$, consisting of all conditional futures $Z_t^z$ accessible for $X_t$. Then one only requires that (6.32) hold on $U$ in order to apply the method. In this case, if for example $Z_t$ is an intrinsic diffusion on $U$, $\mathscr{A}$ may be a differential operator, and the solution may be both unique and obtainable.

We let $M_\lambda^A(t)$ denote the fundamental martingales on $\Omega_Z$ for the counterpart of $X_B(t)$, namely, for $\hat{\rho}(Z_A(t))$, $0 \le t$. Their evaluation can easily be reduced to that of $M_\lambda(t)$, for $\hat{\rho}(Z(t))$, together with the passage times and arrival probabilities for $Z(t)$ on $A$. Indeed, continuing the assumption that $A$ is closed and $A = A^r$, we have, for $z \in M_2$,

$$R_\lambda^Z \hat{\rho}(z) - R_\lambda^{Z_A} \hat{\rho}(z) = E^z(e^{-\lambda T_A}[R_\lambda^Z \hat{\rho}(Z(T_A)) - \lambda^{-1}\hat{\rho}(Z(T_A))]) \tag{6.34}$$

where we used the strong Markov property at $T_A$.

Then, denoting (6.34) by $D_\lambda(z)$, we have easily

$$M_\lambda^A(t) = M_\lambda(t \wedge T_A) - \left[D_\lambda(Z(t \wedge T_A)) - D_\lambda(z) - \int_0^{t \wedge T_A} \lambda D_\lambda(Z(s \wedge T_A))\, ds\right]. \tag{6.35}$$

Since the second factor on the right is a martingale, it can sometimes be written as a stochastic integral by applying Ito's lemma to $D_\lambda(Z(t \wedge T_A))$.

Unfortunately, it is not clear to the author which processes $X_t$ are especially well suited for application of this method. Again, the problem is partly where to begin; there is no systematic way of expressing non-Markovian $X_t$ in general, and the particular possibilities are too numerous to be dealt with. Instead, we shall only present two rather simple illustrations based on Example 4, and leave any deeper applications to the reader as they may arise.

**Example 6** Continuing Example 4, we let $X_t = B_t + I_{[T(1),\infty)}(t)$, $0 \leq t$, and recall that we have

$$\lambda M_\lambda(t) = B_t + \sqrt{2\lambda}\exp(-\sqrt{2\lambda})\int_0^{t \wedge T(1)} (\exp\sqrt{2\lambda}B_s)\,dB_s.$$

Let us apply the above method to calculate passage time distributions (or at least their Laplace transforms) and arrival probabilities for the binary set $B = \{-3, 3\}$. It is easy to see that for this process there is a natural demesne $U$ for $Z_t$ consisting of two sets $S_1$ and $S_2$. For $t < T(1)$ we can identify $Z_t^z$ with the conditional future given $B_t$ and $\{t < T(1)\}$, where $B_t < 1$ may be assumed. For $t \geq T(1)$ we can identify $Z_t^z$ with the future of $(B_{t+}.) + 1$ given $B_t$. These two sets form a (Borel) demesne $U$ for $Z_t$, which suffices to treat $X_t$ in the form $\hat{\rho}(Z_t)$. As above, we let $A = \{z : \hat{\rho}(z) = \pm 3\} \cap U$, and consider $E^z\{e^{-\lambda T_A}; \hat{\rho}(Z(T_A)) = -3\}$. For $z \in S_1$, so that $t < T(1)$, we can identify $z$ with $x = \hat{\rho}(z) < 1$, and write the above as

$[E^x(e^{-\lambda T_B}; T(-3) < T(1)) + E^x(e^{-\lambda T_B}; T(1) < T(-3)$ and

$$B(t) \circ \theta^Z_{T(1)} \text{ reaches } -4 \text{ before } +2)]. \qquad (6.36)$$

The first term in brackets satisfies the equation $\otimes: \frac{1}{2}d^2\varphi/dx^2 = \lambda\varphi$, with the end conditions $\varphi(-3) = 1$, $\varphi(1) = 0$, according to a familiar Brownian motion calculation, since the generator $\mathscr{A}$ of $Z_t$ identifies with $\frac{1}{2}d^2/dx^2$ on $S_1$. The second term in brackets can be expressed, using the strong Markov property, as

$E^x(e^{-\lambda T(1)}e^{-\lambda((T_B)-T(1))}; T(1) < T(-3)$ and $B(t) \circ \theta^Z_{T(1)}$ reaches $-4$ before 2)

$= E^x(e^{-\lambda T(1)}E^1(e^{-\lambda T(-4, 2)}; B(T(-4, 2)) = -4); T(1) < T(-3))$,

where $T(-4, 2)$ is a passage time of $B(t)$ to $\{-4, 2\}$. The inner expectation is the value at 1 of the solution of $\otimes$ with end conditions 1 at $-4$ and 0 at $+2$. Accordingly, it is given by $\sinh\sqrt{2\lambda}/\sinh 6\sqrt{2\lambda}$. Then this term factors out and the other factor has an analogous evaluation as

$$\frac{\sinh\sqrt{2\lambda}(x+3)}{\sinh 4\sqrt{2\lambda}}.$$

Similarly, the first term in brackets is $\sinh\sqrt{2\lambda}(1-x)/\sinh 4\sqrt{2\lambda}$ and (6.36) becomes

$$(\sinh 4\sqrt{2\lambda}\,\sinh 6\sqrt{2\lambda})^{-1}(\sinh \sqrt{2\lambda}(1-x)\sinh 6\sqrt{2\lambda}$$
$$+\sinh\sqrt{2\lambda}(x+3)\sinh\sqrt{2\lambda}) = E^z(e^{-\lambda T_A}; \hat\rho(Z(T_A))=-3)$$
$$\text{for } x=\hat\rho(z),\ z\in S_1. \qquad (6.37)$$

The case when $\hat\rho(Z(T_A))=3$ is a bit shorter, since the process must first reach $+1$, then reach $+3$. By the same method, we obtain

$$E^z(e^{-\lambda T_A}; \hat\rho(Z(T_A))=3)$$
$$=(\sinh 4\sqrt{2\lambda}\,\sinh 6\sqrt{2\lambda})^{-1}(\sinh\sqrt{2\lambda}(x+3)\sinh 5\sqrt{2\lambda}$$
$$\text{for } x=\hat\rho(z),\ z\in S_1. \qquad (6.38)$$

These transforms do not look very easy to invert; however, let us at least check the arrival probabilities, which are obtained simply by letting $\lambda \to 0+$. We have

$$P^z(\hat\rho(Z(T_A))=-3) = (6(1-x)+(x+3))/24 = (9-5x)/24, \qquad (6.39)$$

which holds for $x=\hat\rho(z) < 1$, $z\in S_1$, where we note that this accords with (6.38), whose limit is $(5x+15)/24$. On the other hand, for $z\in S_2$, so that $t\geq T(1)$, we have easily

$$E^z(e^{-\lambda T_A}; \hat\rho(Z(T_A))=3) = (\sinh 6\sqrt{2\lambda})^{-1}\sinh\sqrt{2\lambda}(x+3)$$
$$E^z(e^{-\lambda T_A}; \hat\rho(Z(T_A))=-3) = (\sinh 6\sqrt{2\lambda})^{-1}\sinh\sqrt{2\lambda}(3-x); \quad x=\hat\rho(z), \qquad (6.40)$$

and $P^z(\hat\rho(Z(T_A))=3) = (x+3)/6$, as expected.

It is not very exciting to work out $M_\lambda^A(t)$ from (6.35) in the present example, but we include it for the sake of completeness. We have on $S_1$

$$R_\lambda^Z(\hat\rho(x=-3)) = \lambda^{-1}(-3+E^{-3}e^{-\lambda T(3)}) = \lambda^{-1}(-3+e^{-6\sqrt{2\lambda}}),$$

whence it follows that, for $z\in S_1$,

$$D_\lambda(\hat\rho(z)=x) = \lambda^{-1}e^{-6\sqrt{2\lambda}}(E^x e^{-\lambda T_B}; \hat\rho(Z(T_A))=-3),$$

with the last factor given by (6.37); $x=\hat\rho(z)$. Then, by Ito's lemma in (6.35), the martingale $M_\lambda(t\wedge T_A)-M_\lambda^A(t)$ is given, for $t<T(-3,1)$, by

$$\sqrt{\frac{2}{\lambda}}e^{-6\sqrt{2\lambda}}(\sinh 4\sqrt{2\lambda}\,\sinh 6\sqrt{2\lambda})^{-1}$$
$$\times \int_0^t (-\sinh 6\sqrt{2\lambda}\cosh\sqrt{2\lambda}(1-B_s)+\sinh\sqrt{2\lambda}\cosh\sqrt{2\lambda}(3+B_s))\,dB_s.$$
$$(6.41)$$

It appears difficult to invert $M_\lambda^A(t)$ in $\lambda$ to obtain the PMSIR of $X_t - EX_t$ (by using Lemma 5.9(2)). However, it is by no means impossible, as the following consideration shows. We have already derived the PMSIR for the unstopped process $B_t + I_{[T(1),\infty)}(t)$ in (6.19). Hence to derive it for the present $X_t$ we need only add the change due to stopping at $T_A$, which must also be a martingale stochastic integral with respect to $dB_s$. Thus, we must add the PMSIR of $B(T_A \wedge t) + I_{[T(1),\infty)}(T_A \wedge t) - (B(t) + I_{[T(1),\infty)}(t))$ minus its expectation. But this separates into the representations over the three sets $S_1 = \{B(T_A) = 3\}$, $S_2 = \{B(T_A) = -3; T(-3) > T(1)\}$, and $S_3 = \{B(T_A) = -3, T(-3) < T(1)\}$. On $S_1$ and $S_2$ the representation is just $\int_0^t -I_{((T_A \leq s) \cap S_i)}(s)\, dB(s)$, but over $S_3$ it is given by an expression of the form $\int_0^t -I_{((T_A \leq s) \cap S_3)}(s) F^\#(s)\, dB(s)$, where $F^\#(s)$ is similar to (6.19) except that $t$ is replaced by $t - T_A$, $T(1)$ is replaced by $T(4) \circ \theta_{T_A}^Z$, $B(s)$ is replaced by $(B(s) \circ \theta_{T_A}^Z) + 3$, and we need to replace the factor $\exp(-\sqrt{2\lambda})$ by $\exp(-4\sqrt{2\lambda})$ outside the integral. In short, we can express the previsible integrand explicitly, and this in turn implies an inversion of the transform (6.41).

Continuing with the process $X_t = B_t + I_{[T(1),\infty)}(t)$, let us also indicate an example in which the process $Z_t$ is stopped at a set $A$ which does not correspond to stopping $X_t$ at any set $B$. For instance, let us take $A = \{z : P^z\{X_t \text{ reaches } -2 \text{ before } +3\} \geq \tfrac{1}{2}\}$. Now for $t < T_A \wedge T(1)$,

$P^{Z_t}\{X_t \text{ reaches } -2 \text{ before } +3\}$

$= P^{Z_t}\{B_t \text{ reaches } -2 \text{ before } +1\} + P^{Z_t}\{B_t \text{ reaches } +1 \text{ before } -2\}$

$\quad \times P^2\{B_t \text{ reaches } -2 \text{ before } +3\}$

$= \tfrac{1}{3}(1 - x) + \tfrac{1}{15}(2 + x)$

$= \tfrac{1}{15}(7 - 4x),$

where $x = \hat{\rho}(Z_t)$, while for $T(1) \leq t < T_A$, the same expression becomes $\tfrac{1}{5}(3 - x)$. Setting these probabilities equal to $\tfrac{1}{2}$, it follows that

$$T_A = T(-\tfrac{1}{8})I_{(T(-\tfrac{1}{8}) < T(1))} + (T(1) + T(+\tfrac{1}{2}) \circ \theta_{T(1)}^Z)I_{(T(-\tfrac{1}{8}) > T(1))}. \quad (6.42)$$

Much as in Example 6, it follows that for $t < T_A \wedge T(1)$,

$E^{Z_t}(e^{-\lambda T_A}; T(-\tfrac{1}{8}) < T(1)) = (\sinh \tfrac{9}{8}\sqrt{2\lambda})^{-1} \sinh \sqrt{2\lambda}(1 - B_t),$

$E^{Z_t}(e^{-\lambda T_A}; T(-\tfrac{1}{8}) > T(1)) = (\sinh \tfrac{9}{8}\sqrt{2\lambda})^{-1} \sinh \sqrt{2\lambda}(\tfrac{1}{8} + B_t) \exp -\tfrac{3}{2}\sqrt{2\lambda}.$

On the other hand, for $T(1) \leq t < T_A$, we have $E^{Z_t}(e^{-\lambda T_A}) = \exp -(X_t - \tfrac{1}{2})\sqrt{2\lambda}$; $X_t = \hat{\rho}(Z_t)$. It can be seen that these three transforms determine the passage time distributions and the arrival probabilities of $Z_t$ on the set $A$. For $T(1) \leq t < T_A$, the set $A$ is accessible only at a single point $z_1$, while for $t < T(1) \wedge T_A$ it is accessible at two points $z_0$ and $z_1$.

For representation of the process, let us consider, instead of $Z_A(t)$, the killed process $Z_A^0(t) \doteq Z_t$ for $t < T_A$, and $Z_A^0(t) = \Delta$ for $t \geq T_A$. Setting $\hat{\rho}(\Delta) = 0$, we denote the fundamental martingales for $\hat{\rho}(Z_A^0(t))$ by $M_\lambda^0(t)$ (suppressing $A$). It is easy to see that they are represented as in (6.35) except that $D_\lambda(z)$ ($\doteq$(6.34)) is simplified to $E^z(e^{-\lambda T_A} R_\lambda^Z \hat{\rho}(Z(T_A)))$. We leave the rest as an exercise.

**Exercise 6.11**  By applying Ito's lemma, show that

$$M_\lambda^0(t) = M_\lambda(t \wedge T_A) - (8 \sinh \tfrac{9}{8}\sqrt{2\lambda})^{-1} \sqrt{\frac{2}{\lambda}} \int_0^t H_\lambda(B_s)\, dB_s;$$

$$t \leq T(1) \wedge T(-\tfrac{1}{8}),$$

where

$H_\lambda(x) = (8\exp -\tfrac{9}{8}\sqrt{2\lambda})\cosh\sqrt{2\lambda}(1-x) + (4\exp -\tfrac{3}{2}\sqrt{2\lambda})\cosh\sqrt{2\lambda}(\tfrac{1}{8}+x)$.

For $T(1) \leq t \leq T_A$ one must add to $M_\lambda^0(T(1))$ the term

$$(2\lambda)^{-1/2} \int_{T(1)}^t \exp -\sqrt{2\lambda}(B(s)+\tfrac{1}{2})\, dB_s, \quad \text{and} \quad M_\lambda^0(t) = M_\lambda^0(t \wedge T_A).$$

## Appendix

As a sort of final addendum to the present work, we append here a discussion of how the prediction process relates to the usual approach to stochastic processes, as exemplified by the general theory of processes of Dellacherie and Meyer (1975, 1980, 1987). Let $X_t$, $0 \leq t$, be a measurable process on a complete probability space, with values in the Lusin space $(E, \mathscr{E})$. Thus far, our approach to $X_t$ has been (except in Section 6.1 and a few other places) to represent $X_t$ on the new space $(\Omega_E', \mathscr{F}', P')$ of Definition 1.1, where $P'$ is derived from $P$ by the inverse of the mapping of $\Omega$ into $\Omega_E'$ sending $w$ into the path $X_{(\cdot)}(w)$. Only by this means could we obtain the regularity of conditional probabilities and translation operators necessary to construct the prediction process $Z_t^z$ ($z = P'$), and to go on to the prediction space $(\Omega_Z, \mathscr{L}_\infty^\circ, P^z)$ with its natural translation operators $\theta_t^Z$ when needed.

This seems to be a valid approach in the sense that it makes clear that translation operators of $t$ are a construct on the path space, and do not come ready-made along with $X_t$. But there are also situations in which one wishes to work with the prediction process of $X_t$ directly on $(\Omega, \mathscr{F}, P)$, and in fact it is convenient to do so in the majority of applications of the theory. To do this, one has only to transplant $Z_t^z(S)$, $S \in \mathscr{F}'$, back onto $(\Omega, \mathscr{F}, P)$ in a rather evident way.

**Definition 6.3**  Let $Z_t^X(S, w) = Z_t^z(S, X_{(\cdot)}(w))$, $0 \leq t$, $S \in \mathscr{F}'$, where $X_{(\cdot)}(w)$ is the path of $X$ at $w \in \Omega$, and $z(S) = P\{X_{(\cdot)}(w) \in S\}$. We call $Z^X$ (with whatever further variables are needed) the *prediction process* of the process $X$.

It follows obviously from Theorems 1.14 and 1.9 that $Z_t^X$ is a strong Markov process on $(M_E, \mathscr{M}_E)$ relative to the filtration $\mathscr{F}_{t+}^X$, where $\mathscr{F}_t^X = \{\{X_{(\cdot)}(w) \in S_t\}, S_t \in \mathscr{F}_t'\}$, and that, for $\mathscr{F}_{t+}^X$-stopping times $T < \infty$, $Z_T^X(S) = P\{X_{T+(\cdot)} \in S | \mathscr{F}_{T+}^X\}$. Further, by Theorem 1.16 we know that $\mathscr{F}_{t+}^X$ and $\sigma(Z_s^X, s \leq t)$ are P-equivalent. Of course, if $X_t$ has right-continuous paths then $\mathscr{F}_t^X = \sigma(X_s, s < t)$ and $\mathscr{F}_{t+}^X$ is the usual generated right-continuous filtration, which may be augmented by all P-null sets when needed.

A similar construction obviously exists when one begins with a prescribed filtration $\mathscr{F}_t$ on $(\Omega, \mathscr{F}, P)$, such that $X_t \in \mathscr{F}_t$ and $\mathscr{F}_\infty$ is almost countably generated. As in Section 6.2, one need only introduce a further measurable process $Y_t$ such that $\mathscr{F}_t \equiv \sigma(X_s, Y_s; s \leq t)$, then define the prediction process $Z_t^{X,Y}$ of the pair (with product state space), and finally restrict $Z_t^{X,Y}(S)$ to $S$ depending only on the $X$-coordinate. This defines the *prediction process of $X_t$ relative to the filtration $\mathscr{F}_t$*. It is clear, however, that although this defines the appropriate conditional probabilities relative to $\mathscr{F}_{T+}'$ for $\mathscr{F}_{t+}'$-stopping times $T < \infty$, the Markov properties of $Z_t^{X,Y}$ itself may be lost unless we retain all sets $S$ in the combined $\mathscr{F}'$.

On the other hand, if a right-continuous filtration $\mathscr{G}_t$ contained in $\mathscr{F}_{t+}^X$ is given (automatically it is almost countably generated along with $\mathscr{F}_\infty^X$), and if we define a right-continuous process $W_t$ such that $\mathscr{G}_t \equiv \sigma(W_s, s \leq t)$ then by applying Section 6.2 we may, in a similar way to the above, define the prediction process $G_t^{X,W}$ of $X, W$ relative to the observed filtration of $W$. By Theorem 6.3, $G_t^{X,W}$ will be a strong Markov process relative to $\mathscr{G}_t$, with its own transition function $q^*(t, z, A)$. Thus we can formulate a problem of filtering directly on $(\Omega, \mathscr{F}, P)$.

One other type of problem which is most conveniently viewed on the original probability space is that of comparing several given processes $X_t, Y_t, \ldots$. For convenience of notation, let us suppose that we have only $(X_t, Y_t)$, each with the same state space, $E$, and that they are considered relative to their generated filtrations $\mathscr{F}_{t+}^X$ and $\mathscr{F}_{t+}^Y$. Of course, one can view $(X_t, Y_t)$ as a single process relative to the combined filtration $\mathscr{F}_{t+}^{X,Y}$, but that does not readily yield the separate prediction processes. Instead, we will wish to consider $Z_t^X$ and $Z_t^Y$ simultaneously on $(\Omega, \mathscr{F}, P)$. (On the other hand, to study a modification $Y_t$ of a given process $X_t$, it may be recommended to treat them as a vector process; such a case arose in the appendix to Chapter 3.)

The question which we wish to treat here is that of invariance properties of $Z_t^X$ under various equivalences of filtrations. For any two filtrations $\mathscr{F}_t$ and $\mathscr{G}_t$ on $(\Omega, \mathscr{F}, P)$ we continue to write $\mathscr{F}_t \equiv \mathscr{G}_t$ when $\mathscr{F}_t$ and $\mathscr{G}_t$ differ only by P-null sets for each $t$, i.e. their $L^2$-spaces are the same. For example, it follows from the proof of Theorem 4.3 that whenever processes $X$ and $Y$

have $\mathcal{F}^X_{t+} \equiv \mathcal{F}^Y_{t+}$, the times of discontinuity of $Z^X_t$ and $Z^Y_t$ are also equivalent. Indeed, they are both equivalent to the times of discontinuity of the family of right-continuous martingales of the common filtration. More generally, the entire classification of Chapter 4 according to types of discontinuity reduced to the discontinuities of martingales, and hence does not separate $X$ and $Y$.

It is thus natural to enquire when $Z^X_t$ and $Z^Y_t$ have some stronger form of equivalence. The two forms to be treated here are given by the following definitions.

**Definition 6.4** We say that $Z^X$ and $Z^Y$ are functionally related if, for each $t$, there are functions $f_t$ and $g_t$, $\mathcal{M}_E/\mathcal{M}_E$-measurable, for which $f_t(Z^X_t) = Z^Y_t$, and $g_t(Z^Y_t) = Z^X_t$, $P$-a.s.

**Definition 6.5** We say that $Z^X$ and $Z^Y$ are set related if for each $t$, there are functions $h_t$ and $k_t \colon \mathcal{F}' \to \mathcal{F}'$, such that for $S \in \mathcal{F}'$, $Z^X_t(S) = Z^Y_t(h_t(S))$ and $Z^Y_t(S) = Z^X_t(k_t(S))$, $P$-a.s.

**Remark** We do not know whether one can achieve $g = f^{(-1)}$ in Definition 6.4, or $k_t = h_t^{(-1)}$ in Definition 6.5. Conceivably, $h_t$ may even be chosen to be an isomorphism of $\mathcal{F}'$.

It is not difficult to see that if $Z^X$ and $Z^Y$ are set related they are also functionally related. Indeed, since $\mathcal{F}'$ is countably generated (Definition 1.1), $Z^X_t$ is $P$-a.s. determined by $Z^Y_t(h_t(S))$ for a countable collection of $S$. The determination is effected by a Borel function on the set where the equality holds for each such $S$. It is not as easy to see that set relatedness is a considerably more restrictive concept than function relatedness. This ensues, however, from the following basic theorem.

**Theorem 6.9**

(1) $Z^X$ and $Z^Y$ are functionally related if and only if, for each $t$, $\mathcal{F}^X_{t+} \equiv \mathcal{F}^Y_{t+}$ and $\sigma(Z^X_s, s \geq t) \equiv \sigma(Z^Y_s, s \geq t)$.

(2) $Z^X$ and $Z^Y$ are set related if and only if, for each $t$, $\mathcal{F}^X_{t+} \equiv \mathcal{F}^Y_{t+}$ and $\{X_{t+(\cdot)} \in S, S \in \mathcal{F}'\} \equiv \{Y_{t+(\cdot)} \in S, S \in \mathcal{F}'\}$.

In other words, (1) requires equivalence of the futures of $Z^X$ and $Z^Y$, while (2) requires equivalence of the futures of $X$ and $Y$. It is easy to guess (see also Exercise 6.12 below) that the former does not imply the latter. But it seems less than obvious that, when $\mathcal{F}^X_{t+} \equiv \mathcal{F}^Y_{t+}$, future equivalence of $X$ and $Y$ does imply that of $Z^X$ and $Z^Y$. In any case, before doing the proof a simple example and exercise may serve as an aid to intuition. These are

## Applications and ramifications | 241

stated in the discrete parameter case, but they are trivially extended to a continuous parameter by use of step functions in the obvious way.

**Example 7** Let the parameter set be $\{1, 2, 3\}$, and let $B_1$ and $B_2$ be independent Bernoulli random variables, $P\{B_i = \pm 1\} = \frac{1}{2}$. We define $X_1 = B_1, X_2 = B_2, X_3 = B_1$ and also $Y_1 = B_1, Y_2 = B_1 B_2, Y_3 = B_1$. Recalling that present is contained in both past and future, it is easy to see that $X$ and $Y$ have equivalent pasts and equivalent futures for each $t \in \{1, 2, 3\}$. Clearly $X$ and $Y$ are not, themselves, functionally related at $t = 2$, but $Z_t^X$ and $Z_t^Y$ are even set related, because for $a$ and $b \in \{\pm 1\}$, we have $Z_2^X\{(a, b)\} = Z_2^Y\{(ab, b)\}$.

**Exercise 6.12** Let the parameter set be $\{1, 2, 3\}$ as before, with $X_1, X_2$ and $Y_1, Y_2$ unchanged, but let $X_3 = Y_3$ where the common value has conditional probability $\frac{1}{2}$ for $+1$ or $-1$ if $B_1 = 1$, and has conditional probability $\frac{1}{3}$ for $-1, 0,$ or $+1$ if $B_1 = -1$, irrespective of $B_2$. Show that the pasts of $X$ and $Y$ are equivalent for each $t \in \{1, 2, 3\}$, and also the futures of $Z^X$ and $Z^Y$ are equivalent, but the futures of $X$ and $Y$ are distinct at $t = 2$. Therefore $Z^X$ and $Z^Y$ are functionally, but not set related. Hint: $B_1$ is in the $Z$-futures at $t = 2$, but not in the $X$- or $Y$-futures.

Turning to the proof of Theorem 6.9, part (1) is in fact a simple consequence of the Markov property of $Z^X$ and $Z^Y$. By Theorem 1.16(2) we have $\mathcal{F}_{t+}^X \equiv \sigma(Z_s^X, s \leq t)$, hence the sufficiency is immediate. Conversely, if $\mathcal{F}_{t+}^X \equiv \mathcal{F}_{t+}^Y$, there is a (Borel) function $f_t^-(z_1, z_2, \ldots) \in \times_1^\infty \mathcal{M}_E/\mathcal{M}_E$, and a sequence $s_j^- \leq t$, such that $Z_t^Y = f_t^-(Z_{s_1^-}^X, Z_{s_2^-}^X, \ldots)$, P-a.s. To see this, we can assume, by applying a Kuratowski isomorphism (Theorem 1.1), that $M_E = [0, 1]$, whence the existence of $f_t^-$ is a familiar application of the monotone class theorem. Similarly, there is a $g_t^-$ and $t_j^- \leq t$ for which $Z_t^X = g_t^-(Z_{t_1^-}^Y, Z_{t_2^-}^Y, \ldots)$, P-a.s. The same reasoning applied to the Z-futures yields the existence of $f_t^+$ and $s_j^+ \geq t$, as well as $g_t^+$ and $t_j^+ \geq t$, such that $Z_t^Y = f_t^+(Z_{s_1^+}^X, Z_{s_2^+}^X, \ldots)$ and $Z_t^X = g_t^+(Z_{t_1^+}^Y, Z_{t_2^+}^Y, \ldots)$. However, since $Z_t^Y$ is Markovian, the two expressions for $Z_t^X$ are conditionally independent given $Z_t^Y$. It follows that a conditional distribution of $Z_t^X$ given $Z_t^Y$ (taking, as before, $M_E = [0, 1]$) reduces to a unit step function, P-a.s., and hence we have $\sigma(Z_t^X) \subset \sigma(Z_t^Y)$ up to P-null sets. Hence, as above, we have $Z_t^X = g_t(Z_t^Y)$ for a Borel function $g_t$. The same reasoning with $X$ and $Y$ reversed yields the existence of $f_t$ for Definition 6.4, completing the proof of (1).

The proof of (2) is more subtle. Assuming that $\mathcal{F}_{t+}^X \equiv \mathcal{F}_{t+}^Y$ and $\{X_{t+(\cdot)} \in S, S \in \mathcal{F}'\} \equiv \{Y_{t+(\cdot)} \in S, S \in \mathcal{F}'\}$ (where we note that these are $\sigma$-fields on $(\Omega, \mathcal{F}, P)$) we have $Z_t^X(S) = P\{X_{t+(\cdot)} \in S | \mathcal{F}_{t+}^Y\}$, $S \in \mathcal{F}'$, and analogously with $X$ and $Y$ reversed. By hypothesis, for $S \in \mathcal{F}'$ there is an $S' \in \mathcal{F}'$ with

242 | Foundations of the prediction process

$\{X_{t+(\cdot)} \in S\}$ and $\{Y_{t+(\cdot)} \in S'\}$ $P$-equivalent. Then clearly

$$P\{Z_t^X(S) = Z_t^Y(S')\} = 1,$$

and we set $S' = h_t(S)$. Reversing the roles of $X$ and $Y$ we define $k_t$, proving the necessity.

**Remark** We note, however, the non-uniqueness of $h_t$ and $k_t$: our hypothesis means that the measure algebras of the $X$- and $Y$-futures are isomorphic, but it does not give an obvious set isomorphism.

Assume, conversely, that $Z_t^X$ and $Z_t^Y$ are set related for each $t$. Since this implies functional relatedness, we have $\sigma(Z_s^X, s \le t) \equiv \sigma(Z_s^Y, s \le t)$. According to Theorem 1.16 and its extension to general $E$ in Theorem 2.7, this is equivalent to $\mathscr{F}_{t+}^X \equiv \mathscr{F}_{t+}^Y$.

It remains to show that the $X$- and $Y$-futures are also equivalent. By Theorem 2.7 we have, for bounded continuous $f$ on $E$, and any $t, s > 0$,

$$P\left\{\int_0^s f(X_{t+u})\,du = \int_0^s f(\hat{\rho}_E(Z_{t+u}^X))\,du\right\} = 1.$$

Since $\theta_t^{-1}(\mathscr{F}')$ is generated by such integrals, it follows that

$$\{X_{t+(\cdot)} \in S; S \in \mathscr{F}'\} \equiv \{\hat{\rho}_E(Z_{t+(\cdot)}^X) \in S; S \in \mathscr{F}'\}, \quad (6.43)$$

and since $Z_{t+s}^X$ is right-continuous this implies, as in the proof of Theorem 1.16(2) that

$$\{X_{t+(\cdot)} \in S; S \in \mathscr{F}'\} \equiv \sigma(\hat{\rho}_E(Z_{t+s}^X), \quad 0 \le s). \quad (6.44)$$

We now claim, furthermore, that

$$\{X_{t+(\cdot)} \in S; S \in \mathscr{F}'\} \equiv \sigma(Z_{t+s}^X(S): P\{Z_{t+s}^X(S) = 0 \text{ or } 1\} = 1,$$

$$0 \le s, S \in \mathscr{F}'), \quad (6.45)$$

To see this, we observe by Definition 2.6 that $\tilde{\rho}_E(X_{v+(\cdot)})$ is $\mathscr{F}_{v+}^X/\mathscr{E}$-measurable. Therefore, for $S(A) = \{w' \in \Omega'_E: \tilde{\rho}_E(w') \in A\}$, $A \in \mathscr{E}$, we have $P\{I_A(\tilde{\rho}_E(X_{v+(\cdot)})) = Z_v^X(S(A))\} = 1$, by definition of $Z_v^X$ as conditional probability given $\mathscr{F}_{v+}^X$. On the other hand, by Theorem 2.7 we have $P\{\tilde{\rho}_E(X_{v+(\cdot)}) = \hat{\rho}_E(Z_v^X)\} = 1$. Hence $P\{I_A(\hat{\rho}_E(Z_v^X)) = Z_v^X(S(A))\} = 1$, and as $v$ ranges in $[t, \infty)$ it follows from (6.44) that

$$\{X_{t+(\cdot)} \in S; S \in \mathscr{F}'\} \equiv \sigma(Z_{t+s}^X(S(A)), \quad 0 \le s, A \in \mathscr{E}), \quad (6.46)$$

where $P\{Z_{t+s}^X(S(A)) = 0 \text{ or } 1\} = 1$ as needed for (6.45). Conversely, suppose that $P\{Z_{t+s}^X(S) = 0 \text{ or } 1\} = 1$ for an $s \ge 0$ and $S \in \mathscr{F}'$. Then

$$P\{P(X_{t+s+(\cdot)} \in S | \mathscr{F}_{t+s+}^X) = 0 \text{ or } 1\} = 1,$$

Applications and ramifications | 243

and therefore, up to a $P$-null set

$$\{Z^X_{t+s}(S) = 1\} = \{X_{t+s+(\cdot)} \in S\} = \{X_{t+(\cdot)} \in \theta_s^{-1} S\}. \tag{6.47}$$

This completes the proof of (6.45).

Now to finish Theorem 6.9, we have only to note that if $Z^X$ and $Z^Y$ are set related, then whenever $P\{Z^X_{t+s}(S) = 0 \text{ or } 1\} = 1$ we also have $P\{Z^Y_{t+s}(h_{t+s}(S)) = 0 \text{ or } 1\} = 1$, and conversely with $k_{t+s}$ in place of $h_{t+s}$. Therefore, by (6.45) and its analogue with $Y$ in place of $X$, the $X$-futures and $Y$-futures are equivalent, as was to be shown.

# References

Belayev, Y. (1961). Continuity and Hölder's conditions for sample functions of stationary Gaussian processes. In *Proceedings of the fourth Berkeley Symposium on Mathematical Statistics and Probability*, Vol. 2 (ed. J. Neyman), pp. 23–34. University of California Press, Berkeley.

Blaschke, W. (1916). *Kreis und kugel*. Veit and Comp., Leipzig.

Blumenthal, R. M. and Getoor, R. K. (1968). *Markov processes and potential theory*, Pure and Applied Mathematics, 29. Academic Press, New York and London.

Breiman, L. (1968). *Probability*. Addison-Wesley, Reading.

Chung, K. L. (1982). *Lectures from Markov processes to Brownian motion*, Grundlehren der mathematische Wissenschaften, 249. Springer-Verlag, Berlin.

Cramer, H. (1964). Stochastic processes as curves in Hilbert space. *Theory of Probability and its Applications*, **IX**, 169–79.

Davis, M. H. A. and Varaiya, P. (1974). The multiplicity of a family of $\sigma$-fields. *The Annals of Probability*, **2**, 958–63.

Dellacherie, C. and Meyer, P.-A. (1975), (1980), (1987). *Probabilités et potentiel*, Publications de l'Institut de Mathematiques de l'Universitē de Strasbourg, XV, XVII, XIX. Hermann, Paris.

Doleans, C. (1967). Construction du processus croissant naturel associé à un potential de la classe (D). *Comptes Rendus Académie Science de Paris*, Ser. A, **264**, 600–2.

Doob, J. L. (1949). Heuristic approach to the Kolmogorov–Smirnov theorems. *The Annals of Mathematical Statistics*, **XX**, 393–403.

Doob, J. L. (1953). *Stochastic processes*. Wiley, New York.

Doob, J. L. (1984). *Classical potential theory and its probabilistic counterpart*, Grundlehren der mathematischen wissenschaften, 262. Springer, Berlin.

Feller, W. and McKean, H. P. (1956). A diffusion equivalent to a countable Markov chain. In *Proceedings of the National Academy of Science USA*, **42**, 351–5.

Getoor, R. K. (1975). *Markov processes: Ray processes and right processes*, Lecture Notes in Mathematics, 440. Springer, Berlin.

Getoor, R. K. (1978). Homogeneous potentials. In *Seminaire de Probabilités*, XII (ed. C. Dellacherie, P.-A. Meyer and M. Weil), pp. 398–410. Lecture Notes in Mathematics, 649. Springer, Berlin.

Goswami, A. (1990). On a conjecture of F. B. Knight, two characterization results related to prediction processes. In *Seminaire de Probabilités*, XXIV (ed. J. Azéma, P.-A. Meyer and M. Yor), pp. 480–5. Lecture Notes in Mathematics, 1426. Springer, Berlin.

Hanner, O. (1949). Deterministic and non-deterministic stationary random processes. *Arkiv för Mathematik*, **1**, 161–77.

Hida, T. (1960). Canonical representations of Gaussian processes and their applications. *Memoirs of the College of Science, University of Kyoto, Series A*, **33**, 105–55.

Hille, E. and Phillips, R. S. (1937). *Functional analysis and semigroups*, (revised edn.), American Mathematical Society Colloquium Publications, 31, Providence, RI.

Ito, K. (1942). Differential equations determining a Markoff process. In *Kiyosi Ito, selected papers* (ed. D. W. Strook and S. R. S. Varadhan), pp. 52–75. Springer, Berlin.

Ito, K. (1968). The canonical modification of a stochastic process. In *Kiyosi Ito, selected papers* (ed. D. W. Stroock and S. R. S. Varadhan), pp. 406–80. Springer, Berlin.

Ito, K. and Nisio, M. (1968). On the oscillation function of Gaussian processes. In *Kiyosi Ito, selected papers* (ed. D. W. Stroock and S. R. S. Varadhan), pp. 513–28. Springer, Berlin.

Jain, N. and Monrad, D. (1982). Gaussian quasimartingales. *Zeitschrift für Wahrscheinlichkeitstheorie und verwandte Gebiete*, **59**, 139–59.

Kinney, J. R. (1953). Continuity properties of sample functions of Markov processes. *Transactions of the American Mathematical Society*, **74**, 280–302.

Knight, F. B. (1961). On the regularity of Markov processes. *Illinois Journal of Mathematics*, **5**, 591–613.

Knight, F. B. (1971). A reduction of continuous square-integrable martingales to Brownian motion. In *Conference on Martingales* (ed. H. Dinges and L. Snell), pp. 19–31. Lecture Notes in Mathematics, 190. Springer, Berlin.

Knight, F. B. (1972). On Markov processes with right-deterministic germ fields. *The Annals of Mathematical Statistics*, **43**, 1968–76.

Knight, F. B. (1975). A predictive view of continuous time processes. *The Annals of Probability*, **3**, 573–96.

Knight, F. B. (1979). Prediction processes and an autonomous germ–Markov property. *The Annals of Probability*, **7**, 385–405.

Knight, F. B. (1981a). *Essays on the prediction process*, Lecture Notes Series, 1. Institute of Mathematical Statistics, Hayward, CA.

Knight, F. B. (1981b). *Essentials of Brownian motion and diffusion*, Mathematical Surveys, 18. American Mathematical Society, Providence, RI.

Knight, F. B. (1983a). A post-predictive view of Gaussian processes. *Annales Scientifiques École Normale Supérieure*, **16**, 541–66.

Knight, F. B. (1983b). A transformation from prediction to past of an $L^2$-stochastic processes. In *Seminaire de Probabilités*, XVII (ed. J. Azema and M. Yor), pp. 1–14. Lecture Notes in Mathematics, 986. Springer, Berlin.

Knight, F. B. (1984). On the Ray topology. In *Seminaire de Probabilités*, XVII (ed. J. Azema and M. Yor), pp. 56–69. Lecture Notes in Mathematics, 1059. Springer, Berlin.

Knight, F. B. (1986a). On strict-sense forms of the Hida–Cramer representation. In *Seminar on Stochastic Processes*, 1984 (ed. E. Cinlar, K. L. Chung, and R. K. Getoor), pp. 109–38. Progress in Probability and Statistics, 9. Birkhauser, Boston.

Knight, F. B. (1986b). Poisson representation of strict regular step filtrations. In *Seminaire de Probabilités*, XX (ed. J. Azema and M. Yor), pp. 1–27. Lecture Notes in Mathematics, 1204. Springer, Berlin.

Knight, F. B. (1988). On invertibility of martingale time changes. In *Seminar on Stochastic Processes*, 1987 (ed. E. Cinlar, K. L. Chung, and R. K. Getoor), pp. 193–222. Progress in Probability and Statistics, 15. Birkhauser, Boston.

Kunita, H. and Watanabe, S. (1967). On square-integrable martingales. *Nagoya Mathematical Journal*, **30**, 209–45.

LeJan, Y. (1978). Temps d'arrêt stricts et martingales de saut. *Zeitschrift für Wahrscheinlichkeitstheorie und verwandte Gebiete*, **44**, 213–25.

Lepingle, D., Meyer, P.-A., and Yor, M. (1981). Extrémalité et remplissage de tribus pour certaines martingales purement discontinues. In *Seminaire de Probabilités*, XV (ed. J. Azema and M. Yor), pp. 604–17. Lecture Notes in Mathematics, 850. Springer, Berlin.

# References

Lévy, P. (1948). *Processus stochastiques et mouvement Brownien*. Gauthier-Villars, Paris.

Lévy, P. (1956). A special problem of Brownian motion, and a general theory of Gaussian random functions. In *Proceedings of the third Berkeley Symposium on Mathematical Statistics and Probability*, Vol. 2 (ed. J. Neyman), pp. 133–76. University of California Press, Berkeley.

Lévy, P. (1957). Fonction aléatoires à corrélation linéaire. *Illinois Journal of Mathematics*, **1**, 217–58

McKean, H. P. (1969). *Stochastic integrals*. Probability and Mathematical Statistics (ed. Z. W. Birnbaum and E. Lukacs). Academic Press, New York.

Meyer, P.-A. (1966). *Probability and potentials*. Blaisdell, Waltham, Mass.

Meyer, P.-A. (1967). Intégrales stochastiques III. In *Seminaire de Probabilités*, I (ed. P.-A. Meyer), pp. 118–41. Lectures Notes in Mathematics, 39, Springer, Berlin.

Meyer, P.-A. (1976a). Generation of $\sigma$-fields by step processes. In *Seminaire de Probabilités*, X (ed. P.-A. Meyer), pp. 118–24. Lecture Notes in Mathematics, 511, Springer, Berlin.

Meyer, P.-A. (1976b). La théorie de la prédiction de F. Knight. In *Seminaire de Probabilités*, X (ed. P.-A. Meyer), pp. 86–103.

Meyer, P.-A. (1983). A remark on F. Knight's paper. *Annales Scientifiques École Normale Supérieure*, **16**, 567–9.

Meyer, P.-A. and Yor, M. (1976). Sur la theorie de la prédiction, et le problème de décomposition de tribus $\mathscr{F}_{t+}^{0}$. In *Seminaire de Probabilités* (ed. P.-A. Meyer), pp. 104–17. Lecture Notes in Mathematics, 511, Springer, Berlin.

Phillips, R. S. (1940). A characterization of Euclidean spaces. *Bulletin of the American Mathematical Society*, **46**, 930–3.

Ray, D. B. (1959). Resolvents, transition functions, and strongly Markovian processes. *Annals of Mathematics*, **70**, 43–72.

Revuz, D. and Yor, M. (1991). *Continuous martingales and Brownian motion*. Grundlehren der mathematischen Wissenschaften, 293. Springer, Berlin.

Sharpe, M. (1988). *General theory of Markov processes*. Pure and Applied Mathematics, 133. Academic Press, Boston.

Stroock, D. and Yor, M. (1980). On extremal solutions of martingale problems. *Annales Scientifiques Éole Normale Supérieure*, **13**, 95–164.

Walsh, J. B. and Meyer, P.-A. (1971). Quelques applications des résolvantes de Ray. *Inventiones Mathematicae*, **14**, 143–66.

Widder, D. V. (1946). *The Laplace transform*, Princeton Mathematical Series, 18 (ed. M. Morse and A. W. Tucker). Princeton University Press, Princeton.

Yor, M. (1977). Sur les théories du filtrage et de la prédiction. In *Seminaire de Probabilités*, XI (ed. C. Dellacherie, P.-A. Meyer, M. Weil), pp. 257–97. Lecture Notes in Mathematics, 581. Springer, Berlin.

# Index

absorbed process 233
additive functional 109
announcable stopping time 27
arrival (hitting) probabilities 233
Ascoli and Arzela, theorem of 21

Banach space 12
Bessel's inequality 10
Blaschke theorem 14
Blumenthal, R. M., 0-1 law 39
branch point 38
Brownian motion 1, 83, 111
  arrested 2
  reflecting 60

conditional distributions 241
conditional expectation 11
conditional probabilities, regular 21

decomposition of Doob and Meyer 218, 220
demesne, of the prediction process 64
  complete, Borel 64
Dellacherie, C. and Meyer, P.-A., theorems on previsibility and announcibility 27
diffusion 151, 200
  intrinsic 149, 200
Doob, J. L., representation theory of 8; *see also* supermartingale

essentially countably generated 7
excessive functions 67
expectations, process of 104

Feller, W., process and semigroup 70
Feller and McKean's example 174
filtering, incomplete information 212
Fourier transform 225

Gaussian
  family of random variables 93, 97
  Markov process 132
  process 93
  stationary semimartingale 223

general theory of processes 212
Getoor, R. K.
  proposition on potentials 73
  theorem on Ray processes 76
  Ray space 87

Hilbert space 12
hitting probabilities, *see* arrival probablities
Hunt, G. A.
  lemma 25
  process 68

index of multiplicity
  strict sense 182, 192
  wide sense 122, 181
infinitely divisible law 59
infinitesimal
  generator, weak 234
  operator, extended 175
inversion formula, Laplace 90

jump process 160
  intrinsic 157, 159
  well-ordered 161
jump prediction process 196

killed process 238
Knight, F. B., theorem of 170, 228
Kolmogorov, A. N.
 Chapman-K. identity 31
  extension theorem 52
Kunita, H. and Watanabe, S., martingales of 145
Kuratowski, C., isomorphism 17

Laplace transform
  continuity theorem 23
  inversion formula 90
Lebesgue measure 16
Lévy, P.
  process 59
  proper canonical representation 123, 132
linear martingale representation (LMR) 168, 192
Lusin space 16

## 248 Index

Markov
  process 51
  property
    strong 25, 37, 63, 215
    moderate 37, 63, 215
    intrinsic 31
  transition function 51
measurable process 17, 92, 188
Meyer, P.-A.
  0–1 Law 38
  theorem on supermartingales 42
  orthogonal decomposition 231
  see also decomposition
modification
  of the future 229
  of a Markov process 57, 58
  standard 52
multiplicity, see index of multiplicity

optional process 24
Ornstein–Uhlenbeck velocity process 133, 138, 204, 222

passage time distributions 233
Poisson process 29, 170, 210
prediction
  demesne, see demesne
  process 23
    functionally related 240
    set related 240
    of $X_t$ 238
    relative to a general filtration 212
    relative to $\mathscr{F}'_{E_1}$ 214
  space 63
  topology 46
previsible (predictable)
  martingale stochastic integral representation (PMSIR) 169, 192
  process 24
  quadratic variation 173
  $\sigma$-field 183
proper canonical, see Lévy, P.
pseudo-path topology 44
pseudo-trajectory 16

quasi-left-continuity 68

Ray, D.
  compactification 81
  process 71
  resolvent 69
  space 80

theorem 70, 82
  topology 74
Rayification 87
regularization of a Markov process 57, 62, 72; see also modification
regular point 234
representation of filtrations problem 169, 170, 228
representation theory, see Doob, J. L.
resolvent
  identity 70, 234
    for Kunita–Watanabe martingales 146
  of $Z$ 42
  of $X$ 52
Riesz–Fisher theorem 10
Riesz representation theorem 21, 82
right
  process 67
  semigroup 67

section theorems 28
semimartingale
  special 219
  stationary Gaussian 223
  totally integrable 219
Sharpe, M., theorem on right processes 163
sojourn measures 44
square field operator (carré du champs) 175
stationary
  covariance 133
  Gaussian process 133
  process 135
  wide sense 136
Stone–Weierstrass theorem 23, 33, 113
stopping time 5, 25
strict sense concept 165

times of discontinuity 29, 151, 240
  of a Gaussian process 111
  foreseeable 152
  totally unforeseeable 152
  well ordered 161
  see also jump processes
transition function, see Markov transition function
triangular covariance 139, 209, 222

weak-* convergence 21, 46
wide sense
  concept 165
  martingale 111

Yor, M., theorem on filtering 213